MECHANICAL ENGINEERING THEORY AND APPLICATIONS

FAULT DETECTION

CLASSIFICATION, TECHNIQUES AND ROLE IN INDUSTRIAL SYSTEMS

MECHANICAL ENGINEERING THEORY AND APPLICATIONS

Additional books in this series can be found on Nova's website under the Series tab.

Additional e-books in this series can be found on Nova's website under the e-book tab.

ENGINEERING TOOLS, TECHNIQUES AND TABLES

Additional books in this series can be found on Nova's website under the Series tab.

Additional e-books in this series can be found on Nova's website under the e-book tab.

MECHANICAL ENGINEERING THEORY AND APPLICATIONS

FAULT DETECTION

CLASSIFICATION, TECHNIQUES AND ROLE IN INDUSTRIAL SYSTEMS

FAUSTO PEDRO GARCÍA MÁRQUEZ, PH.D.
AND
MAYORKINOS PAPAELIAS, PH.D.
EDITORS

nova publishers
New York

For permission to use material from this book please contact us:
Telephone 631-231-7269; Fax 631-231-8175
Web Site: http://www.novapublishers.com

Library of Congress Cataloging-in-Publication Data

Fault detection : classification, techniques, and role in industrial systems / editors, Fausto Pedro Garcma Marquez and Mayorkinos Papaelias (Birmingham University, UK).
pages cm
Includes bibliographical references and index.
ISBN: 978-1-62808-999-8 (hardcover)
1. Fault location (Engineering) 2. Defects. 3. System failures (Engineering)--Prevention. I. Garcma Marquez, Fausto Pedro, editor of compilation. II. Papaelias, Mayorkinos, editor of compilation.
TA169.6.F358 2013
620'.00452--dc23

2013033670

Published by Nova Science Publishers, Inc. † New York

CONTENTS

PREFACE

Fault Detection: Classification, Techniques and Role in Industrial Systems synthesises the principles of fault detection, i.e. the book provides an interface between the main disciplines of methods/technologies and the industrial systems. It is complementary to other sub-disciplines such as maintenance, engineering, safety, risk analysis, etc.

This book is intended for engineers, economists, technical consultants, researchers, etc. who are involved in the advancement of fault detection, or who employ fault detection as part of their work. The authors of this volume describe their pioneering work in the area or provide material for case studies successfully applying the fault detection discipline in real life cases.

The first chapter describes a case study on wind turbines (WT). Wind energy is the fastest growing renewable energy source of power production in the world today. There is a constant need to reduce the costs of operating and maintaining wind turbines. Condition Monitoring (CM) is a tool commonly employed for the early detection of faults/failures so as to minimise downtime and maximize productivity. This chapter provides a review of the state-of-the-art in the CM of WTs, describing the different sensory signals and signal processing methods.

The second chapter is focused in improving the efficiency of inspection activities in a rail network and ensuring an acceptable level of safety for rolling stock operations, where rails are inspected at regular intervals for structural defects, profile irregularities, excessive wear, missing fastenings, failed sleepers, and abnormal variations in rail gauge. Rails are inspected either manually or with hi-rail vehicles and test trains using traditional non-destructive evaluation techniques such as vision (including automated systems), magnetic particle inspection, magnetic flux leakage, ultrasonic and eddy current testing. The accuracy of the inspection process is of critical importance in minimising the risk of catastrophic failures, improving maintenance scheduling and optimising rolling stock operations. The inspection procedures currently employed by rail infrastructure managers are not adequately efficient as they depend on the mobilisation of a significant amount of resources, both in terms of special test equipment and expert personnel. Moreover, the inspection process needs to be carried out in segments either at night or when the line is closed to avoid disruption of normal rolling stock operations. Rail infrastructure managers due to various reasons have not benefited sufficiently from recent breakthroughs and innovation in inspection technologies that other industrial sectors are already taking advantage of. This chapter analyses the efficiency of current inspection strategies and discusses a plausible way of implementing substantial improvements without the requirement for a fundamental shift in the existing understanding of the overall inspection process and the ways it is related to preventive and reactive maintenance procedures.

Complex industrial systems such as gas turbine engines have many heterogeneous subcomponents (e.g., thermal, mechanical, hydraulic and electrical) with complex electro-mechanical interactions are studied in chapter three. Today, schedule-based policies are followed throughout the industry for complex system maintenance. However, due to manufacturing and usage variations, degradation profile of every system is different. Therefore, a scheduled maintenance policy in most cases becomes overly conservative or a serious safety concern. This calls for reliable and cost-effective health management systems for complex industrial systems. However, reliable first principle based modelling of such systems is often very expensive (memory, computation and cost) and hence may become inappropriate for condition monitoring, fault detection, diagnostics and prognostics. This book chapter describes a novel data-driven framework to obtain abstract inexpensive models of complex systems from multiple sensor observations to enable condition monitoring, diagnostics and supervisory control. The health management algorithms are formulated in the setting of symbolic dynamic filtering (SDF) that has been recently reported in literature. The underlying concept of SDF is built upon the principles of symbolic dynamics, statistical pattern recognition, probabilistic graphical models and information theory. SDF was initially developed for statistically stationary data and then it was successfully extended to analysis of transient data. The tool-chain involves data space abstraction for analyzing mixed data (continuous, discrete and categorical) without significant loss of information; spatiotemporal pattern discovery & fusion built on concepts of Markov and Bayesian networks; and health status classification using machine learning techniques. The book chapter presents application of the generic data-driven tool in heterogeneous application areas, such as aircraft gas turbine engines, coal gasification systems, nuclear power plants, electric motors, fatigue damage problems and aviation data anomaly.

Chapter four presents the importance role in maintenance system of the machine condition monitoring with fault diagnosis. It reduces maintenance costs, improves productivity and efficiency and increases machine availability. This chapter presents an expert system for online and offline induction motor fault detection based on vibration analysis and support vector machines (SVM). Vibration analysis is used because of its easy measurability, high accuracy and reliability. Vibration signals of few types of healthy and faulty induction motors are collected and characteristic features, as indicators of fault presence, are calculated, in both time and frequency domain. The most frequent faults present at induction motors are considered, such as bearing wear, static eccentricity and broken rotor bar. Obtained feature sets were then used for training of support vector machines classifiers, a type of artificial intelligence classification technique which can determine whether some of considered faults is present or not. The performance of SVM classifiers is improved using General Particle Swarm Optimization (GPSO) algorithm for SVM parameter tuning. This expert system can function online, when vibration sensors are connected to the motor and fault detection is carried out simultaneously, or offline, when collected vibration signal is recorded and processed afterwards. The system was tested and validated on a number of healthy and faulty motors in industrial facility for sunflower oil processing during the maintenance period. Obtained results show that this system can detect faults in early stages with high accuracy and reliability, which could provide malfunction and failure prevention and improve overall performance and efficiency of industrial systems.

Recent events have highlighted the criticality of "common mode" failures on the Air Data System (ADS) and that physical redundancy is insufficient when all the redundant sensors are subjected to the same environmental conditions. Chapter five evaluates the analytical redundancy approaches to the Sensor Failure Accommodation (SFA) of the ADS - model free based on an Unscented Kalman Filter (UKF) a Multilayer Feedforward (MLP) and the Extended Minimal Resource Allocating Network (EMRAN) and a model based approach, where the angle of attack equation is rearranged as a quadratic equation in V and solved to provide airspeed estimates. For Sensor Failure Detection and Identification (SFDI), three approaches are evaluated - Cumulative Sum (CUSUM), Floating Limiter and a sequential Generalized Likelihood Ratio Test (GLRT) for a Step and a Ramp Bias failures on ADS – using flight data from a small UAV flown at the WVU Flight Control Systems Laboratory (FCSL).

Chapter six presents a new approach to the H_∞ fault detection and diagnosis (FDD) problem for a class of descriptor systems. A new structure via an elegant construction with adjustable parameters, which can simultaneously design the control input signal and the FDD signal, is proposed. The state and fault signal can be also estimated at the same time. The design procedure is based on LMI techniques.

Chapter seven analyses the intelligent signal processing methods role in identifying 'novelty' in machine behaviour, thereby facilitating greater operational availability and early detection of emerging faults. Complex industrial systems often capture a high volume of data, with high dimensionality, and therefore clustering techniques have emerged as primary candidates for machine fault detection and classification due to their pattern-matching and data compression capabilities. To address well-known 'sensitivity' problems associated with classical clustering techniques, however, a real-coded Genetic Algorithm (GA) is now considered that is integrated into a traditional K-means clustering methodology. The resultant hybridisation provides an enhanced search algorithm by incorporating the local search capabilities rendered by hill-climbing optimisation and the global search ability provided by a GA. Performance of the proposed algorithm is compared with other clustering techniques using two artificial data sets, and a benchmark iris classification problem. Results show that, in all cases, the proposed algorithm outperforms its counterparts in terms of global search capability and robustness. Moreover, the scalability of the proposed algorithm to higher-dimensional problems featuring a greater number of data points, is also considered, and validated using an application to provide a reduced data-representation of measurements taken from industrial gas turbine units, and for fault detection based on rundown vibration signatures.

Finally, chapter eight describes how a suitable design of a fault diagnosis system can be successfully applied on industrial mobile robots for increasing their reliability and safety, improving their operability and maximize their performance for different tasks and different environments. Structural analysis is applied to the nonlinear model of an industrial mobile robot for residual generation. An adaptive/thresholding decision algorithm is described for developing the residual evaluation module. This includes a threshold test based on the minimum and maximum values of the generated residuals to decide if a fault has occurred or not. Moving thresholds are updated using a sliding window of a fixed number of sequential sample readings. With this solution, probability density functions of residuals are unknown and need not be estimated. Experimental validation of the proposed scheme is performed on a differential drive industrial mobile robot, and it is presented in the second part of the chapter.

In: Fault Detection
Editors: F. P. G. Márquez and M. Papaelias
ISBN: 978-1-62808-999-8

Chapter 1

CONDITION MONITORING OF WIND TURBINES: SENSORY SIGNALS AND SIGNAL PROCESSING METHODS

***Fausto Pedro García Márquez*[1*]*, Jesús María Pinar Pérez*[2]**
***and Raúl Ruiz de la Hermosa González-Carrato*[2]**
[1]Ingenium Research Group, Universidad Castilla-La Mancha, Ciudad Real, Spain
[2]Ingenium, Centro Universitario de Estudios Financieros, Madrid, Spain

ABSTRACT

Wind Turbines (WT) are one of the fastest growing sources of power production in the world today and there is a constant need to reduce the costs of operating and maintaining them. Condition monitoring (CM) is a tool commonly employed for the early detection of faults/failures so as to minimise downtime and maximize productivity. This chapter provides a review of the state-of-the-art in the CM of WTs, describing the different sensory signals and signal processing methods.

Keywords: Fault Detection and Diagnosis, Condition Monitoring, Maintenance Management, Wind Turbines

ACRONYMS

AE: Acoustic emission
AI: Artificial intelligence
CBM: Condition based maintenance
CM: Condition monitoring
CMS: Condition monitoring system
FDD: Fault detection and diagnosis

---

* Faustopedro.garcia@uclm.es.

FFT: Fast Fourier transform
HMM: Hidden Markov model
O&M: Operation and maintenance
PM: Preventive maintenance
RMS: Root mean square
RCM: Reliability centred maintenance
SCADA: Supervisory command and data acquisition
SPM: Shock pulse method
TSA: Time Synchronous Averaging
UT: Ultrasonic testing
WEC: Wind energy converter
WT: Wind turbine

1. Introduction

In 2011, world wind energy capacity was 237 gigawatts (GW), having been more than doubling every three years, and one forecast (by WWEA [1]) is for world wind energy capacity to rise to at least 1,000 GW by 2020. World wind energy capacity has been increasing year by year and is expected to continue increasing. But to make wind power competitive with other sources of energy, availability, reliability and the life of turbines will all need to be improved. As the wind energy sector grows, business economics will demand increasingly careful management of costs. For a 20-year life, the operations and maintenance (O&M) costs of 750kW turbines might account for about 25% - 30% of the overall energy generation cost [2] or 75%–90% of the investment costs [3].

Furthermore, one projection in 2002 was that the O&M costs for 2 MW turbines (which together with 2.5 and 3 MW turbines have since become the workhorses of the wind power industry) "might be 12% less than an equivalent project of 750kW machines"[4].But new wind farms typically have higher capacity and comprise more machines. Tavner et al. [5] suggest that larger turbines fail more frequently and thus require more maintenance. Reducing inspection and maintenance costs has thus become increasingly important as wind turbine size and numbers have continued to rise.

Of course, some WT components fail earlier than expected and, because unscheduled downtime can be so costly, condition monitoring systems (CMS) are employed to improve WT availability and reduce the O&M costs. However there is a degree of uncertainty about the appropriateness of applying specific maintenance policies to WT components. This chapter discusses the applicability of various maintenance strategies to WT condition monitoring, collects the available techniques and reviews the sensory signals and signal processing methods in the literature.

2. Wind Turbines

Most WT machines are three-blade units comprising the major components illustrated in Figure 1. Driven by the wind, the blades and rotor transmit energy via the main shaft through the gearbox to the generator, the main shaft being supported by the bearings, and the gearbox being such that the generator speed is as near as possible to optimal for the generation of electricity.

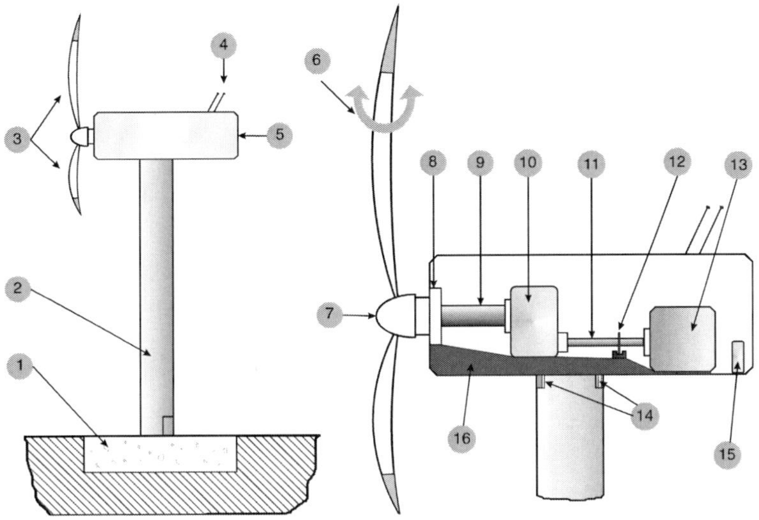

Figure 1. Components of the WT where: 1-Base/Foundations; 2-Tower; 3-Blades; 4-Meteorological unit (vane and anemometry); 5-Nacelle; 6-Pitch; 7-Hub; 8-Main bearing; 9-Low speed (main) shaft; 10-Gearbox; 11- High speed shaft; 12-Brake system; 13-Generator; 14-Yaw system, 15-Converter, 16-Bedplate. N.B. Drive train = 9+11.

Alignment with the direction of the wind is controlled by a yaw system and the housing (or "nacelle") is mounted at the top of a tower. The capacity power per turbine has increased dramatically over the past thirty years from 50 kW machines in late 1990s to 7.5 MW at the present, while 8-12 MW turbines are currently under development (see Figure 2).

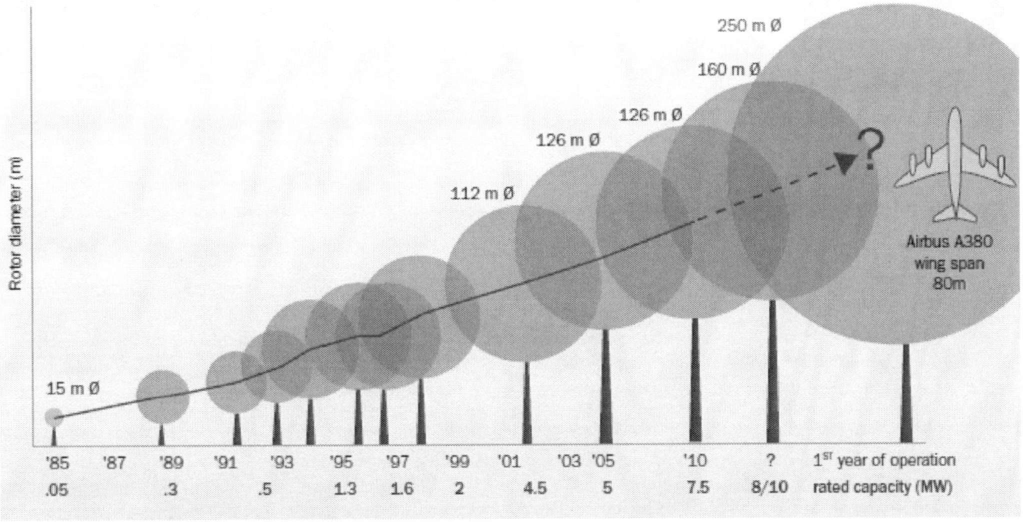

Figure 2. WT sizes and power [47].

Over the same period, tower heights, rotor diameters and overall weights of turbines have almost quadrupled in size and capacity, increasing the complexity of construction, operation, maintenance and inspection procedures [47]. Many technologies have been developed and implemented in WTs to increase their capacities. The most common configuration is the horizontal axis WT with three blades, for which different combinations of rotational speed, power control, drive train configuration and generator can be used. Nowadays, the 3 blades, power control by pitch system and variable rotational speed configuration is the most installed WT around the world.

Some defects such as leaking and corrosion can be detected by visual inspection; discolouration of component surfaces may indicate slight temperature variations or deteriorating condition, and the sound coming from the bearings can also indicate physical condition [7]. However, many of the most typical failures like cracking and roughness on the surfaces of the blades, electric short circuits in the generator, and overheating of the gearbox all demand a more sophisticated approach to maintenance.

3. MAINTENANCE THEORY

Maintenance is required to make sure that the components continue to perform the functions for which they were designed. The basic objectives of the maintenance activity are to deploy the minimum resources required to ensure that components perform their intended functions properly, to ensure system reliability and to recover from breakdowns [8].

3.1. Corrective, Scheduled and Condition Based Maintenance

Classical theory sees maintenance as either corrective or preventive. The former(also known as unscheduled or failure based maintenance) is carried out when turbines break down and when faults are detected or failures occur in any of the components. Immediate refurbishment or replacement of parts may be necessary [9] and unscheduled downtime will result. Corrective maintenance is therefore the most expensive of strategies and wind farm operators will hope to resort to it as little as possible. The various stages are shown in Figure 3.

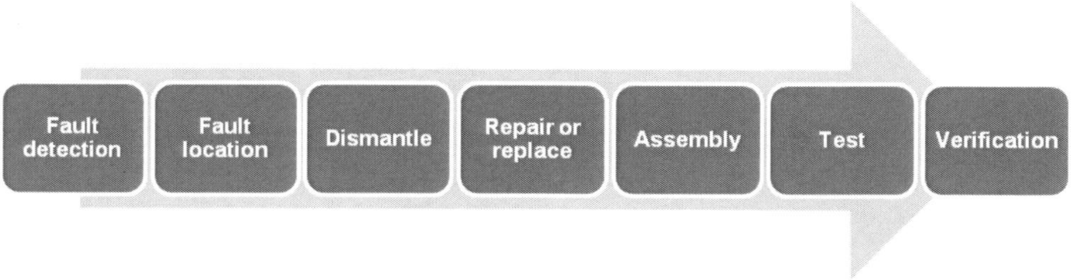

Figure 3. Stages of a corrective maintenance task.

By contrast, the objective behind preventive maintenance (PM) is to either repair or replace components before they fail as shown in Figure 4. This has most straightforwardly been achieved by scheduled maintenance, also known as time based (or planned) maintenance and involving repair or replacement at regular time intervals as recommended by the supplier and regardless of condition. Scheduled maintenance activities in WT include the changing of oil and filters, and the tightening and torqueing of bolts.

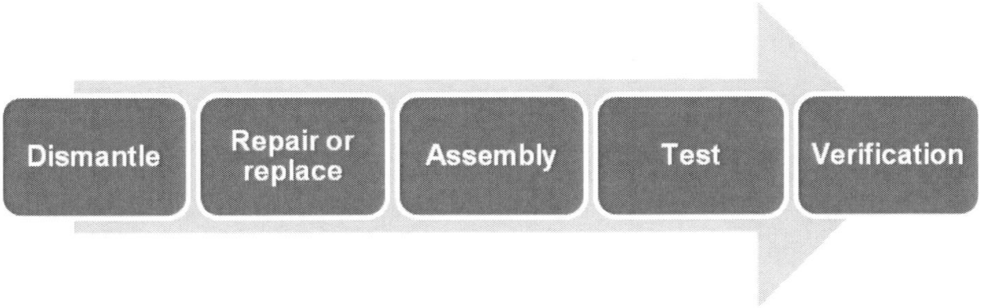

Figure 4. Stages of a preventive maintenance task.

But reducing failures in this way comes at the cost of completing maintenance tasks more frequently than absolutely necessary and not exhausting the full life of the various components already in-service. An alternative is to mitigate against major component failure and system breakdown with condition based maintenance (CBM) in which continuous monitoring and inspection techniques are employed to detect incipient faults early, and to determine any necessary maintenance tasks ahead of failure [10].

This involves acquisition, monitoring, processing, analysis and interpretation of data and selection of optimal maintenance actions [11] and is achieved using condition monitoring systems (see Figure 5) [12].

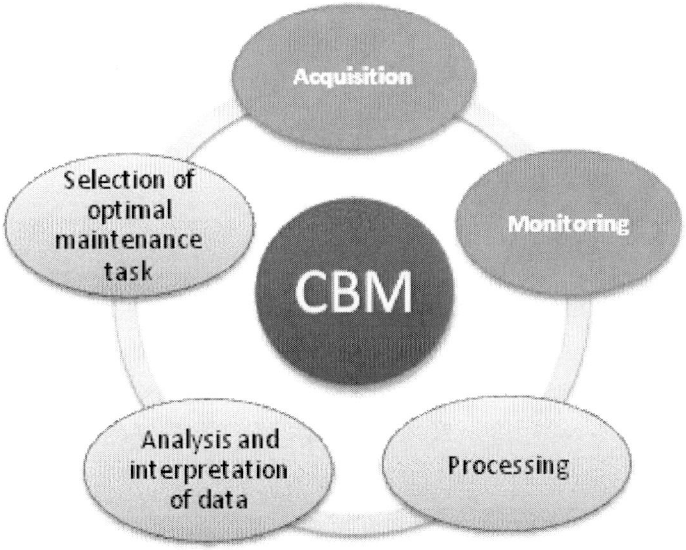

Figure 5. Items of CBM, where CMS involves acquisition and monitoring.

CBM has been shown to minimise the costs of maintenance, improve operational safety, and reduce the quantity and severity of in-service system failures. Byon and Ding [13] have demonstrated its applicability to WTs, and McMillan and Ault [14] have used Monte Carlo simulation to evaluate its cost effectiveness when applied to WTs. CBM is now the most widely employed strategy in the WT industry.

3.2. Reliability Centred Maintenance

The state-of-the-art way of deciding upon maintenance strategy in the WT industry is reliability centred maintenance (RCM), which has been formally defined as "*a process used to determine what must be done to ensure that any physical asset continues to do whatever its users want it to do in its present operating context*" [15]. It involves maintaining system functions, identifying failure modes, prioritizing functions, identifying PM requirements and selecting the most appropriate maintenance tasks with the objective of managing system failure risk effectively; operational and maintenance policies are optimised such that the overall maintenance task is reduced [9]. RCM has been recognized and accepted in many industrial fields, such as steel plants, railway networks, ship maintenance and other industries. RCM in the wind turbine industry is addressed by Andrawus et al. [16].

4. Condition Monitoring of Wind Turbines

On the basis that a significant change is indicative of a developing failure, condition monitoring systems (CMS) [8] comprise combinations of sensors and signal processing equipment that provide continuous indications of component (and hence wind turbine) condition based on techniques including vibration analysis, acoustics, oil analysis, strain measurement and thermography (see Figure 6). On WTs they are used to monitor the status of critical operating major components such as the blades, gearbox, generator, main bearings and tower. Monitoring may be online (and hence provide instantaneous feedback of condition) or offline (data being collected at regular time intervals using measurement systems that are not integrated with the equipment) [17].

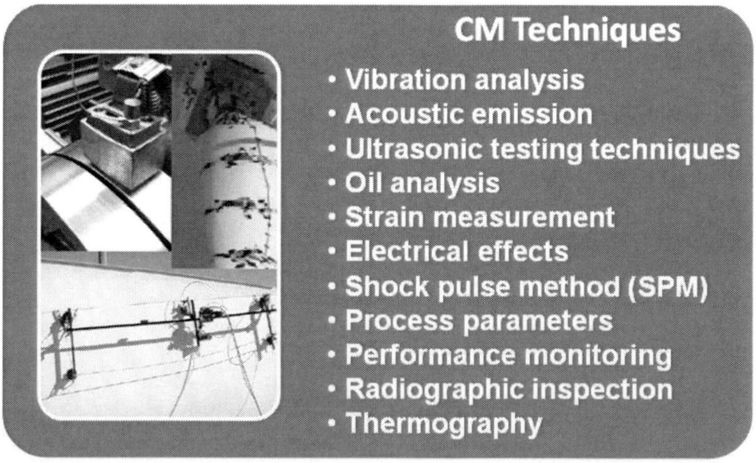

Figure 6. CM techniques for WTs.

With good data acquisition and appropriate signal processing, faults can thus be detected while components are operational and appropriate actions can be planned in time to prevent damage or failure of components. Maintenance tasks can be planned and scheduled more efficiently, resulting in increased reliability, availability, maintainability and safety (RAMS) whilst down-time, maintenance and operational costs are reduced. CM techniques are thus used throughout the industry and benefits are "especially shown for offshore wind farm(s)" [18] because of not only the high costs of O&M at sea but also the typically larger turbines.

5. Sensory Signals and Signal Processing Methods

Regardless of the technique, the capability of a CMS relies upon two basic elements: the number and type of sensors, and the associated signal processing and simplification methods utilized to extract important information from the various signals. An electronic measurement system will acquire the data and then process and distribute them to an observer or other technical control system. Data acquisition will involve measuring the required variables (e.g. current, voltage, temperature, speed) and turning them into electronic signals as shows Figure 7.

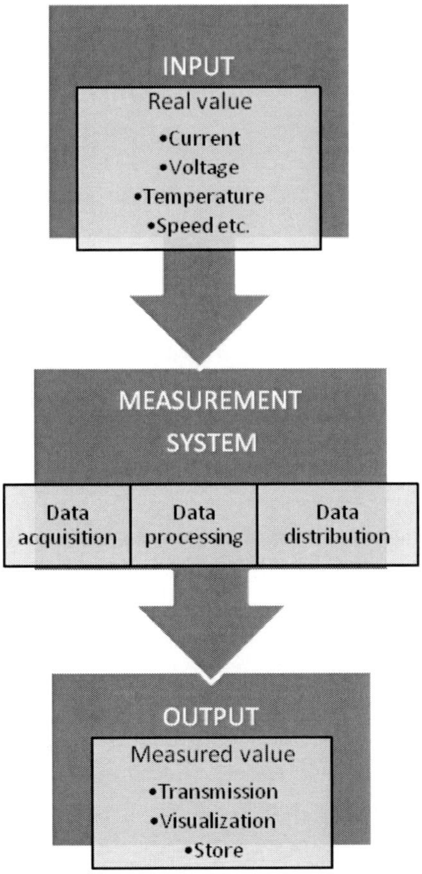

Figure 7. Measurement system process.

To do so effectively will involve judicious choice and placement of the right type and number of sensors; conditioning (performing basic operations including amplification, filtering, linearization, and finally modulation/demodulation) may be necessary to reduce the susceptibility of the signals to interference.

Optimization techniques may then be employed in the processing of the signals by a digital signal processor (DSP), involving not only the processing itself but also sorting and manipulation as necessary. Subsequent distribution will be to either a screen, computer, storage device or other system. There are several options, including e.g. Ethernet networks with TCP/IP protocol together with WLAN for a WT for communicating with either a Farm Server or a supervisory command and data acquisition (SCADA) system.

The latter is a particular computer-based system that allows local and remote control of the functions of a WT, gathering data from the wind farm and analysing them in order to report operational performance, and hence ensure efficient operation. SCADA uses various signal processing methods, the most relevant to WT being those covered below as shows Figure 8.

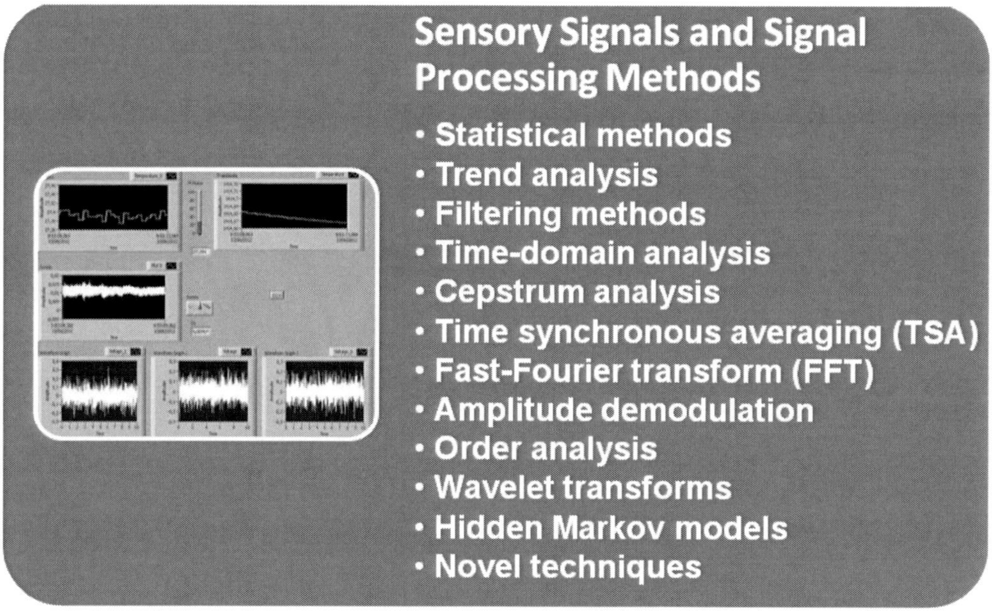

Figure 8. Sensory signals and signal processing methods for WTs.

5.1. Statistical Methods

One common application of statistical algorithms for the purposes of CM is to analyse the data signals from the various sensors in WTs. Common statistical measures such as root mean square (RMS) and peak amplitude are widely used for the diagnosis of failures but more advanced features are also being developed. Other important statistical parameters are the maximum value, minimum value, mean, peak to peak, standard deviation, shape factor, crest factor, impulse factor, definite integral, energy ratio and kurtosis.

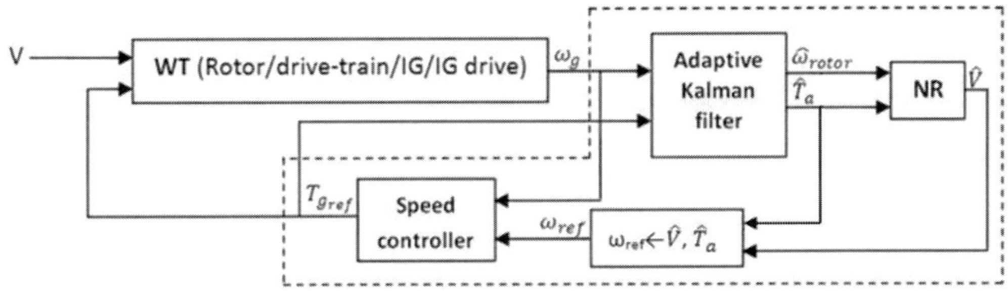

Figure 9. Proposed control scheme of a variable speed stall regulated wind turbine [45].

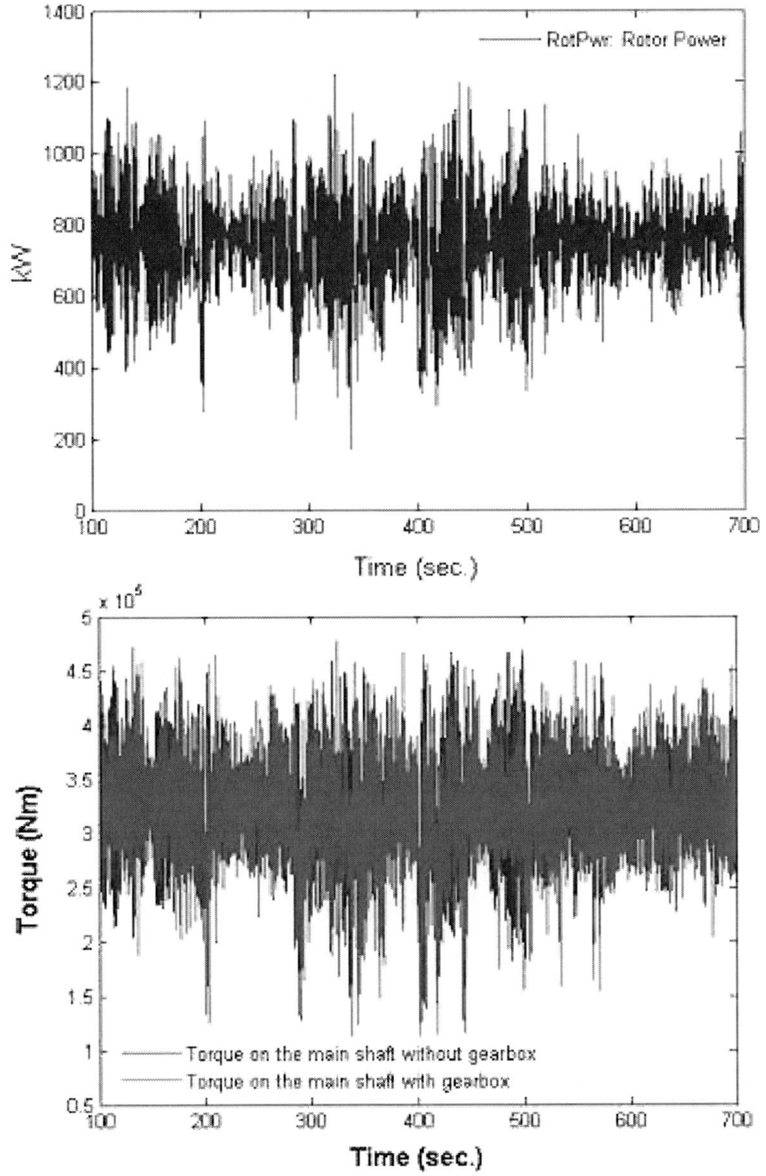

Figure 10. Time domain analysis for rotor power (left) and torque of the main shaft (right) of a WT [44].

5.2. Trend Analysis

When applied to WT, trend analysis refers to the concept of collecting data from the various sensors and looking for trends. It requires particular algorithms, and applications include the monitoring of pitch mechanisms although most common is its use on power output patterns from WT generators. Note that trend estimation is a different technique specifically for forecasting.

5.3. Filtering Methods

Sometimes there are redundant data that contain information that is not useful and must be eliminated so as to avoid compromising the computations. For example, the vibrations from a nacelle will have to be filtered whilst measuring the vibration of the gearbox inside [22].

A method using least median squares that filters in parallel with estimating the power curve is presented in [23], and filtering with and without a classical statistical method (based on standard deviation) is described and compared in [24]. The drawback of filtering whilst monitoring trends is that the parameters have to be adjusted to take account of varying operating conditions. Figure 9 shows a control scheme of a WT using a Kalman filter.

5.4. Time-Domain Analysis

Analysis in the time domain is a way of monitoring WT faults like resistive and inductive imbalances between the rotor and the stator phases, and turn-to-turn faults in the rotor windings of the generator. Variations in current signals and trends are typically used for vibration analysis, oil analysis and AE. Figure 10 shows the time domain analysis of two different variables.

5.5. Cepstrum Analysis

The power cepstrum is a time based approach defined as “the inverse Fourier transform of the logarithmic power spectrum” [26]. The cepstrum is well suited for applications to equipment diagnostics in a WT. Gear vibration spectra commonly show sidebands of meshing frequency and its harmonics arising from the modulation of tooth meshing waveform [19] (see Figure 11).

For gearboxes in good condition, the sideband level generally remains constant with time. Therefore, changes in the number and amplitude of the sidebands normally indicate deterioration. The presence of several families of sidebands and other components can complicate the distinction and evaluation of the sideband spacing [27].

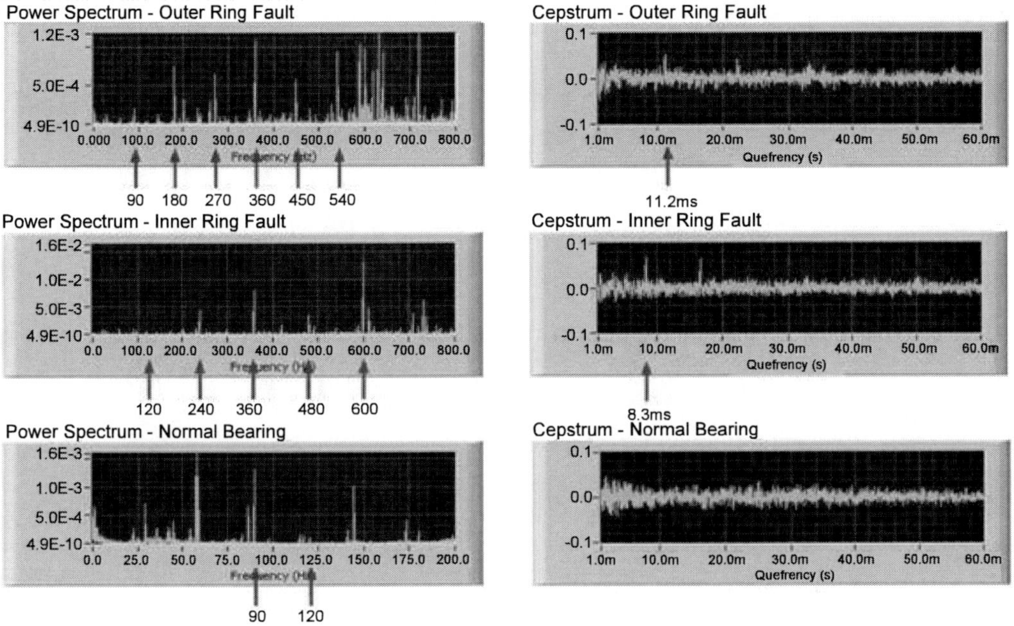

Figure 11. Power spectra and cepstrums of a bearing with an outer ring fault (top), an inner ring fault (middle), and a no fault (bottom) [46].

5.6. Time Synchronous Averaging (TSA)

TSA, also called time domain averaging, is a signal processing technique that serves as the basis for many gear fault detection algorithms. The method is for identifying a rotating bearing defect by measuring the vibration of a rotating bearing and obtaining a waveform signal. It may also help in identifying the source of vibration in WT gearboxes [28].

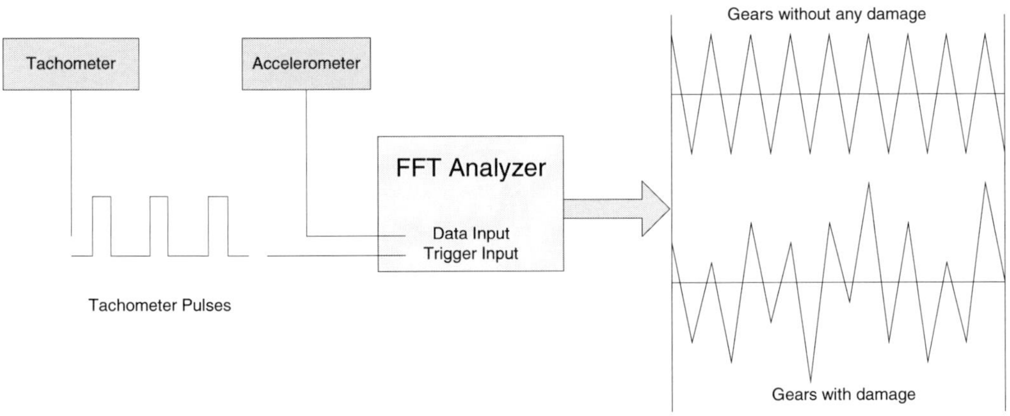

Figure 12. Tachometer and accelerometer signals measured by a signal analyzer with TSA.

For instance, a cracked gear tooth in a WT gearbox that meshes once per revolution produces a highly periodic vibration signature which can be very weak. TSA can be used to highlight the vibration signal features taking place over a given period, and can work with non-periodic signal components, filtering the noise of the time-domain signal. Once the average time signal is achieved, it is possible to compute the FFT (as below). A review of other TSA algorithms is presented in [29]. The result from tachometer and accelerometer signals measured by a signal analyzer with TSA is shown in Figure 12.

5.7. Fast-Fourier Transform (FFT)

The FFT algorithm is used for the conversion of a digital signal from the time domain into one in the frequency domain (see Figure 13). Particular frequency ranges correspond to particular states (e.g. fault-free, defective bearing), the ranges reflecting the rotational speed of the main shaft and the shape and size of the element concerned. All bearing elements generate vibration at specific frequencies (referred to as fault frequencies) and hence FFT finds most usage in gearbox monitoring [6].

FFT is also used for bearings wherein, if damage starts to develop, the shape of the vibration distribution deviates from the nominal Gauss-shaped curve [7]. "The advantage of frequency domain analysis over time-domain analysis is its ability to identify and isolate certain frequency components of interest" [30].

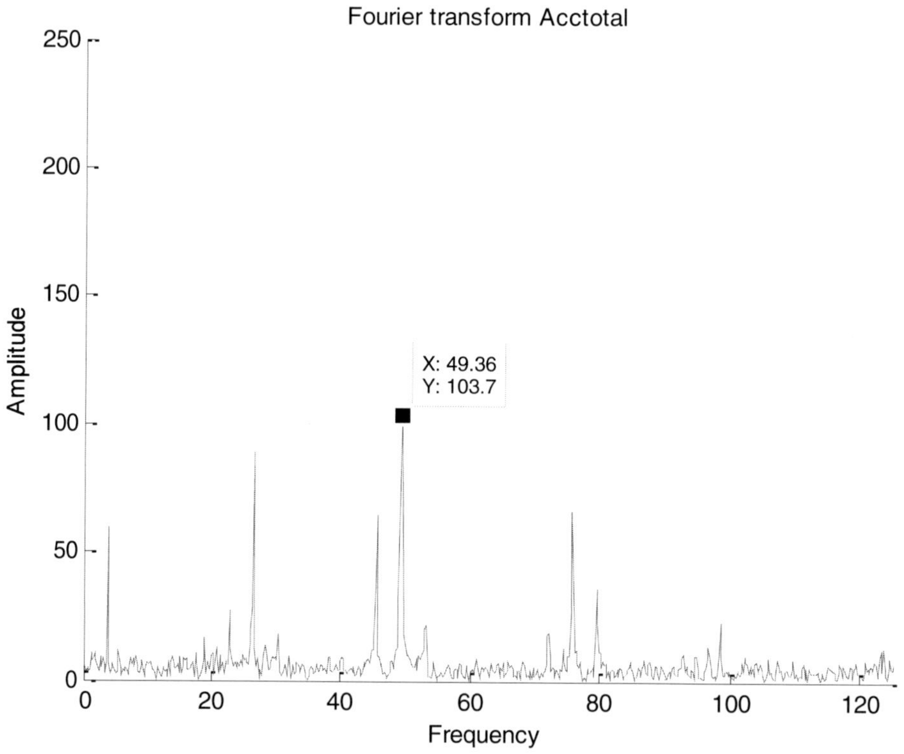

Figure 13. FFT of a vibration signal.

5.8. Amplitude Demodulation

This approach can extract very low-amplitude and low-frequency periodic signals that might be masked by other higher energy vibrations as in WT gearboxes [25]. While the raw spectrum can be useful for monitoring gear mesh frequencies, the envelope spectrum provides superior sensitivity to bearing defect frequencies in WT applications [22].

With its high sensitivity, demodulation has been proven to be good for evaluating defects that produce impacting, e.g. rolling contacts in bearings and tooth-to-tooth contacts in the gear meshes [20].

It also helps to reduce the complexity of the analysis, the main advantage being that it provides excellent visibility of bearing defects frequencies without the interference of gear mesh frequencies in the same spectrum. Despite the advantages for WT gearboxes, some barriers to implementation remain. The selection of transducer location is crucial to guarantee the results and, if the demodulation process is based on low-pass filtering, the original amplitude of the original defect signals is not preserved.

5.9. Order Analysis

FFT is useful for studying oscillations in constant speed wind energy converters (WEC) when applied to time series signals. However, this algorithm is not suitable for variable speed WEC where a different algorithm based on rotational angles is required, i.e. so-called order analysis, particularly well-suited for rotor imbalances and aerodynamic asymmetries [19]. Torsional oscillation of the nacelle generates a signal which is out of phase with any transverse oscillations, so they can be separated and analyzed individually [31]. Interpolation and production of the order spectrum then leads to the computation of the FFT. Such analysis may be used for monitoring the overall rotor condition including surface roughness, mass imbalance and aerodynamic asymmetry.

5.10. Wavelet Transforms

Wavelet transformation is a time-frequency technique similar to short time Fourier transform (STFT) but more appropriate for non-stationary signals. It provides a time-frequency 3D map of the signal being analyzed in [32] and involves decomposing it into a set of sub-signals or levels with different frequencies [33], where each of them has an energy rate associated from the original signal (see Figure 14). It is applied to WTs in order to monitor the vibration level caused by misalignment, bearing and other problems, and can be used as a general sign or indication of a faulty WT. Wavelet transforms have been applied to waveform data analysis in fault detection and diagnostics of various WT parts including gears, bearings, and other mechanical systems. Assessment of the effectiveness and reliability of wavelet transforms and comparison with other vibration analysis techniques has been completed by Dalpiaz and Rivola [34]. Baydar and Ball [35] successfully applied wavelet transforms to vibration signals and acoustic signals [21]. FFT (and indeed RMS) have also been used to estimate the maximum amplitude of the wavelet coefficients which helps in discriminating between normal and damaged components [37].

5.11. Hidden Markov Models

Hidden Markov models (HMM) have successfully been applied to the classification of patterns in trend analysis [38] and CM [36]. Atlas et al. developed a method to predict wear accurately [39], when applying a method for monitoring of milling processes with HMM."Ocak and Loparo [40] presented the application of HMM in bearing fault detection" [36] and the dynamic statistical characteristics that exist in the current observations of vibration signals in the machine have been modelled using HMM utilizing a Markov chain (see Figure 15).

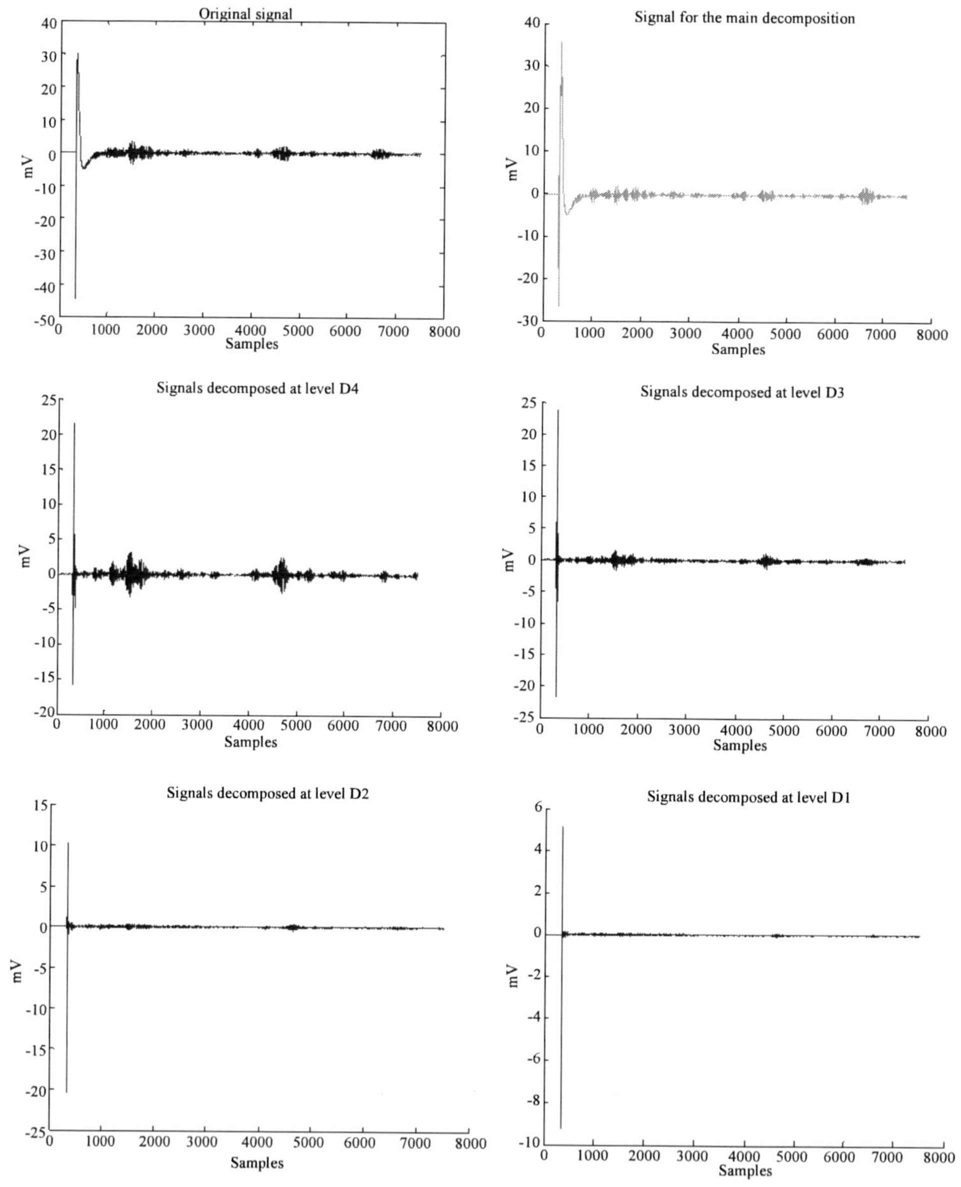

Figure 14. Wavelet decompositions.

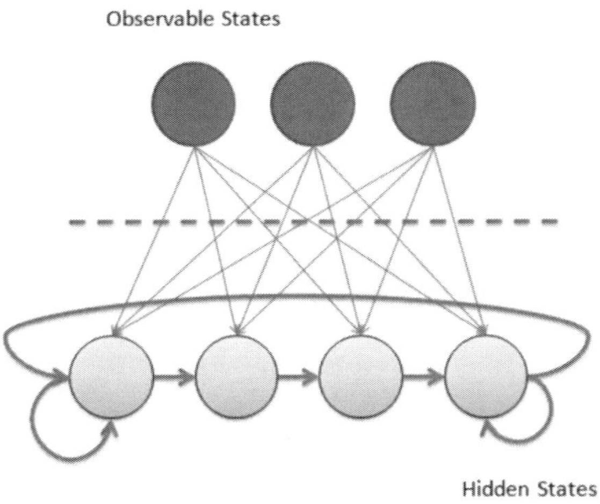

Figure 15. Hidden Markov Model [48].

5.12. Novel Techniques

Fault detection and diagnosis (FDD) is a sophisticated adaptation of CMS that incorporates 'intelligent' algorithms suitable for early detection of incipient faults providing an insight into the corresponding level of criticality [41]. FDD methods can be model-based or non-model based depending upon the way process knowledge is incorporated within the signal processing unit; a more specific classification is shown in Figure 16 [19].

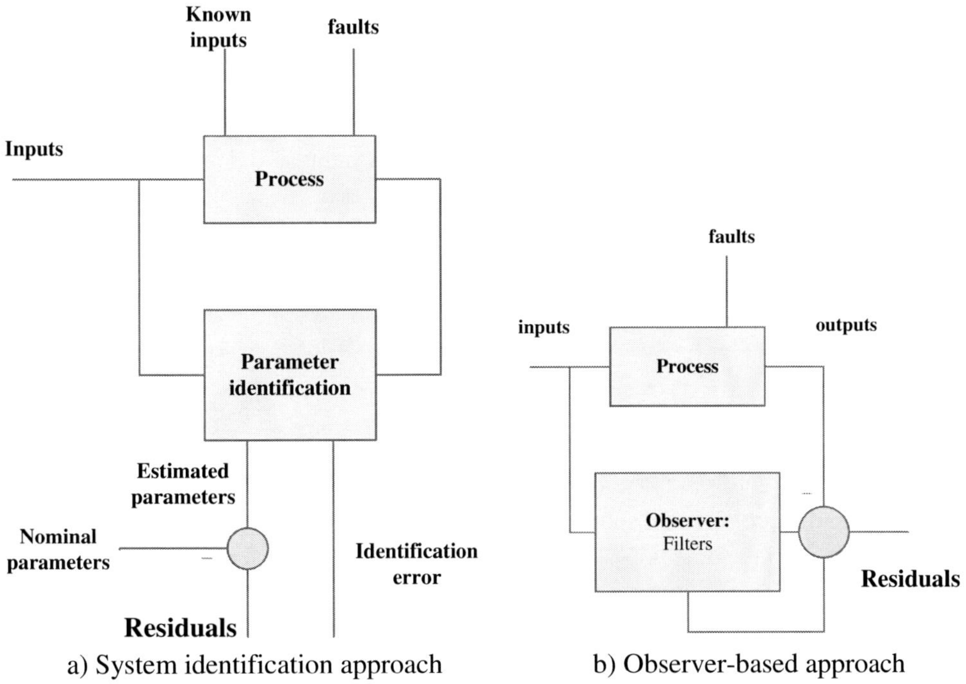

Figure 16. Continued on next page.

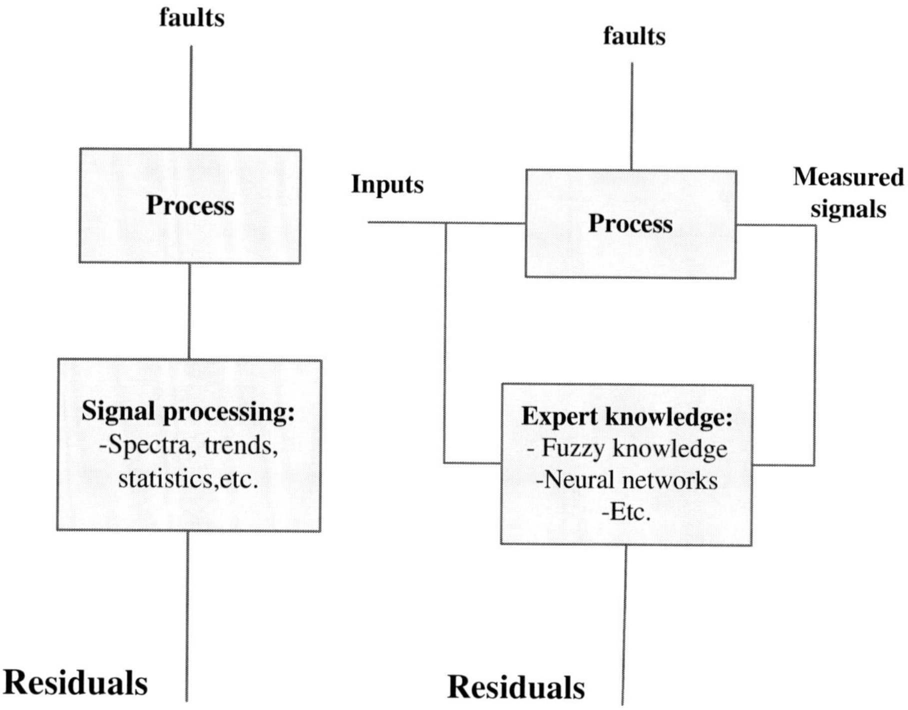

c) Signal analysis approach

d) Expert systems and artificial

Figure16. Different fault detection methods.

Artificial intelligence (AI) is essentially employed to reproduce human reasoning as accurately as possible, the reasoning process being based on the behaviour of the system, and written down in terms of rules. The dynamic nature of the environments in which WTs operate has led to the emergence of predictive maintenance plans that take qualitative account of the environment as well as its actual effect on the condition of the components.

Expert systems or so-called rule-based diagnostic systems "detect and identify incipient faults in accordance with the rules representing the relation of each possible fault with the actual monitored equipment condition" [42]. They have to fulfil two capabilities so as to be effective: the acquisition and integration of new knowledge, and also the explanation of their reasoning [43].

CONCLUSION

Wind Turbines (WT) are one of the fastest growing sources of power production in the world today and there is a constant need to reduce the costs of operating and maintaining them. Condition monitoring (CM) is a tool commonly employed for the early detection of faults/failures so as to minimise downtime and maximize productivity. This chapter provides a review of the state-of-the-art in the CM of WTs, describing the different sensory signals and signal processing methods.

The main obstacles facing the designers of condition monitoring systems for wind turbines clearly continue to be:

- selection of the number and type of sensors;
- selection of effective signal processing methods associated withthe selected sensors; and
- design of an effective fusion model (i.e., the combination of sensors and signal processing methods which give an improved performance);

References

[1] Gsänger S, Pitteloud J D. World wind energy report 2011. Word wind energy association (WWEA); 2012. Available at www.wwindea.org (06-2012).

[2] Milborrow D. 2006. *Operation and Maintenance Costs Compared and Revealed*, Wind Stats, Vol. 19(3), pp. 3.

[3] Vachon W. 2002. Long-Term O&M Costs of Wind Turbines Based on Failure Rates and Repair Costs," in *Proceedings WINDPOWER, American Wind Energy Association Annual Conference,* Portland, Oregon, June 2-5.

[4] Walford, C. (2006) *Wind Turbine Reliability: Understanding and Minimizing Wind Turbine Operation and Maintenance Costs.* Sandia Report, SAND2006-1100. Sandia National Laboratories, Albuquerque, New Mexico 87185 and Livermore, California 94550.

[5] Tavner, P. J., Spinato, F., Bussel, G J W van, and Koutoulakos, E. 2008. *Reliability of Different Wind Turbine Concepts with Relevance to Offshore Application.* European Wind Energy Conference, Brussels. April 2008.

[6] de Novaes Pires G., Alencar E. and Kraj A., *Remote Conditioning Monitoring System For A Hybrid Wind Diesel System-Application At Fernando de Naronha Island*, Brasil. http://www.ontario-sea.org (19-07-10).

[7] Igarashi T. and Yabe S. 1983. Studies on the Vibration and Sound of Defective Roller Bearings (First report: sound of ball bearing with one defect), Bull, *JSME* Vol. 26(220), pp. 1791-1798.

[8] Knezevic, J. 1993. *Reliability, Maintainability and Supportability Engineering: A Probabilistic Approach*, McGraw Hill.

[9] Ben-Daya M. S. and Duffuaa, A. R. 2009. *Handbook of Maintenance Management and Engineering.* Springer Verlag London Limited.

[10] Pedregal D. J., Garcia Marquez F. P., Roberts C. 2009. An Algorithmic Approach for Maintenance Management. *Annals of Operations Research*, Vol. 166, pp. 109-124.

[11] Campbell, J. D. and Jardine, A.K.S. 2001.*Maintenance Excellence: Optimizing Equipment Life-cycle Decisions,* Marcel Dekker, New York.

[12] Garcia Marquez F.P., Pedregal D.J. and Roberts C. 2010. Time Series Methods Applied to Failure Prediction and Detection. *Reliability Engineering & System Safety,* Vol. 95(6), pp. 698-703.

[13] Byon E. and Ding Y. 2010. "Season-Dependent Condition-Based Maintenance for a Wind Turbine Using a Partially Observed Markov Decision Process" *IEEE Transactions on Power Systems,* Vol. 25(4), pp. 1823-1834.

[14] McMillan D. and Ault G. W. 2008. *Condition Monitoring Benefit for Onshore Wind Turbines: Sensitivity to Operational Parameters.* IET Renewable Power Generation, Vol. 2(1), pp. 60–72.

[15] Moubray J. 1997. *Reliability-Centered Maintenance*, New York: Industrial Press.

[16] Andrawus, J. A., Watson, J., Kishk, M., Adam, A. 2009. The Selection of Suitable Maintenance Strategy for Wind Turbines, *Wind Engineering,* Vol. 30(6), pp. 471-486.

[17] Scarf PA. 2007. *A* Framework for Condition Monitoring and Condition Based Maintenance. *Quality Technology &. Quantitative Management,* Vol. 4(2), pp. 301–12.

[18] Nilsson, J. and Bertling, L. 2007. Maintenance Management of Wind Power Systems using Condition Monitoring Systems – Life Cycle Costs Analysis for Two Case Studies, *IEE Transactions on Energy Conversion* Vol. 22(1).

[19] Caselitz, P., Giebhardt, J. and Mevenkamp, M. 1994. On-Line Fault Detection and Prediction in Wind Energy Converters. *Proceedings of EWEC'94, Thessaloniki, Greece,* pp. 623–627.

[20] McFadden, P.D. 1996. Detecting Fatigue Cracks in Gears by Amplitude and Phase Demodulation of the Meshing Vibration, *Journal of Vibration, Acoustics, Stress and Reliability in Design,* Vol. 108, pp. 165-170.

[21] Miller A.J. 1999. *A New Wavelet Basis for the Decomposition of Gear Motion Error Signals and Its Application to Gearbox Diagnostics*, Master of Science Thesis, The Pennsylvania State University.

[22] Tandon N. and Nakra B. C. 1992. Comparison of Vibration and Acoustic Measurement Techniques for the Condition Monitoring of Rolling Element Bearings. *Tribology International,* Vol. 25(3), pp. 205-212.

[23] Llombart-Estopiña, A. 2008. Improving the Operation and Maintenance of Wind Farms: Determination of Wind Turbines Performance. *Proceedings of ICREPQ.*

[24] Giebhardt, J. 2007. *Condition Monitoring for Wind Turbines. State of the Art Report as Deliverable 7.1.1*, WP7 Condition monitoring. UPWIND.

[25] Cheng J., Yang Y., Yu D. 2010. The Envelope Order Spectrum Based on Generalized Demodulation Time–Frequency Analysis and Its Application to Gear Fault Diagnosis. *Mechanical Systems and Signal Processing*, Vol. 24, pp.508–521.

[26] Wismer N.J., 1994. *Gearbox analysis using cepstrum analysis and comb liftering,* Application Note Brüel & Kjaer, Denmark.

[27] Zheng G.T. and Wang W.J. 2001. A New Cepstral Analysis Procedure of Recovering Excitations for Transient Components of Vibration Signals and Applications to Rotating Machinery Condition Monitoring. 2001. *ASME Journal of Vibration and Acoustics,* Vol. 123, pp 222–229.

[28] Combet, L., Gelman L., 2007. An automated methodology for performing time synchronous averaging of a gearbox signal without speed sensor, *Mechanical Systems and Signal Processing,* Volume 21(6), pp. 2590-2606.

[29] Bechhoefer E., Kingsley M, 2009.*A Review of Time Synchronous Average Algorithms.* In: *Annual Conference of the Prognostics and Health Management Society*, San Diego, CA.

[30] Jardine, A.K.S., Lin, D. and, Banjevic, D. 2006. A review on machinery diagnostics and prognostics implementing condition-based maintenance, *Mechanical Systems and Signal Processing*, Vol. 20 (7), pp.1483-1510.

[31] Caselitz P., Giebhardt J. 2005. Rotor Condition Monitoring for Improved Operational Safety of Offshore Wind Energy Converters, Transactions of the ASME. *Journal of Solar Energy Engineering,* Vol. 127(2), pp. 253-261.

[32] Yang W.X., Tavner P.J. and Wilkinson M. 2008. *Condition Monitoring and Fault Diagnosis of a WT with a Synchronous Generator using Wavelet Transforms*. The 4th International Conference on Power Electronics, Machines and Drives 2008 (PEMD2008), York, UK, pp. 6-10.

[33] Strang G. and Nguyen T. 1997. *Wavelets and Filter Banks*, Wellesley-Cambridge Press.

[34] Dalpiaz G. and Rivola A. 1997. Condition Monitoring and Diagnostics in Automatic Machines: Comparison of Vibration Analysis Techniques, *Mechanical Systems and Signal Processing,* Vol.11, pp. 53–73.

[35] Baydar N. and Ball A. 2003. Detection of Gear Failures via Vibration And Acoustic Signals Using Wavelet Transform, *Mechanical Systems and Signal Processing*, Vol. 17, pp. 787–804.

[36] Miaoa, Q. and Makis, V. 2007. Condition Monitoring and Classification of Rotating Machinery Using Wavelets and Hidden Markov Models, M*echanical Systems and Signal Processing*, Vol. 21, pp. 840–855.

[37] Xiang, J. and Watson, S.J. 2008. *Practical Condition Monitoring Techniques for Offshore Wind Turbines*. In: EWEC 2008 Conference Proceedings, Brussels, Belgium, PO. 195, pp1-8.

[38] Kwon, K.C. and Kim, J.H. 1999. Accident Identification in Nuclear Power Plants Using Hidden Markov Models, *Engineering Applications of Artificial Intelligence*, Vol. 12, pp. 491–501.

[39] Atlas, L., Ostendorf, M., Bernard G. D. 2000. Hidden Markov Models for Monitoring Machining Tool-Wear, *IEEE. ICASSP* 2000, Vol. 6, pp. 3887–3890.

[40] Ocak H., Loparo K. A. 2001. A New Bearing Fault Detection and Diagnosis Scheme Based on Hidden Markov Modeling of Vibration Signals, *IEEE ICASSP* 2001, Vol. 5, pp. 3141–3144.

[41] Garcia Marquez F. P., Roberts C. and Tobias A. 2010. Railway Point Mechanisms: Condition Monitoring and Fault Detection. Proceedings of the Institution of Mechanical Engineers, Part F, *Journal of Rail and Rapid Transit.* Professional Engineering Publishing, Vol. 224(1), pp. 35-44.

[42] Yam, R. C., M., Tse, P. W., Li, L., Tu, P. 2001. Intelligent Predictive Decision Support System for Condition-Based Maintenance, *The International Journal of Advanced Manufacturing Technology.* Vol. 17. (5), pp. 383-391.

[43] Garcia M.C., Sanz-Bobi M. A. and del Pico J. 2006. *SIMAP:* Intelligent System for Predictive Maintenance Application to the Health Condition Monitoring of a Windturbine Gearbox, *Computers in Industry*, Vol. 57(6), pp. 552–568.

[44] Wenbin Dong, Yihan Xing, Torgeir Moan, Zhen Gao, 2013. Time domain-based gear contact fatigue analysis of a wind turbine drive train under dynamic conditions, *International Journal of Fatigue*. Vol. 48, pp. 133-146.

[45] Dimitris Bourlis. A Complete Control Scheme for Variable Speed Stall Regulated Wind Turbines. *Fundamental and Advanced Topics in Wind Power*, book edited by Rupp Chapter 14. Carriveau, Intech. ISBN 978-953-307-508-2, DOI: 10.5772/17569. Published: July 5, 2011

[46] Vibration Analysis and Signal Processing in Lab VIEW. www.ni.com (April-2013).

[47] UPWIND Final Publication. March 2011. Up Wind - European project funded under the EU's Sixth Framework Programme (FP6). http://www.upwind.eu/ (June 2012).

[48] Crouse, M.S.; Nowak, R.D.; Baraniuk, R.G. 1998. Wavelet-based statistical signal processing using hidden Markov models. *Signal Processing, IEEE Transactions on,* Vol. 46(4), pp.886-902. DOI: 10.1109/78.668544.

In: Fault Detection
Editors: F. P. G. Márquez and M. Papaelias
ISBN: 978-1-62808-999-8

Chapter 2

AN INTEGRATED STRATEGY FOR EFFICIENT NON-DESTRUCTIVE EVALUATION OF RAILS

M. Ph. Papaelias*
Centre of Rail Research and Education,
The University of Birmingham,
Birmingham, UK

ABSTRACT

To improve the efficiency of maintenance activities in a rail network and ensure an acceptable level of safety for rolling stock operations, rails are inspected at regular intervals for structural defects, profile irregularities, excessive wear, missing fastenings, failed sleepers, and abnormal variations in rail gauge. Rails are inspected either manually or with hi-rail vehicles and test trains using traditional non-destructive evaluation techniques such as vision (including automated systems), magnetic particle inspection, magnetic flux leakage, ultrasonic and eddy current testing. The accuracy of the inspection process is of critical importance in minimising the risk of catastrophic failures, improving maintenance scheduling and optimising rolling stock operations. The inspection procedures currently employed by rail infrastructure managers are not adequately efficient as they depend on the mobilisation of a significant amount of resources, both in terms of special test equipment and expert personnel. Moreover, the inspection process needs to be carried out in segments either at night or when the line is closed to avoid disruption of normal rolling stock operations. Rail infrastructure managers due to various reasons have not benefited sufficiently from recent breakthroughs and innovation in inspection technologies that other industrial sectors are already taking advantage of. This study analyses the efficiency of current inspection strategies and discusses a plausible way of implementing substantial improvements without the requirement for a fundamental shift in the existing understanding of the overall inspection process and the ways it is related to preventive and reactive maintenance procedures.

Keywords: Rail; inspection; strategy; maintenance; defects; RCF; non-destructive evaluation

---

* Contact E-mail and Telephone: m.papaelias@bham.ac.uk - ++44 (0) 121 414 2626.

1. Introduction

Technical improvements in the production process of rails combined with strict quality control procedures based on automated vision, eddy current (EC) and ultrasonic testing (UT) systems have enabled a significant reduction in the occurrence of manufacturing-related defects [1-2]. Quality control inspection is arbitrarily used to detect hot marks, protrusions, scratches, rolled in scale, seams, cold marks, microstructural damage, internal cracking and large inclusions before the rails are installed on the network. Thus, the vast majority of defects found on the rail network nowadays have initiated and grown while in-service.

In service damage can develop in the head, web or foot of the rail. In modern rail networks the majority of the defects tend to occur more often on the rail head, with Rolling Contact Fatigue (RCF) being a particularly common surface defect. Rail head defects can be distinguished to those having internal origin, such as progressive transverse cracking or kidney-shaped fatigue cracks, horizontal cracking with or without transverse cracking of the rail head, horizontal cracking beneath the gauge corner, and longitudinal-vertical cracking, and those having surface origin, such as RCF damage (including gauge corner cracking, head checks, squats, shelling, and corrugation), wheelburns, and indentures. Rail web and rail foot defects include longitudinal and vertical cracking, cracking occurring at fishbolt holes or other holes found in the web (star-cracking), transverse fatigue cracking, and rail foot corrosion. Profile irregularities and low levels of conicity that develop with time due to wear of the rail head can also be considered as rail defects [3-5].

Internal defects in rail welds such as shrinkage cavities, microporosities, inclusions and a coarse dendrite microstructure that may develop during the solidification process adversely influence the structural integrity and fatigue performance of a weld by acting as crack initiation points [6]. Occasionally weld breaks can occur without warning during cold weather due to rail contraction. As a result, rail welds can fail very rapidly with no indication of fatigue cracking. On the other hand, during summer welded rails can buckle if the stress levels exceed a certain level.

2. State-of-the-Art Rail Inspection Technology

Rail tracks are inspected either visually by appropriately trained personnel walking along the tracks and noting down any defects or missing components observed, or by using traditional inspection techniques such as magnetic particle inspection (MPI), UT, MFL, EC testing and automated vision. Occasionally, radiography has been employed for the assessment of welds but it has never been regarded as a standard rail inspection procedure mainly due to the complex health and safety issues associated with it [7-8].

Dry or wet MPI is applied in rail sections which are already suspected to contain surface or near-surface defects. Magnetic particles are applied on the surface of the rail head and then a permanent magnet or electromagnet is used to magnetise the rail section under inspection. If a surface crack is present then the magnetic flux leaks in that area causing the fine magnetic particles to be attracted in the region of the crack creating a visible indication. If the crack is not surface-breaking but it is still very near to the surface (1 or 2mm deep) then some magnetic flux will still leak in that area creating a fuzzy pattern which may be difficult to identify as an indication of a near-surface defect being present sometimes. Deeper cracks are not detectable with MPI and the technique does not give any information regarding the depth of the cracks detected.

Rail UT is carried out manually using walking sticks and portable instruments or at speed with hi-rail vehicles and special test trains. Manual UT inspection is a slow process which requires experienced personnel to carry out the tests and interpret the signals while the line is closed to normal rolling stock traffic. The results produced are fairly accurate, providing an approximation regarding the size of the defects detected. Therefore the rail industry relies heavily on this type of inspection despite the inefficient logistics associated with this process. The main problem in manual UT inspection of rails is that shallow (near-surface defects) that develop in the zone 2.5-6mm from the rail head surface can be easily missed. In addition, smaller surface defects can block the interrogating ultrasonic beam from reaching deeper and more significant defects leading to an underestimation of the actual structural condition of a rail section [7-8].

High-speed conventional ultrasonic systems have limited ability in detecting RCF type defects. This is due to the compromise that needs to be made in adjusting the threshold level and time window of such systems in order to avoid the generation of too many false alarms without missing larger defects. Increasing inspection speed also leads to lower resolution, i.e. at a speed of 100km/h state-of-the-art systems have a resolution of around 6mm, meaning that moderately sized defects can be missed. The signal-to-noise ratio decreases with speed and therefore certain defects can be underestimated or even ignored completely if the signal associated with them does not exceed the threshold level set. For that reason although certain UT systems are capable of operating at speeds up to 100km/h, inspection is usually carried out at lower speeds. Depending on the degree of confidence required during the inspection, speeds can be as low as 15km/h. Usually hi-rail vehicles operate at 15km/h while their maximum possible speed does not exceed 30km/h. Test trains can operate at speeds up to 100km/h with inspection operations usually being carried out at a speed range between 40-80km/h [4, 7-8]. The maximum inspection speed possible is influenced by the design of the deployment mechanism of the UT transducers, i.e. sled or wheel carriers. Sled carriers allow in general higher speeds and better resolution in comparison to wheel probes as the ultrasonic beam needs to travel a smaller distance from the transducer to the rail and back. However, the control of sled carriers is more complicated and can be influenced by corrugation. As an alternative to sleds, rubber tractors have been considered and seem to offer several advantages during inspection at speed in terms of probe control particularly in lines with several joints as well as the overall wear of the deployment mechanism.

UT systems deployed by hi-rail vehicles or test trains can be complemented by MFL or more recently eddy current sensors to increase the probability of detection of surface and near-surface transverse defects, such as RCF cracking [7-12]. MFL sensors are adversely affected by the speed of inspection which usually should not exceed 35km/h. This is due to the fact that MFL requires the adequate magnetisation of the rail head so in the presence of a discontinuity sufficient flux can leak in order to be detected by the sensors of the system. With increasing speed the flux leakage in the presence of a given defect will be lower and thus the overall detection capability of the system will decrease. Although, MFL sensors increase the reliability of the inspection process itself, they also reduce the maximum inspection speed possible when integrated with high-speed UT systems. Moreover, MFL sensors have poor sensitivity in detecting longitudinal defects. This is due to the fact that magnetic flux lines run parallel to these defects without sufficient flux leaking to be detectable by the sensor. Eddy current sensors offer increased sensitivity in the detection of surface-breaking defects at high inspection speed but cannot detect near-surface defects that are more than couple of millimetres away from the surface of the rail head. They are also strongly affected by lift-off variations which can result in missed defects or false alarms. Satisfactory lift-off control can be difficult depending on the quality of the surface of the track (e.g. presence of corrugation and profile variations). Eddy current probes perform better than MFL sensors at higher inspection speed and are more sensitive to all types of surface defects but severe grinding marks can make the detection of smaller RCF cracks difficult. Inspection speeds up to 100km/h have been reported as possible with state-of-the-art eddy current systems used to supplement UT inspection [9-12]. Manual eddy current systems are used to inspect the rail head for RCF defects but this practice is not as common as manual UT inspection.

Automated vision systems are used as an integral part of rail track inspection for the measurement of the rail head profile, percentage of wear, rail gauge, corrugation, failed sleepers, missing components and depending on the speed of inspection for RCF detection [13]. Certain vision systems are portable, whilst others are deployed at high speed by test trains.

On some occasions high speed vision systems can be installed on a temporary basis on passenger trains to complete a single or several inspections of a certain route. Depending on the mode of inspection speeds up to 320km/h are possible. However, if RCF defect detection is required then the speed of inspection cannot exceed 30km/h in order to maintain adequate image resolution.

Heavy grinding marks can make the visual detection of RCF extremely difficult or even impossible regardless of the speed. Recent developments have resulted in automated vision systems being integrated with UT sensors to allow inspection of internal defects at the same time with the visual assessment of the rails [14-15].

Rail welds due to their large grain microstructure cannot be reliably inspected manually or at speed with conventional UT systems [16]. The high frequency interrogating ultrasonic beam experiences significant attenuation due to grain boundary scattering resulting in poor penetration and very low signal-to-noise ratios. For that reason defects such as centreline shrinkage in the foot of aluminothermic welds are not detectable with UT and alternative means of inspection are required. Rail welds upon completion of the welding process can be inspected radiographically to detect such defects [17]. However, the use of radiography is not commonplace in the rail industry and it is almost never applied for subsequent evaluations of a weld during its lifetime.

Residual stresses in rails can be evaluated to give an early indication of the onset of cracking and thus preventive remedial action such as rail grinding can be undertaken. A Magnetic Anisotropy and Permeability System (MAPS^{TM}) has been developed by MAPS Technology (formerly a division of ESR Technology and part of AEA) in the UK which can be used to measure the residual stresses in the rail crown as well as evaluate the Stress Free Temperature (SFT) of a rail section [18]. MAPS^{TM} is a portable system which can be deployed in the field manually to perform an on-the-spot inspection quickly without the need of closing down the line of interest.

3. Rail Inspection Technology Undergoing Pilot Trials or under Development

Alternative rail inspection techniques, in addition to those mentioned already in the previous section, have undergone pilot trials by the industry to evaluate their capabilities in the field or are currently under development. Such techniques include ultrasonic phased arrays, Alternating Current Field Measurement (ACFM) sensors, guided wave systems, laser ultrasonics, Electromagnetic Acoustic Transducers (EMATs) and acoustic emission [7, 19-46]. The aforementioned techniques are currently under different stages of development with some having reached the stage of commercialisation already.

Ultrasonic phased arrays have been used in pilot studies for the assessment of rail welds as an alternative means to radiography [20-24]. Ultrasonic phased arrays have the ability of focusing the interrogating beam apart from transmitting more energy inside the weld. However, they suffer from the same attenuation mechanism that influences conventional UT transducers and therefore the quality of the results of such inspections is still questionable. Phased arrays have also been used for the accurate quantification of defects including RCF cracking, with promising results [23]. Commercial phased array systems are readily available and are widely used by other industrial sectors. Therefore it is not the characteristics of the systems themselves but the inspection procedure and relevant standards which require further development for widespread application of ultrasonic phased arrays within the rail industry.

Portable Alternating Current Field Measurement (ACFM) inspection systems have been used for the evaluation of RCF defects during trials in the UK and Portugal. The operation of ACFM sensors is based on the induction of a uniform alternating electric current field near the surface of the rail head. In the presence of a defect such as rolling contact fatigue cracks, the current field will be disturbed travelling around the edges and under the defect giving rise to variations in the electromagnetic field produced by the induced field as shown in figure 1.

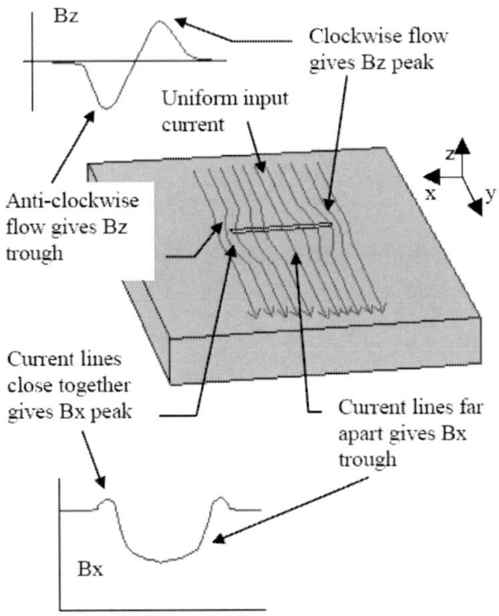

Figure 1. Definition of field directions and co-ordinate system used in ACFM [Taken from Ref. 26].

These variations can be quantified and then related to the surface length and depth of the defect detected. The ACFM walking sticks incorporate an array of probes which is shaped to conform to the shape of the head of the rail. This allows the application of the ACFM system in both new and worn rails. The inspection across the rail head is carried out by sequentially scanning across the group of sensors enabling the uninterrupted inspection of the rail. A customised software package incorporating the appropriate defect sizing algorithms has been developed that enables the automated sizing of the RCF cracks that are detected with the walking stick. The ACFM array can detect and size gauge corner cracks and head checks smaller than 2mm in depth at inspection speeds 2-3km/h [25-27]. A high speed ACFM array system is under construction by the consortium of the INTERAIL FP7 project for inspection of the rail head surface at speeds up to 80km/h [15, 29-30]. The system is intended to undergo field trials in Portugal in 2012. The photographs in Figures 2 and 3 show tests carried out in the laboratory and in the field using ACFM sensors. A portable Alternating Current Stress Measurement (ACSM) system has also been developed as a spin-off from ACFM to evaluate residual stresses in the rail crown. The ACSM system is currently undergoing pilot trials in the UK by TSC. The ACSM technique involves inducing currents into the metal surface and taking measurement of the magnetic fields produced above the surface being inspected. Small changes in the strength and direction of the magnetic field can be related to both the type (i.e. tensile or compressive) and magnitude of the residual stresses present locally [26].

Figure 2. Slow speed ACFM and EMAT pilot trials on a Network Rail's line carried out by researchers from the Universities of Birmingham and Warwick as part of the demonstration activities within the INNOTRACK FP6 project.

Guided wave systems have undergone pilot trials in the UK and North America for the evaluation of welded rail sections several metres long, including level crossings [30-35]. Guided wave systems use much lower frequencies (60-280kHz) in comparison to conventional UT which involves frequencies in the MHz range. Thus the ultrasonic waves produced can travel several metres along the rail and be less affected by attenuation due to the microstructure of the welds. Such systems are sensitive to transverse defects and need to be big enough to reflect back a sufficient proportion of the transmitted energy otherwise they cannot be detected. For example, a transverse rail head defect in order to be detectable with guided waves needs to have a size equal to at least 5-10% the overall cross-section of the rail head.

Figure 3. Laboratory based high speed ACFM test configuration at Birmingham University. The interrogating ACFM probe can be seen at the right hand side of the photograph. Tests are carried on a special set of rails with artificially induced surface defects using spark erosion.

A hi-rail vehicle-based non-contact laser ultrasonics system has been tested by Tecnogamma S.p.A. (now part of the MER MEC Group of companies in Italy) in collaboration with the Transportation Technology Centre Inc. (TTCI) in North America [36]. In this technique lasers are used in the place of piezoelectric transducers to emit pulses which strike the surface of the rail from a distance generating ultrasonic waves either due to thermal or ablation effects. When the reflected ultrasonic waves reach back to the surface of the sample, the resulting surface displacement is measured with a laser ultrasonic receiver based on an adaptive interferometer or an air coupled ultrasonic transducer. The developed system has not been made available commercially but its further development is under consideration once further funding becomes available.

The applicability of EMATs for rail inspection was tested using a hi-rail vehicle for deployment purposes by Tektrend (now part of NDT Olympus) between 2000 and 2001 [37-38]. Since then a high speed EMAT guided wave system based on a pitch-catch concept (shown in figure 4) has been under development by UK researchers from the Universities of Warwick and Birmingham, and the I-RAIL consortium which is funded by the European Commission under Framework 7 [39-46]. The prototype I-RAIL EMAT system is expected to undergo demonstration trials in Portugal before the end of 2012. The system will be deployed for field trials at speeds up to 100km/h using a special test train [46].

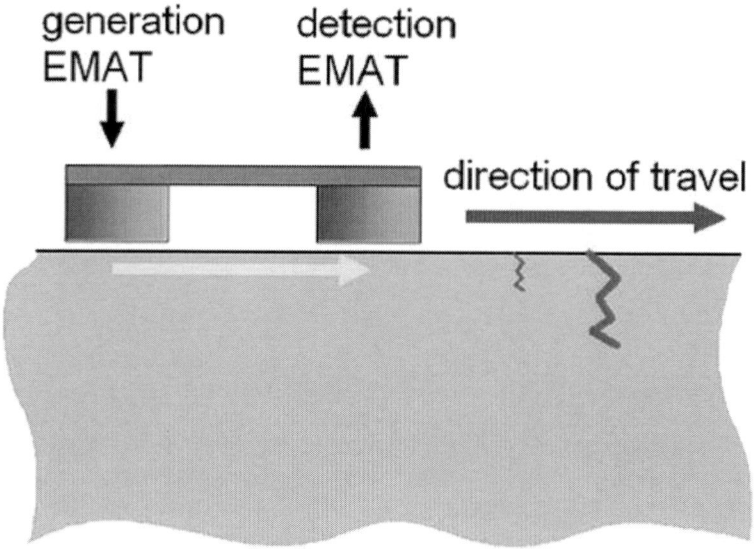

Figure 4. Schematic showing the pitch-catch EMAT concept. Detection signal is dominated by the deepest defect [Taken from reference 45].

Acoustic emission (AE) techniques have been tested in the laboratory and in the field regarding their suitability for the detection of surface rail defects at speed by researchers at AEA Technology Rail (now DeltaRail) and Cranfield University [46-47]. AE has also been under investigation by researchers in Birmingham University and elsewhere to evaluate the applicability of the technique for structural health condition monitoring of rail sections where structural defects have already been identified by other means and where their further growth requires attention.

A portable high-frequency vibration analysis system has been under development by APT Rail in Belgium in order to identify rail foot corrosion in grooved rails. The system has been undergoing test trials at STIB's network in Brussels as part of the INTERAIL FP7 project [15]. The system uses a small impact mechanism which is used to excite the rail section under evaluation. A dynamic sensor records the response and then compares it with the response obtained for a calibration rail section which is known to be in good condition. If rail foot corrosion has occurred then the response recorded will be different from that obtained for the calibration sample indicating the presence of the defect. High-frequency vibration analysis is also under investigation for temporary or permanent condition monitoring purposes of rail sections that are of interest for continuous evaluation.

4. State-of-the-Art Rail Inspection Strategy

Rail infrastructure managers follow different rail track inspection strategies which are dependent on the technical and human resources available to them as well as the operational 'sweating' of the rail network that they are responsible. The complexity of the rail inspection strategy used can vary significantly between different infrastructure managers depending on the actual requirements of the network.

Urban rail networks, i.e. rail lines with heavy usage and low rolling stock speeds, are mostly inspected using manual techniques based on vision and UT although in certain larger networks special test trains are also employed (e.g. Seoul Metro, etc.). Metro lines with heavy use are inspected visually every two weeks whilst UT takes place every 6 or 12 months depending on the amount of traffic. Line sections with low usage are also visually inspected every two weeks but UT inspection is not carried out. Tram networks on the other hand are only visually inspected every few weeks [15].

Small national rail networks with low to moderate levels of rolling stock traffic and average train speeds tend to deploy small teams of inspection engineers who are responsible for inspecting the rail tracks visually as well as with manual UT and portable optical equipment.

Depending on the rolling stock traffic and speed certain lines are inspected twice or more times in a year whilst lines with lower traffic are only inspected once every one or two years. The frequency of inspections can be altered if the number of severe defects found on the network is higher than that deemed as an acceptable level of risk by the infrastructure manager [49]. In addition, the development of risk assessment tools allows the systematic evaluation of hypothetical scenarios and the potential benefits of alternative rail testing strategies [50].

The use of test trains deploying automated vision systems to assess the profile quality, rail head wear and rail gauge is also common and supplements the data acquired manually by the inspection teams working on the network throughout the year. On networks with low density rolling stock traffic line possession is not absolutely necessary in order to carry out manual inspection as there is ample time between train passages. Also test trains can be deployed without interfering too much with normal rolling stock operations.

Medium and large sized national rail networks regardless of the rolling stock traffic density are in general inspected using manual as well as automated inspection systems deployed by hi-rail vehicles and test trains. Infrastructure managers responsible for such networks may adjust the inspection frequency with respect to a number of factors including the rate of detected defects, the target defect density per track km, traffic tonnage accumulation between two inspections and probability of detection (POD). Weibull distribution is used to estimate the rate at which internal rail defects will develop in order to assess the operational risk [49].

Rail networks with heavy rolling stock traffic incorporating high-speed lines employ more complex strategies with inspections being carried out at intervals which may vary from every 1 week up to 24 months depending on the operational criticality of the line and the section in question [15]. High-speed lines are inspected visually as often as once a month while critical areas such as points may be tested up to once a week. UT inspection is usually carried out manually or using test trains between every two to six months depending on the accumulating tonnage over a certain period. Less heavily used lines are inspected every six, twelve or twenty-four months both visually and ultrasonically. Ultrasonic inspection is usually supplemented by MFL or EC sensors to improve the detectability of surface and near-surface defects.

Infrastructure managers may have their own in-house inspection teams and equipment which they manage directly or they may subcontract inspection to external providers [15]. This usually depends on the human resources with adequate expertise in inspection technology available within an individual infrastructure manager. Certain external inspection providers may only sell their services whilst others may sell both inspection services and equipment in whichever format requested by the infrastructure managers. In certain cases however, where the inspection equipment provider is a relatively small company with limited human resources available, it is common practice to sell the equipment and then undertake the adequate training of personnel from the infrastructure managing company. Small equipment providers always tend to provide expert consultancy services particularly in designing an optimum inspection procedure and in data assessment.

In general, inspection techniques currently used by the rail industry are based on mature technologies that despite their well-known deficiencies have until now enabled an acceptable level of operational reliability. Nonetheless, their cost-effectiveness has come under fire in recent years by the infrastructure managers whilst the ever increasing traffic density, rolling stock speeds and axle loads means that maintaining the existing level of reliability becomes increasingly harder. Moreover, the cost of inspection both direct and indirect is growing and inspection planning is becoming a very complex task.

Rail inspection is of critical importance for the safety of operations in any rail network since it is the last line of defence against possible accidents. However, with increasing trends in traffic density, speed and axle loads, inspection under its current form and given the technological stagnation over a relatively long time period in the past may be gradually evolving to a necessary evil which impedes growth and decreases cost efficiency for certain rail networks while only a marginally acceptable level of safety is achieved.

The rail industry has globally recognised the need for measurable and valid innovation in rail track inspection procedures and technologies in order to improve existing safety standards and optimise cost efficiency. The development of reliable inspection procedures based on mature and extensively validated technologies is of particular interest to the rail industry. Nonetheless, rail infrastructure managers are unanimously open in adopting alternative new inspection technologies as long as their capabilities can be proven in the field and realistically offer distinctly better results than existing techniques.

Each inspection technique has its own advantages but also technical problems and limitations and for that reason it is necessary to have a clear understanding of what it can and cannot do. For new technologies which are entering or seeking to enter the active rail inspection portfolio of rail infrastructure managers it is of paramount importance that appropriate standards are developed. However, it is also equally important that the staff members asked to use them have developed sufficient expertise on them before they take them to the field.

A fundamental shift to inspection techniques where completely new standards are required and on which existing staff members have no basic expertise is highly undesirable for the rail industry.

Current international practice has recently focused in the combination of certain fundamental inspection techniques (i.e. UT, EC, MFL and automated vision) with preventive maintenance procedures, such as rail head grinding, in order to optimise the trade-off between maintenance costs and structural reliability [12]. Still, there is room for substantial improvements that could be achieved in the short to medium term based on existing inspection technology and in the long term with the adaptation of novel technologies.

5. An Integrated Approach to Rail Inspection

The rail industry has remained until now largely attached to conventional inspection techniques and procedures that have been proven to contribute in achieving at least acceptable levels of reliability despite their shortcomings. The procedures commonly used, with the exception of automated vision and UT, provide qualitative information regarding the actual condition of the rail track.

Several infrastructure managers carry out visual inspections separately from UT or other types of inspection. This is mainly due to the different schedules that are used for visual inspection and also the higher inspection speeds that can be achieved from automated vision systems. Nonetheless, visual inspection only provides information for very specific visible defects and features and therefore it is not of any help in the internal evaluation of rails. Some infrastructure companies have integrated Ground Penetrating Radar (GPR) systems with automated vision to enhance the amount of information gained in a single inspection run with respect to the condition of the substructure of the rail track.

In certain cases UT hi-rail vehicles and test trains have been upgraded by incorporating apart from the UT probes, MFL or EC sensors to increase the POD of surface and near-surface breaking defects of the system. More recently the integration of automated vision systems together with UT, EC, MFL or ACFM sensors has been put forward based on a modular operation concept. The integration of inspection systems with grinding test trains has also been developed in order to carry out the inspection simultaneously with the rail head grinding whilst evaluating the quality of the grinding itself [12].

Integration of conventional inspection techniques between them as well as with other maintenance processes offers significant advantages as it minimises the need for multiple possessions and requires less human effort, hence more limited availability of expert personnel. Moreover integration contributes to the improvement of cost-efficiency of the inspection process.

The integration of conventional techniques is an intermediate technological step forward in comparison to current common practice with short to medium term positive implications in terms of reliability and cost efficiency. It should be stressed here that the type of inspection systems and techniques being combined and the nature of integration itself have significant impact on the efficiency of the final integrated system. Automated vision systems and conventional UT probes are irreplaceable at the moment as there are no realistic alternatives that can be employed by the rail industry. Both techniques are continuously undergoing further development in order to provide improved inspection tools to rail infrastructure managers.

In the case of high-speed UT inspection, better results can be obtained at higher speeds if a sled carrier or tractor-based deployment mechanism is used rather than wheel carriers since the interrogating beam needs less time to travel from the transducer down the rail and back. However, control of the probe is more difficult at least in the case of sleds carriers. The use of novel data acquisition boards with sampling rates up to 100MHz allow better inspection resolution at higher speeds than those possible in the past.

The addition of EC sensors offers distinct advantages to an integrated inspection system since surface and near-surface defects are more likely to be detected by EC testing rather than UT. However, EC sensors require excellent lift-off control in order to function properly adding to the complexity of the deployment mechanism and only provide qualitative information regarding the size of detected defects. ACFM probes are an alternative to EC sensors which also requires good lift-off control but it is not as sensitive to slight lift-off variation as EC sensors. Moreover, ACFM probes can be used to get a good approximation regarding the size of the defects detected. One of the main problems in ACFM inspection is when RCF cracks are closely spaced it may be difficult to resolve them, hence making reliable quantification very difficult. However, better resolution may be possible with new ACFM arrays currently being tested within the INTERAIL project.

EMAT guided wave systems based on a pitch-catch concept such as the one developed within the I-RAIL project can be used for the evaluation of RCF defects in the place of EC or ACFM sensors at high-speeds. Since the ultrasonic wave generated by the excitation EMAT will travel at a speed of 3000m/s along the rail towards the sensing EMATs, the system is practically not concerned about the speed of the test train since it will be always significantly lower (a few tens of metres per second). Although, EMAT guided wave systems require excellent lift-off control like EC sensors they can be very accurate in quantifying the depth of RCF defects using amplitude and frequency analysis of the measured data [40-47]. EMAT tests at high sampling rate will generate a significant amount of information which is difficult to fully analyse in real time. Therefore, post processing is required to obtain accurate results from the raw data. Another drawback of the technique is that it has a dead zone of 2mm deep from the surface of the rail head and thus any RCF shallower than 2mm remains undetected. On the other hand RCF cracks that are less than 2mm deep are not considered to be severe enough to be of concern to train safety based on existing standards.

Each of the above techniques has its own advantages and disadvantages and is designed to detect certain types of rail track damage and wear. Integration of automated vision with UT and EC or ACFM or EMAT sensors permits the full assessment of the structural condition of the rail track in a single run, resulting in a noteworthy reduction of inspection times while POD for certain defects is improved. Furthermore, data convolution during post-processing can increase the amount of useful information gained from the individual systems minimising the likelihood of false defect indications and increasing qualitative and quantitative accuracy for certain defects.

It is important to remember that each inspection technique has each own physical limitations with respect to the maximum speed it can be performed at. For example in UT, ultrasonic waves need to be emitted from the transducer, down the rail and back. The ultrasonic waves travel at a finite speed in the rail and hence above a certain inspection velocity the reflected waves will not have sufficient time to return back to the transducer. Thus, this is the absolute limit of conventional UT techniques unless far more complex probe designs are employed. EC, ACFM and EMAT techniques can be carried out at higher speeds than UT but do not offer any feedback regarding deeper defects in the rail. Similarly automated vision tests can be carried out at higher speeds but can only be used for the inspection of certain visible features only as discussed earlier. Thus test trains incorporating several inspection systems should be able to carry out modular operations when and if required since the maximum inspection velocity of an integrated system will always be determined by the UT module's limits which will remain an irreplaceable component of rail inspection.

The use of alternative techniques for manual inspection is more straightforward to implement and can provide increased accuracy. Ultrasonic phased arrays offer distinct advantages over existing conventional UT equipment used in manual inspection. Although ultrasonic phased arrays are more expensive they have better POD for most types of defects and can be used for the accurate quantification of the defects detected. They also perform better in rail weld inspection although their performance is not as good as desired due to attenuation caused by the large grain microstructure. Portable EC equipment, ACFM arrays and pitch-catch EMATs can provide better quantification results for RCF defects detected in comparison to MPI while the manual inspection process is speeded up.

The advent of digital X-ray radiography offers new possibilities and advantages in rail weld evaluation over traditional film radiography based on gamma rays. Rail welds have been traditionally a weak link in the rail network which has always been very difficult to evaluate accurately. Portable digital X-ray equipment can give very accurate information regarding the actual state of suspect rail welds.

In addition, high-frequency vibration analysis equipment under development within the INTERAIL project can assist in the detection of damage in grooved rails, switches and crossings. Guided wave systems can also be used for the evaluation of level crossings in one off measurement.

It is important that the performance of high-speed rail inspection systems is complemented through the application of advanced defect verification and evaluation techniques based on ACFM, ultrasonic phased arrays, high-frequency vibration analysis, pitch-catch EMAT testing, guided waves and digital radiography as they offer distinct benefits over existing state-of-the-art manual inspection procedures. The simplified schematic in figure 5 showing the overall INTERAIL inspection platform concept is a useful example of what improvements are possible if an integrated inspection strategy is adopted by infrastructure manager. In the INTERAIL approach each of the techniques complements the other and therefore increases the versatility of the overall inspection process.

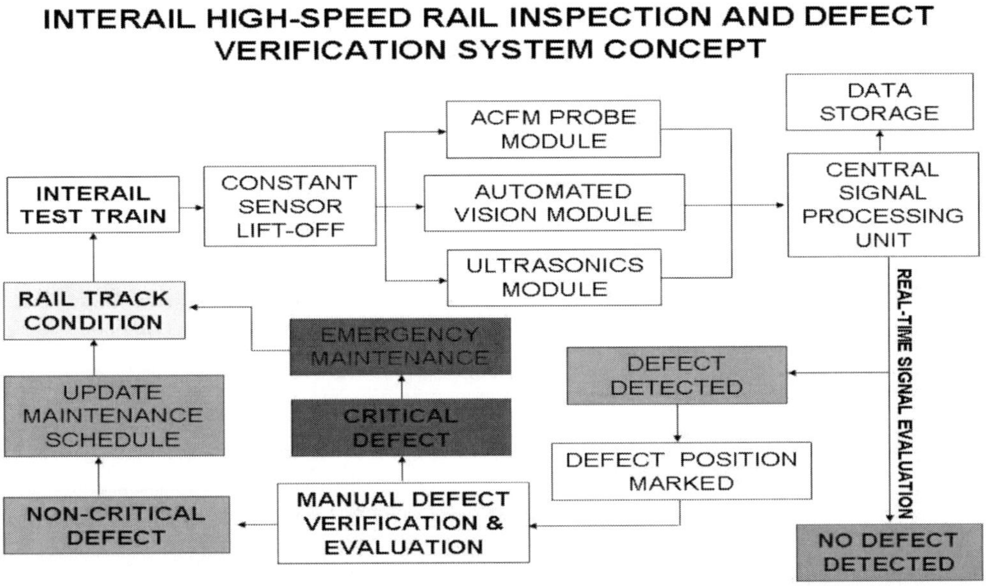

Figure 5. Simplified schematic showing the overall concept of the INTERAIL high-speed rail inspection and defect verification platform.

CONCLUSION

Rail infrastructure managers agree that inspection results influence profoundly their maintenance planning as well as their strategy for future inspection scheduling. Although, the rail industry has until now remained attached to traditional based inspection techniques which have been proven to offer an acceptable level of reliability the vast majority of infrastructure managers are prepared to adopt new technologies and procedures within a reasonable time frame. In most cases this time-frame has been indicated to be in the range of twelve and thirty-six months after proper certification has been obtained.

Increasing traffic density, rolling stock speeds and axle loads are resulting in more complicated inspection schedules while the cost-efficiency of the non-destructive evaluation process itself is gradually decreasing and the desirable level of operational reliability is growing more difficult to maintain. Integrated inspection strategies coupled with measurable innovation in inspection technology can lead to significant improvements in operational cost efficiency and reliability without the requirement for a fundamental shift in the existing understanding of the inspection process and standards. The rail industry has already acknowledged this and is moving steadily towards the adoption of integrated inspection procedures based on existing commercial systems available in the market. INTERAIL, a European FP7 collaborative research project which is currently underway is looking into demonstrating the value of integrated inspection, coupled with the application of novel non-destructive evaluation techniques for both manual and automated inspection of the rail track under actual conditions. Several other projects which are currently on-going, such as RAILECT, I-RAIL and MONITORAIL also supported by the European Commission through FP7, are looking into demonstrating alternative inspection technologies in the field.

It is undeniable that substantial improvements in rail non-destructive evaluation can be achieved when appropriate techniques are integrated together in a way so as to complement the capabilities of each other. Current developments in the rail industry already point to this direction, with conventional ultrasonic probes being integrated with other types of inspection systems.

Table 1. Summary of rail inspection techniques

NDT Technique	Systems Available	Defects Detected	Performance
Ultrasonics	Manual and high-speed systems (up to 100 km/h)	Surface defects, rail head internal defects, rail web and foot defects	Reliable manual inspection but can miss rail foot defects. At high speed can miss smaller surface defects as well as internal defects particularly at the rail foot
Magnetic Flux Leakage	High-speed systems (up to 35 km/h)	Surface defects and near surface internal rail head defects	Reliable in detecting surface defects and shallow internal rail head defects although cannot detect cracks smaller than <4mm. MFL performance deteriorates at higher speeds

NDT Technique	Systems Available	Defects Detected	Performance
Eddy current	Manual and high-speed systems (up to 100 km/h)	Surface and near-surface internal defects	Reliable in detecting surface breaking defects. Adversely affected by grinding marks and lift-off variations
Automated Visual Inspection	Manual and high speed systems (up to 320 km/h depending on the mode of inspection)	Surface breaking defects, rail head profile, corrugation, missing parts, defective ballast	Reliable in detecting corrugation, rail head profile missing parts and defective ballast at high speeds. Cannot reliably detect surface breaking defects above a certain speed due to resolution deterioration. Cannot assess the rail for internal defects
Electromagnetic Acoustic Transducers	Low speed hi-rail vehicle (<10km/h)	Surface defects, rail head, web and foot internal defects	Reliable for surface and internal defects. Can miss rail foot defects. Adversely affected by lift-off variations
Long range Ultrasonics	Manual systems and low-speed hi-rail vehicle systems (<10 km/h)	Surface defects, rail head internal defects, rail web and foot defects	Reliable in detecting large transverse defects (>5% of the overall cross-section). Difficult to deploy dynamically. More appropriate for static testing.
Alternating Current Field Measurement	Manual systems (hi-speed system under development)	Surface breaking defects	Reliable in detecting and quantifying surface breaking defects. Cannot detect sub-surface defects. Very good tolerance to lift-off variations
Acoustic Emission	Experimental static and high-speed testing	Condition monitoring, rail breaks, wheel burns, squats, wet spots, worn rail profiles	Limited to experimental work only at the moment.
High-frequency vibration	Experimental	Grooved rails and crossings.	Limited experiments – Capable of assessing the quality of the rail with static tests. Needs calibration and trending.
Radiography	Static	Applicable for rail welds only	Limited to pilot trials.
Ultrasonic phased arrays	Static or walking speed tests	Applicable for quantification of rail defects	Limited to pilot trials.

ACKNOWLEDGMENTS

The author would like to thank the European Commission for partially funding the results of this work through the INTERAIL project (www.interailproject.eu) under the 7th Framework Programme, contract reference number SCP8-GA-2009-234040.

REFERENCES

[1] Kristan, J., Sawley, K. Material property characterisation and specifications for improved rail steels – TTCI R&D, *Railway Track and Structures*, October 2002.

[2] Aglan, H. A., Fateh, M. Fatigue damage tolerance of bainitic and pearlitic rail steels, *International Journal of Damage Mechanics*, 2006, 15, 393-410.

[3] Cannon, D. F. An international cross reference of rail defects – 2nd edition, *UIC Rail Defect Management Report*, June 2003.

[4] Cannon, D. F., Edel, K. –O., Grassie, S. L., Sawley, K. Rail defects: an overview, Fatigue Fract. *Engng Mater. Struct.*, 2003, 26, 865-887.

[5] Grassie, S. L. Rolling Contact Fatigue on the British railway system: treatment, *Wear*, 2005, 258, 1310-1318.

[6] Lonsdale, C. P., Luke S.T. An investigation of thermite rail weld composition and properties, *Proceedings of the International Conference of Advances in Welding Technology*, Columbus, Ohio, USA, 1996.

[7] Papaelias, M. Ph., Roberts, C., Davis, C. L. A review on non-destructive evaluation of rails: state-of-the-art and future development, *Proc.IMechE Part F: J. Rail and Rapid Transit,* vol. 222, 2008.

[8] Clark. R. Rail flaw detection: overview and needs for future developments, *NDT & E International*, 2004, 37, 111-118.

[9] Thomas, H. –M., Junger, M., Hintze, H. Krull R., Rühe, S. Pioneering inspection of railroad rails with eddy currents, *Proceedings of the 15th World Conference on Non-Destructive Testing,* Rome, Italy, 2000.

[10] Pohl, R., Erhard, A., Montag, H.-J., Thomas, H.-M., Wüstenberg, H. NDT techniques for railroad wheel and gauge corner inspection, *NDT & E Int.,* 2004, 37, 89-94.

[11] Junger, M., Thomas, H.-M., Krull, R., Rühe, S. The potential of eddy current technology regarding railroad inspection and its implementation, in the *Proceedings of the 16th World Conference on Non-Destructive Testing*, Montreal, Canada, August-September 2004.

[12] Thomas, H.-M., Heckel, T., Hanspach, G. Advantage of a combined ultrasonic and eddy current examination for railway inspection trains, in the *Proceedings of ECNDT 2006,* Berlin, Germany, 2006.

[13] The MER MEC track surface defects detection system, www.mermec.com.

[14] The diagnostic complex "INTEGRAL", http://www.tvema.com/ru/productList_2060.html.

[15] INTERAIL integrated rail inspection platform, www.interailproject.eu.

[16] Clark, R., Singh, S. The inspection of thermite welds in railroad rail – a perennial problem, *Insight,* 2003, 45, 387-393.

[17] Eklund, B., Crocker, B. *Inspecting welds in rail, Inspection Trends Magazine*, March 2005.

[18] Buttle, D. J., Dalzell, W., Thayer, P. J. Non-destructive residual stress measurement in rail heads and rolling contact fatigue, *Insight,* 2002, 44 (6), 364-368.

[19] Utrata, D. Exploring enhanced rail flaw detection using ultrasonic phased array inspection, *Review of Quantitative Non-destructive Evaluation*, Ed. D. O. Thompson and D. E. Chimenti, 2002, 21, 1813-1818.

[20] Utrata, D., Clark, R. Groundwork for rail flaw detection using ultrasonic phased array inspection, *Review of Quantitative Nondestructive Evaluation*, Ed. D. O. Thompson and D. E. Chimenti, 2003, 22, 799-805.

[21] Garcia, G., Zhang, J. Application of ultrasonic phased arrays for rail flaw inspection, *TTCI Report for the U.S. Department of Transportation*, July 2006.

[22] Advances in phased array ultrasonic technology applications, *ISBN* 0-9735933-4-2, published by Olympus NDT.

[23] Breidif, P., Papaelias, M. Ph., Potter, M. Operation evaluation of and inspection demonstrator, FP6 INNOTRACK Internal Report, Project no. *TIP5-CT*-2006-031415O, 2009.

[24] Development of an ultrasonic technique for the volumetric examination of rail welds, The RAILECT project, www.railect.com.

[25] Howitt, M. Bombardier brings ACFM into the rail industry, *Insight*, 2002, 44 (6).

[26] Lugg, M., Topp, D. Recent developments and applications of the ACFM inspection method and ACSM stress measurement method, *Proceedings of ECNDT 2006,* Berlin, Germany, 2006.

[27] Papaelias, M. Ph., Lugg, M., Smith, M., Roberts, C., Davis, C. L. Detection and quantification of rolling contact fatigue cracks in rails using ACFM technology, *Proceedings of the BINDT 2007* Conference, Glasgow, UK, 2007.

[28] Papaelias, M. Ph., Roberts, C., Davis, C. L., Blakeley, B., Lugg, M. High speed inspection of Rolling Contact Fatigue in Rails using ACFM sensors, *Journal of the British Institute of Non-destructive Testing* (Insight), Issue 7, July 2009.

[29] Papaelias, M. Ph., Lugg, M., Roberts, C., Davis, C. L. High-Speed inspection of rails using ACFM technology', Journal of NDT&E International, 2009. 42, pp. 328-335.

[30] Rose, J. L., Afioli, M. J., Song, W.-J. Application and potential of guided wave rail inspection, *Insight*, 2002, 44 (6), 353-358.

[31] Hayashi, T., Song, W.-J., Rose, J. L. Guided wave dispersion curves for a bar with an arbitrary cross-section, a rod and rail example, *Ultrasonics*, 2003, 41,175-183.

[32] Cawley, P., Wilcox, P., Alleyne, D. N., Pavlakovic, B., Evans, M., Vine, K., Lowe, M. J. S. Long range inspection of rail using guided waves – field experience, *Proceedings of the 16th World Conference on Non-Destructive Testing*, Montreal, Canada, August-September 2004.

[33] Rose, J. L., Lee, C. M., Hay, T. R., Cho, Y., Park, I. K. Rail inspection with guided waves, 12th A-PCNDT 2006 – Asia-Pacific Conference on NDT, 5th-10th November 2006, Auckland, New Zealand.

[34] Gharaibeh, Y., Sanderson, R., Mudge, P., Ennaceur, C., Balachandran, W. Investigation of the behaviour of selected ultrasonic guided wave modes to inspect rails for long-range testing and monitoring, *Proc. Of the IMECHE, Part F: J. Rail and Rapid Transit*, May 2011, Vol. 225, no. 3, pp. 331-324.

[35] Long range condition monitoring of rails – MONITORAIL, www.monitorail.eu.

[36] TTCI 12th Annual Research Review CD-ROM, Pueblo, Colorado, USA, 2007.

[37] Chahbaz, A., Development of a mobile inspection system for rail integrity assessment, Tektrend International Report for the Transportation Development Centre, Montreal, Quebec, Canada, June 2000.

[38] Chabhaz, A., Brassard, M., Pelletier, A. Mobile inspection system for rail integrity assessment, in the *Proceedings of 15th World Conference of Non-Destructive Testing*, Roma, Italy, 2000.

[39] Dixon, S., Edwards, R. S., Jian, X. Inspection of the railtrack head surfaces using electromagnetic acoustic transducers, *Insight*, 2004, 46, 326-330.

[40] Edwards, R. S., Jian, X., Dixon, S. Rail defect detection using ultrasonic surface waves, *Review of Quantitative Nondestructive Evaluation*, Ed. D. O. Thompson and D. E. Chimenti, 2006, Vol. 25, 1601-1608.

[41] Edwards, R. S., Dixon, S. Jian, X. Characterisation of defect in the railhead using ultrasonic surface waves, *NDT & E Int.*, 2006, 39, 468-475.

[42] Edwards, R. S., Dixon, S., Jian, X. Depth gauging of defects using low frequency wideband Rayleigh waves, *Ultrasonic*s, 2006, 44, 93-98.

[43] Edwards, R. S., Fan Y., Papaelias, M., Dixon, S., Davis, C. L., Roberts, C. Ultrasonic detection of surface-breaking railhead defects, in the P*roceedings of RQNDE 2007 Conference,* Colorado, USA, July 2007.

[44] Fan, Y., Dixon, S., Edwards, R. S., Jian, X. Ultrasonic surface wave propagation and interaction with surface defects on rail track head, *NDT & E Int.*, 2007, 40, 471-477.

[45] Edwards, R. S., Papaelias, M., Holmes, C., Fan, Y., Dixon, S. Ultrasonic detection of surface-breaking railhead defects, *Journal of British Institute of NDT, Insight*, Vol. 50, No. 7 July 2008.

[46] I-RAIL high-speed EMAT system, www.i-railproject.eu.

[47] Cawser, S. J., Hardy, A. E. J., Wright, C. E. Acoustic track monitoring - Phase 1: Initial data gathering analysis, *AEA Report for RSSB*, December 2002.

[48] Bruzelius, K. Mba, D. An initial investigation on the potential applicability of Acoustic Emission to rail track fault detection, *NDT & E Int.*, 2004, 37 (7), 507-516.

[49] Jeong, D. Y. Analytical modelling of rail defects and its application to rail defect management, *UIC/WEC Joint Research Project Report on Rail Defect Management*, January 2003.

[50] Jeong, D. Y., Gordon, J. E. Evaluation of rail test frequencies using risk analysis, *Proceedings of the 2009 ASME Joint Rail Conference*, March 3-5, 2009, Pueblo, Colorado, USA.

In: Fault Detection
Editors: F. P. G. Márquez and M. Papaelias

ISBN: 978-1-62808-999-8

Chapter 3

DATA-ENABLED HEALTH MANAGEMENT OF COMPLEX INDUSTRIAL SYSTEMS*

***Soumik Sarkar*†, *Soumalya Sarkar*‡ *and Asok Ray*§**
Mechanical Engineering Department,
The Pennsylvania State University, University Park, US

Abstract

Complex industrial systems, such as gas turbine engines, have many heterogeneous subsystems (e.g., thermal, mechanical, hydraulic and electrical) with complex thermal-hydraulic, electro-mechanical, and electronic interactions. Schedule-based policies are largely followed in the industry for maintenance of complex systems. Since the degradation profiles of individual systems are usually different due to manufacturing and usage variations, a scheduled maintenance policy in most cases becomes either overly conservative or a source of serious safety concern. This calls for development of reliable and cost-effective health management for complex industrial systems. However, reliable first-principle modeling of such systems may often be very expensive (e.g., in terms of computational memory, execution time, and cost) and hence may become inappropriate for condition monitoring, fault detection, diagnostics and prognostics.

This chapter describes a data-driven framework to obtain abstract models of complex systems from multiple sensor observations to enable condition monitoring, diagnostics and supervisory control. The algorithms are formulated in the setting of symbolic dynamic filtering (SDF) that has been recently reported in literature. The underlying concept of SDF is built upon the principles of symbolic dynamics, statistical pattern recognition, probabilistic graphical models and information theory. SDF was initially developed for statistically quasi-stationary data and then it was successfully extended to analysis of transient data. The tool-chain involves data space abstraction for analyzing mixed data (i.e., continuous, discrete and categorical) without significant loss of information. Spatiotemporal pattern discovery & fusion algorithms have been built upon Markovian & Bayesian network concepts and health status classifications

*This work has been supported in part by the U.S. Army Research laboratory and the U.S. Army Research Office under Grant No. W911NF-07-1-0376 and by NASA under Cooperative Agreement No. NNX07AK49A.
†Currently with United Technologies Research Center, E-mail address: sarkars@utrc.utc.com
‡E-mail address: soumalya.still.dreaming@gmail.com
§E-mail address: axr2@psu.edu

by using machine learning techniques. The chapter also presents applications of the generic data-driven tool in heterogeneous areas, such as fault detection in aircraft gas turbine engines, fatigue damage prediction in structural materials, and health monitoring in nuclear power plants & ship-board auxiliary systems.

1. Introduction

With the advent of advanced sensing and computational resources, opportunities for data-driven technologies have become abundant in most of the industries dealing with complex human-engineered systems. Need for high-fidelity condition monitoring, fault detection, diagnostics and prognostics also grew due to the increasing inter-connectedness of such electro-mechanical systems. Apart from the physical components, their cyber counterparts added another dimension from the perspective of complexity, uncertainty and disturbance. Although model-based approaches have shown their merits to utilize the understanding of physical phenomena to detect and isolate faults in industrial systems, data-driven methods are receiving increased attention to address the underlying issues, such as scalability, robustness and cost of deployment. To this end, this chapter summarizes the decade-long development of a statistical time-series analysis tool called the Symbolic Dynamic Filtering (SDF) that has been shown to be extremely effective for numerous industrial systems, some of which will be discussed here.

The chapter is organized in six sections including the present one which briefly introduces the notion of nonlinear time series analysis and presents the two-time-scale problem formulation for anomaly detection using symbolic dynamic filtering. Section 1.1.2 provides a brief overview of symbolic dynamics and encoding of time series data. Before describing different aspects of the SDF tool-chain, Section 2 introduces the industrial applications that will be used to illustrate the capabilities and performance of the tool. This applications include Gas Turbine engines, Fatigue damage problems, Nuclear power plants and Ship-board auxiliary systems. The first step of the tool-chain is data space abstraction where raw time-series can be transformed into a more suitable domain via wavelet or Hilbert transform. After the pre-processing step time-series data is spatially abstracted via a discretization process known as partitioning. Details and varieties of the abstraction procedure are presented in Section 3. While the standard steady-state symbolic modeling is discussed at length in 1.1, a recent development to encode short-length time-series motifs after partitioning is discussed in Section 4. This section also highlights the classification scheme of such motifs. Today's industrial systems almost always are monitored by multiple sensors of heterogeneous modality and it is getting increasingly obvious to design health management systems with information fusion capability. However, most of the state-of-the art fusion technologies deployed today fuse information at the decision layer after extracting condition indicators from individual sensors. This strategy suffers from acute loss of information that may prevent early detection and prognostics. To this end, SDF has a unique advantage especially due to the data abstraction step in the ability of seamless fusion of multi-sensor information at the data and feature layers and Section 5 presents this capability in detail. Finally, Section 6 summarizes and concludes the chapter with recommendations for future research directions.

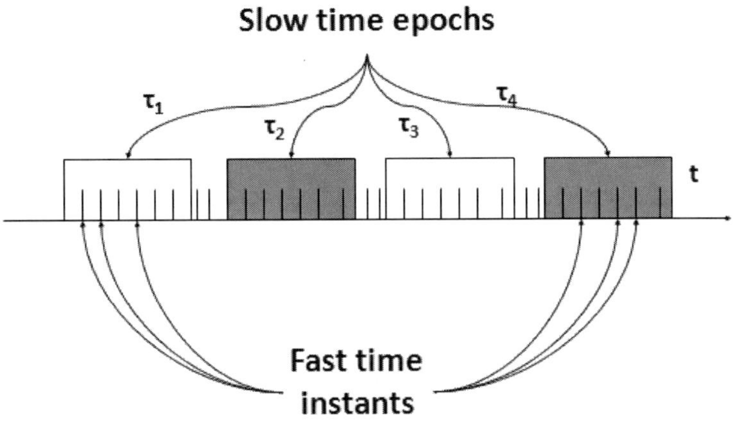

Figure 1. Pictorial view of the two time scales: (i) *Slow time scale* of system evolution and (ii) *Fast time scale* for data acquisition and signal conditioning.

1.1. General Framework of Symbolic Dynamic Filtering (SDF)

Symbolic Dynamic Filtering (SDF) is built upon the concepts from multiple disciplines including Statistical Mechanics [1, 2], Symbolic Dynamics [3], Statistical Pattern Recognition [4], and Information Theory [5]. It is shown to be a fast, efficient and computationally inexpensive pattern recognition tool. These qualities are particularly important requirements for autonomous cyber-physical system applications. It has been shown by laboratory experimentation [6] that this method of pattern identification yields superior performance in terms of early detection of anomalies and robustness to measurement noise in comparison with other existing techniques such as Principal Component Analysis, (PCA) and Artificial Neural Networks (ANN).

1.1.1. A Brief Overview

The Symbolic Dynamic Filtering (SDF) technique is built upon the concept of two time-scales while analyzing time-series data from a given dynamical system. The *fast time scale* is related to the response time of the system dynamics. Over the span of a given time series data sequence, the dynamic behavior of the system is assumed to remain invariant, i.e., the process is quasi-stationary on the fast time scale. On the other hand, the *slow time scale* is related to the time span over which the critical parameters of the system may change and exhibit non-stationary dynamics. The concept of two time scales is illustrated in Fig. 1, where a long time span in the fast scale could be a tiny time interval in the slow scale.

1.1.2. Symbolic Dynamics Encoding and State Machine

The basic concepts of *Symbolic Dynamics* include:

1. Encoding (possibly nonlinear) system dynamics from observed time series data for generation of symbol sequences.

2. Construction of probabilistic finite state automata ($PFSA$) from symbol sequences for generation of pattern vectors for representation of the environment's dynamical characteristics.

The continuously-varying finite-dimensional model of a dynamical system is usually formulated in the setting of an initial value problem as:

$$\frac{d\mathbf{x}(t)}{dt} = f(\mathbf{x}(t), \theta(t_s)); \; \mathbf{x}(0) = \mathbf{x}_0, \tag{1}$$

where $t \in [0, \infty)$ denotes the (fast-scale) time; $\mathbf{x} \in \mathbb{R}^n$ is the state vector in the phase space; and $\theta \in \mathbb{R}^\ell$ is the (possibly anomalous) parameter vector varying in (slow-scale) time t_s. Let $\Omega \subset \mathbb{R}^n$ be a compact (i.e., closed and bounded) region, within which the trajectory of the dynamical system, governed by Eq. (1), is circumscribed. The region Ω is partitioned into a finite number of (mutually exclusive and exhaustive) cells, so as to obtain a coordinate grid. Let the cell, visited by the trajectory at a time instant, be denoted as a random variable taking a symbol value from the alphabet Σ. An orbit of the dynamical system is described by the time series data as: $\mathbb{O} \equiv \{\mathbf{x}_0, \mathbf{x}_1, \mathbf{x}_2, \cdots\}$, where each $\mathbf{x}_i \in \Omega$, which passes through or touches one of the cells of the partition. Each initial state $\mathbf{x}_0 \in \Omega$ generates a sequence of symbols defined by a mapping from the phase space into the symbol space as:

$$\mathbf{x}_0 \rightarrow \sigma_{i_0}\sigma_{i_1}\sigma_{i_2}\cdots \tag{2}$$

where each σ_{i_k}, $k = 0, 1, \cdots$ takes a symbol from the alphabet Σ. The mapping in Eq. (2) is called *Symbolic Dynamics* as it attributes a legal (i.e., physically admissible) symbol sequence to the system dynamics starting from an initial state. The partition is called a generating partition of the phase space Ω if every legal (i.e., physically admissible) symbol sequence uniquely determines a specific initial condition $\mathbf{x}_0$. In other words, every (semi-infinite) symbol sequence uniquely identifies one continuous space orbit [7].

In physics and signal processing literature, similar procedure is known as coarse-graining or quantization. However, in both cases, typically very fine and uniform quantization cells are considered to reduce loss of information. On the other hand, literature of dynamical systems in mathematics presents the concept of partitioning in the context of symbolic dynamics. Unlike coarse-graining or quantization, partitioning typically involves nonuniform and much coarser representation of the phase space, which entails a low order modeling of complex dynamical systems. Moreover, in case of a generating partition, there is no loss of information. Although symbolic dynamics is subjected to (possible) loss of information resulting from granular imprecision of partitioning boxes, the essential robust features (e.g., periodicity and chaotic behavior of an orbit) are expected to be preserved in the symbol sequences through an appropriate partitioning of the phase space [8].

Thus partitioning a finite region of the phase space leads to mapping from the partitioned space into the symbol alphabet [9]. This represents a spatial and temporal discretization of the system dynamics defined by the trajectories. Although the theory of phase-space partitioning is well developed for one-dimensional mappings [7], very few results are known for two and higher dimensional systems. Furthermore, the state trajectory of the system variables may be unknown in case of systems for which a model as in Eq. (1) is not known or is difficult to obtain. As such, as an alternative, the time series data set of selected observable outputs can be used for partitioning and symbolic dynamic encoding. In general, the time series data can be generated from the available sensors and/or from analytically derived model variables. After partitioning, the symbol sequence is converted into a probabilistic finite-state machine, that is considered as the compressed abstract model of the system.

1.1.3. Phase Space Partitioning

Several partitioning techniques have been reported in literature for symbol generation [10][11], primarily based on symbolic false nearest neighbors ($SFNN$). These techniques rely on partitioning the phase space and may become cumbersome and extremely computation-intensive if the dimension of the phase space is large. Moreover, if the time series data is noise-corrupted, then the symbolic false neighbors would rapidly grow in number and require a large symbol alphabet to capture the pertinent information on the system dynamics. Therefore, symbolic sequences as representations of the system dynamics should be generated by alternative methods because phase-space partitioning might prove to be a difficult task in the case of high dimensions and presence of noise. The wavelet or Hilbert transforms largely alleviate the difficulties of phase-space partitioning and are particularly effective with noisy data from high-dimensional dynamical systems [12, 13]. A further theoretical development has been made to generalize Hilbert transform [14] that might prove to be a better choice as a pre-processing technique.

In both of these approaches, time series data are first converted to wavelet domain or analytic signal domain, and the transformed space is then partitioned with alphabet size $|\Sigma|$ into segments. The choice of $|\Sigma|$ depends on specific experiments, noise level and also the available computation power. A large *alphabet* may be noise-sensitive while a small alphabet could miss the details of signal dynamics [12]. The partitioning can be done such that the regions are partitioned uniformly (equal width) or the regions with more information are partitioned finer and those with sparse information are partitioned coarser. This is achieved by maximizing the Shannon entropy [5], which is defined as:

$$S = -\sum_{i=1}^{|\Sigma|} p_i \, log(p_i) \tag{3}$$

where p_i is the probability of a data point to be in the ith partition segment. In this case, the size of the cells is smaller for regions with higher density of data points to ensures an unbiased partition such that each cell is allocated equal number of visits at the nominal condition. Uniform probability distribution, i.e., $p_i = \frac{1}{|\Sigma|}$ for $i = 1, 2, \ldots |\Sigma|$, is a consequence of maximum entropy partitioning [12]. In the illustrative example of Fig. 2, the partitioning contains 4 cells (i.e., line intervals in this case).

Apart from this maximum entropy (equal frequency) or uniform (equal width) partitioning schemes, partitioning can be optimized based on other objective functions. This will be discussed in details later. Once the partitioning is done with alphabet size $|\Sigma|$ at a reference slow time scale condition, it is kept constant for all slow time epochs, i.e., the structure of the partition is fixed at the reference condition. Therefore, the partitioning structure generated at the reference condition serve as the reference frame for computation of pattern vectors from time series data at subsequent slow time epochs.

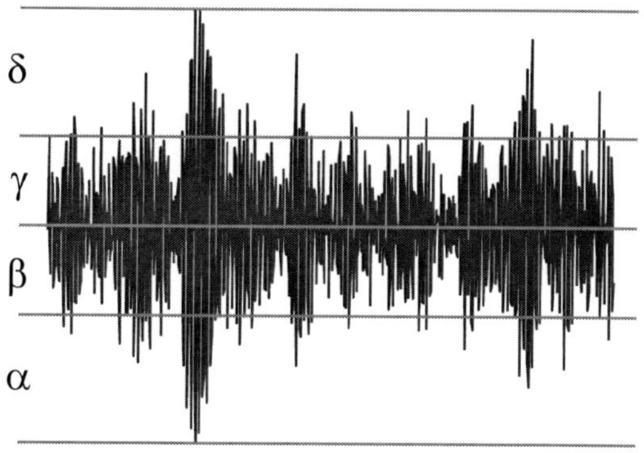

Figure 2. An Example of Partitioning.

1.1.4. Probabilistic Finite State Automata ($PFSA$) Construction

Once the symbol sequence is obtained, the next step is the construction of a Probabilistic Finite State Automaton ($PFSA$) (as shown in Fig. 3) and calculation of the respective state probability vector. The partitioning (see Fig. 2) is performed at the reference slow time epoch. A $PFSA$ is then constructed, where the states of the machine are defined corresponding to a given *alphabet* set Σ and window length D. This explicit state construction method helps defining abstract states for systems without considerable domain knowledge. The alphabet size $|\Sigma|$ is the total number of partition segments while the window length D is the length of consecutive symbol words [15], which are chosen as all possible words of length D from the symbol sequence. Each state belongs to an equivalence class of symbol words of length D, which is characterized by a word of length D at the leading edge. Therefore, the number n of such equivalence classes (i.e., states) is less than or equal to the total permutations of the alphabet symbols within words of length D. That is, $n \leq |\Sigma|^D$; some of the states may be forbidden with zero probability of occurrence. For example, if $\Sigma = \{0, 1\}$, i.e., $|\Sigma| = 2$ and if $D = 2$, then the number of states is $n \leq |\Sigma|^D = 4$; and the possible states are $00, 01, 10$, and 11.

The choice of $|\Sigma|$ and D depends on specific applications and the noise level in the time series data as well as on the available computation power and memory availability. As stated earlier, a large *alphabet* may be noise-sensitive and a small alphabet could miss the details of signal dynamics. Similarly, while a larger value of D is more sensitive to signal distortion, it would create a much larger number of states requiring more computation power and increased length of the data sets. For the analysis of data sets in this research, the window length is set to D=1; consequently, the set of states Q is equivalent to the symbol alphabet Σ. With the selection of the parameters D=1 and $|\Sigma|$=8, the $PFSA$ has only 8 states, which yields very fast computation and low memory requirements; hence, the algorithm is capable of early detection of anomalies and incipient faults. However, other applications, such as two-dimensional image processing, may require larger values of the parameter D and hence possibly larger number of states in the $PFSA$.

Using the symbol sequence generated from the time series data, the state machine is constructed on the principle of sliding block codes [3]. The window of length D on the

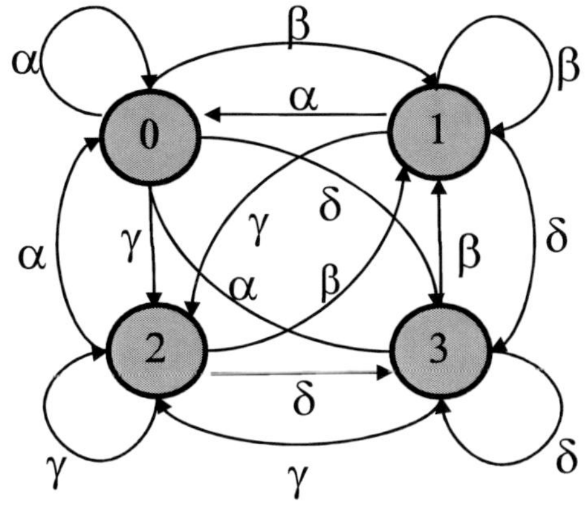

Figure 3. Example of Finite State Automaton.

symbol sequence $\ldots\sigma_{i_1}\ \sigma_{i_2}\ldots\sigma_{i_k}\ldots$ is shifted to the right by one symbol, such that it retains the last (D-1) symbols of the previous state and appends it with the new symbol σ_{i_ℓ} at the end. The symbolic permutation in the current window gives rise to a new state. The $PFSA$ constructed in this fashion is called the D-Markov machine [15], because of its Markov properties.

Definition 1.1. *A symbolic stationary process is called D-Markov if the probability of the next symbol depends only on the previous D symbols, i.e.,*
$P\left(\sigma_{i_0}|\sigma_{i_{-1}}....\sigma_{i_{-D}}\sigma_{i_{-D-1}}....\right) = P\left(\sigma_{i_0}|\sigma_{i_{-1}}....\sigma_{i_{-D}}\right).$

The finite state machine constructed above has D-Markov properties because the probability of occurrence of symbol σ_{i_ℓ} on a particular state depends only on the configuration of that state, i.e., the previous D symbols. Once the alphabet size $|\Sigma|$ and word length D are determined at the nominal condition (i.e., time epoch t_0), they are kept constant for all other slow time epochs. That is, the partitioning and the state machine structure generated at the reference condition serve as the reference frame for data analysis at subsequent slow time epochs.

The states of the machine are marked with the corresponding symbolic word permutation and the edges joining the states indicate the occurrence of a symbol σ_{i_ℓ}. The occurrence of a symbol at a state may keep the machine in the same state or move it to a new state.

Definition 1.2. *The probability of transitions from state q_j to state q_k belonging to the set Q of states under a transition $\delta : Q \times \Sigma \rightarrow Q$ is defined as*

$$\pi_{jk} = P\left(\sigma \in \Sigma \mid \delta(q_j, \sigma) \rightarrow q_k\right); \sum_k \pi_{jk} = 1; \qquad (4)$$

Thus, for a D-Markov machine, the irreducible stochastic matrix $\mathbf{\Pi} \equiv [\pi_{ij}]$ describes all transition probabilities between states such that it has at most $|\Sigma|^{D+1}$ nonzero entries. The left eigenvector $\mathbf{p}$ corresponding to the unit eigenvalue of $\mathbf{\Pi}$ is the state probability vector under the (fast time scale) stationary condition of the dynamical system [15].

On a given symbol sequence $....\sigma_{i_1}\sigma_{i_2}...\sigma_{i_l}....$ generated from the time series data collected at a slow time epoch, a window of length D is moved by keeping a count of occurrences of word sequences $\sigma_{i_1}\cdots\sigma_{i_D}\sigma_{i_{D+1}}$ and $\sigma_{i_1}\cdots\sigma_{i_D}$ which are respectively denoted by $N(\sigma_{i_1}\cdots\sigma_{i_D}\sigma_{i_{D+1}})$ and $N(\sigma_{i_1}\cdots\sigma_{i_D})$. Note that if $N(\sigma_{i_1}\cdots\sigma_{i_D}) = 0$, then the state $q \equiv \sigma_{i_1}\cdots\sigma_{i_D} \in Q$ has zero probability of occurrence. For $N(\sigma_{i_1}\cdots\sigma_{i_D}) \neq 0$, the transitions probabilities are then obtained by these frequency counts as follows:

$$\pi_{jk} \equiv P(q_k|q_j) = \frac{P(q_k, q_j)}{P(q_j)} = \frac{P(\sigma_{i_1}\cdots\sigma_{i_D}\sigma)}{P(\sigma_{i_1}\cdots\sigma_{i_D})}$$
$$\Rightarrow \pi_{jk} \approx \frac{N(\sigma_{i_1}\cdots\sigma_{i_D}\sigma)}{N(\sigma_{i_1}\cdots\sigma_{i_D})} \tag{5}$$

where the corresponding states are denoted by $q_j \equiv \sigma_{i_1}\sigma_{i_2}\cdots\sigma_{i_D}$ and $q_k \equiv \sigma_{i_2}\cdots\sigma_{i_D}\sigma$.

The time series data at the reference condition, set as a benchmark, generates the *state transition matrix* $\mathbf{\Pi}$ that, in turn, is used to obtain the *state probability vector* $\mathbf{q}$ whose elements are the stationary probabilities of the state vector, where $\mathbf{q}$ is the left eigenvector of $\mathbf{\Pi}$ corresponding to the (unique) unit eigenvalue. The state probability vector $\mathbf{p}$ is obtained from time series data at a (possibly) faulty condition. The partitioning of time series data and the state machine structure should be the same in both cases but the respective state transition matrices could be different.

Pattern changes may take place in dynamical systems over slow time epochs due to various reasons. The pattern changes are quantified as deviations from the reference pattern (i.e., the probability distribution at the reference condition). To identify variation of a single parameter in the dynamical system, the resulting anomalies (i.e., deviations of the evolving patterns from the reference pattern) are characterized by a scalar-valued function, called *anomaly measure* μ. The anomaly measures are obtained as:

$$\mu \equiv d\left(\mathbf{p}, \mathbf{q}\right) \tag{6}$$

where the $d(\bullet, \bullet)$ is an appropriately defined distance function. An extension of this method for identification of multiple parameters in dynamical systems has also been made in [16]. Instead of using the low dimensional state probability vector as a pattern, the entire Perron-Frobenius operator (state transition matrix) may be used as the pattern to reduce loss of information.

1.1.5. Stopping Rule for Determining Symbol Sequence Length

A stopping rule is necessary to find a lower bound on the length of symbol sequence required for parameter identification of the stochastic matrix $\mathbf{\Pi}$. The stopping rule [17] is based on the properties of irreducible stochastic matrices [18]. The state transition matrix, constructed at the rth iteration (i.e., from a symbol sequence of length r), is denoted as $\mathbf{\Pi}(r)$ that is an $n \times n$ irreducible stochastic matrix under stationary conditions. Similarly, the state probability vector $\mathbf{p}(r) \equiv [p_1(r)\ p_2(r)\ \cdots\ p_n(r)]$ is obtained as

$$p_i(r) = \frac{r_i}{\sum_{j=1}^{n} r_i} \tag{7}$$

where r_i is the number of symbols in the ith state such that $\sum_{i=1}^{n} r_i = r$ for a symbol sequence of length r. The stopping rule makes use of the Perron-Frobenius Theorem [18] to establish a relation between the vector $\mathbf{p}(r)$ and the matrix $\mathbf{\Pi}(r)$. Since the matrix $\mathbf{\Pi}(r)$ is stochastic and irreducible, there exists a unique eigenvalue $\lambda = 1$ and the corresponding left eigenvector $\mathbf{p}(r)$ (normalized to unity in the sense of absolute sum). The left eigenvector $\mathbf{p}(r)$ represents the state probability vector, provided that the matrix parameters have converged after a sufficiently large number of iterations. That is, under the hypothetical arbitrarily long sequences, the following condition is assumed to hold.

$$\mathbf{p}(r+1) = \mathbf{p}(r)\mathbf{\Pi}(r) \;\; \Rightarrow \;\; \mathbf{p}(r) = \mathbf{p}(r)\mathbf{\Pi}(r) \;\; \text{as } r \to \infty \tag{8}$$

Following Eq. (7), the absolute error between successive iterations is obtained such that

$$\| \left(\mathbf{p}(r) - \mathbf{p}(r+1)\right) \|_\infty = \| \mathbf{p}(r) \left(\mathbf{I} - \mathbf{\Pi}(r)\right) \|_\infty \leq \frac{1}{r} \tag{9}$$

where $\| \bullet \|_\infty$ is the max norm of the finite-dimensional vector $\bullet$.

To calculate the stopping point r_{stop}, a tolerance of η $(0 < \eta \ll 1)$ is specified for the relative error such that:

$$\frac{\| \left(\mathbf{p}(r) \; - \; \mathbf{p}(r+1)\right) \|_\infty}{\| \left(\mathbf{p}(r)\right) \|_\infty} \leq \eta \;\;\; \forall \, r \geq r_{stop} \tag{10}$$

The objective is to obtain the least conservative estimate for r_{stop} such that the dominant elements of the probability vector have smaller relative errors than the remaining elements. Since the minimum possible value of $\| \left(\mathbf{p}(r)\right) \|_\infty$ for all r is $\frac{1}{n}$, where n is the dimension of $\mathbf{p}(r)$, the least of most conservative values of the stopping point is obtained from Eqs. (9) and (10) as:

$$r_{stop} \equiv int\left(\frac{n}{\eta}\right) \tag{11}$$

where $int(\bullet)$ is the integer part of the real number $\bullet$.

More advanced forms of stopping rules are reported by Wen and Ray [19, 20], where the parameter η in Eq. 10 is statistically determined.

2. Overview of Industrial System Testbeds

This section briefly introduces the industrial applications that will be used in the sequel to validate the effectiveness of the SDF tool. The applications include aircraft gas turbine engines, fatigue damage problems, nuclear power plants and ship-board auxiliary systems.

2.1. Aircraft Gas Turbine Engines

This section presents the C-MAPSS test bed along with the fault injection scheme. The C-MAPSS simulation test bed [21] was developed at NASA for a typical commercial-scale two-spool turbofan engine and its control system. Figure 4 shows the schematic diagram of a commercial aircraft gas turbine engine used in the C-MAPSS simulation test bed. The engine under consideration produces a thrust of approximately 400,000 N and is designed

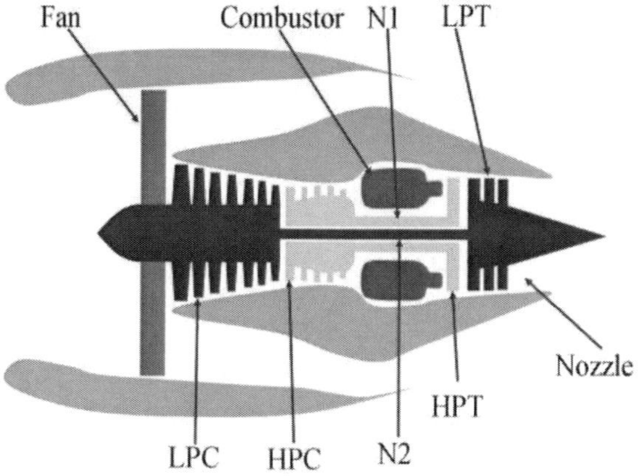

Figure 4. Gas turbine engine schematic [21].

for operation at altitude (A) from the sea level (i.e., 0 m) up to 12,200 m, Mach number (M) from 0 to 0.90, and temperatures from approximately -50°C to 50°C. The throttle resolving angle (TRA) can be set to any value in the range between $0°$ at the minimum power level and $100°$ at the maximum power level. The gas turbine engine system consists of five major rotating components, namely, fan (F), low pressure compressor (LPC), high pressure compressor (HPC), high pressure turbine (HPT), and low pressure turbine (LPT), as seen in Figure 4. Apart from the rotating components, three actuators are modeled in the simulation test bed, namely, Variable Stator Vane (VSV), Variable Bleed Valve (VBV), and Fuel Pump that controls the fuel flow rate (Wf).

Given the inputs of TRA, A and M, the interactively controlled component models in the simulation test bed compute nonlinear dynamics of real-time turbofan engine operation. A gain-scheduled control system is incorporated in the engine system, which consists of speed controllers and limit regulators for engine components. Out of the different types

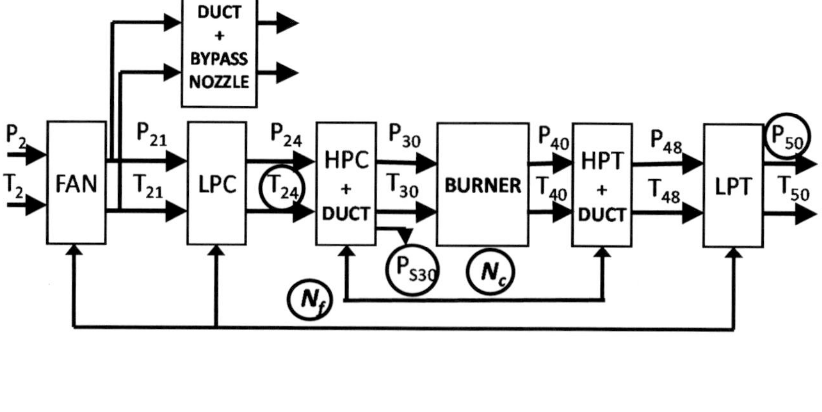

Figure 5. Schematic diagram of the C-MAPSS engine model with Sensors.

Table 1. Sensor Suite for the Engine System

Sensors	Description
T_{24}	LPC exit/ HPC inlet temperature
Ps_{30}	HPC exit static pressure
T_{48}	HPT exit temperature
P_{50}	LPT exit pressure
N_f	Fan spool speed
N_c	Core spool speed

of sensors (e.g., pressure, temperature, and shaft speed) used in the C-MAPSS simulation test bed, Table 1 lists those sensors that are commonly adopted in the Instrumentation & Control system of commercial aircraft engines, as seen in Figure 5.

In the current configuration of the C-MAPSS simulation test bed, there are 13 component level health parameter inputs, namely, efficiency parameters (ψ), flow parameters (ζ) and pressure ratio modifiers, that simulate the effects of faults and/or degradation in the engine components. Ten, out of these 13 health parameters, are selected to modify efficiency (η) and flow (ϕ) that are defined as:

- $\eta \triangleq$ Ratio of actual enthalpy and ideal enthalpy changes.
- $\phi \triangleq$ Ratio of rotor tip and axial fluid flow velocities.

For the engine's five rotating components F, LPC, HPC, LPT, and HPT, the ten respective efficiency and flow health parameters are: (ψ_F, ζ_F), $(\psi_{LPC}, \zeta_{LPC})$, $(\psi_{HPC}, \zeta_{HPC})$, $((\psi_{HPT}, \zeta_{HPT})$, and $(\psi_{LPT}, \zeta_{LPT})$. An engine component C is considered to be in nominal condition if both ψ_C and ζ_C are equal to 1 and fault can be injected in the component C by reducing the values of ψ_C and/or ζ_C. For example, $\psi_{HPC} = 0.98$ signifies a 2% relative loss in efficiency of HPC. Actuator faults can also be injected through the scale shift parameters for the three actuators, VSV, VBV and Wf.

2.2. Fatigue Damage in Polycrystalline Alloys

Fatigue damage is one of the most commonly encountered sources of structural degradation of mechanical structures, made of polycrystalline alloys. Therefore, analytical tools for online fatigue damage detection are critical for a wide range of engineering applications. The process of fatigue damage is broadly classified into two phases: crack initiation and crack propagation. The damage mechanism of these two phases are significantly different and similar feature extraction policies may not work effectively to classify different damage levels in these two phases. For example, damage evolution in the crack initiation phase is much slower, resulting in smaller change in ultrasonic signals as compared to that in the crack propagation phase. The phase transition from crack initiation to crack propagation occurs when several small micro-cracks coalesce together to develop a single large crack that propagates under the oscillating load. Several crack propagation models have been developed based on the inherent stochastic nature of fatigue damage evolution for prediction of the remaining useful life. Due to stochastic nature of material microstuctures and

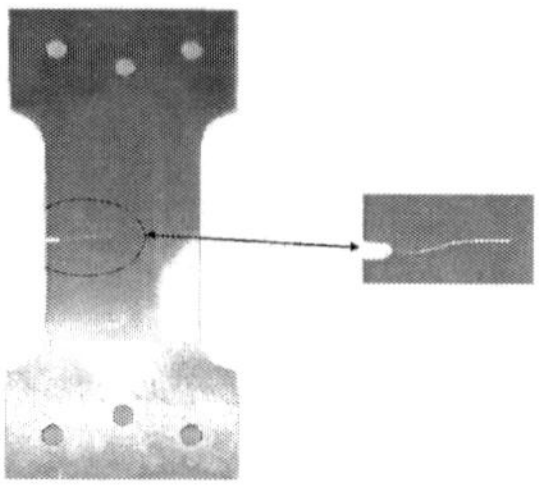

Figure 6. Cracked specimen with a side notch.

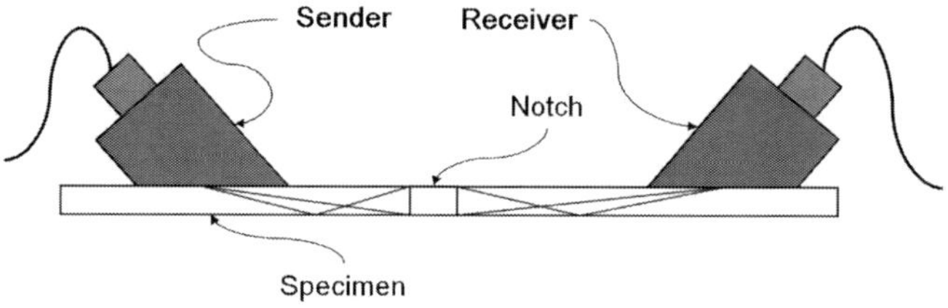

Figure 7. Ultrasonic flaw detection scheme.

operating conditions, a physics-based model would require the knowledge of the parameters associated with the material and geometry of the component. These parameters are often randomly varying and may not be accurately predicted a priori. Crack propagation rate is a function of crack length and, after a certain crack length, it becomes unstable.

Experimental Apparatus and Test Procedure: An experimental apparatus is designed to study the fatigue damage growth in mechanical structures. The apparatus consists of an MTS 831.10 Elastomer Test System that is integrated with Olympus BX Series microscope with a long working-distance objective. A camera, mounted on the microscope, takes images with a resolution of 2 micron per pixel at a distance of 20mm. The ultrasonic sensing device is triggered at a frequency of 5 MHz at each peak of the fluctuating load. Various components of the apparatus communicate over a TCP/IP network and post sensor, microscope and fatigue test data on the network in real time. This data can used by analysis algorithms for anomaly detection and health monitoring of the specimens in real-time.

Figure 6 shows a side-notched 7075-T6 aluminum alloy specimen used in the fatigue damage test apparatus. Each specimen is $3\ mm$ thick and $50\ mm$ wide, and has a slot of $1.58\ mm \times 4.57\ mm$ on one edge. The notch is made to increase the stress concentration factor that localizes crack initiation and propagation under the fluctuating load. Fatigue tests were conducted at a constant amplitude sinusoidal load for low-cycle fatigue, where the maximum and minimum loads were kept constant at 87MPa and 4.85MPa, respectively. For low cycle fatigue studied in this study, the stress amplitude at the crack tip is sufficiently high to observe the elasto-plastic behavior in the specimens under cyclic loading. A significant amount of internal damage caused by multiple small cracks, dislocations and microstructural defects alters the ultrasonic impedance, which results in signal distortion and attenuation at the receiver end.

The optical images were collected automatically at every 200 cycles by the optical microscope which is always focussed in the crack tip. As soon as crack is visible by the microscope, crack length is noted down after every 200 cycles. Ultrasonic waves with a frequency of 5 MHz were triggered at each peak of the sinusoidal load to generate data points in each cycle. Since the ultrasonic frequency is much higher than the load frequency, data acquisition was done for a very short interval in the time scale of load cycling. Therefore, it can be implied that ultrasonic data were collected at the peak of each sinusoidal load cycle, where the stress is maximum and the crack is open causing maximum attenuation of the ultrasonic waves. The slow time epochs for data analysis were chosen to be 1000 load cycles (i.e., $\sim$80 sec) apart. To generate training and test data sample multiple experiments are conducted on different specimen. For each specimen all ultrasonic signals are labeled with crack length.

2.3. Nuclear Power Plants

The International Reactor Innovative & Secure (IRIS) simulator of nuclear power plants is based on the design of a next-generation nuclear reactor. It is a modular pressurized water reactor (PWR) with an integral configuration of all primary system components. Figure 8 shows the layout of the primary side of the IRIS system that is offered in configurations of single or multiple modules, each having a power rating of $1000MWt$ (about $335MWe$) [22]. The nominal reactor core inlet and outlet temperatures are $557.6°F$ ($292°C$) and $626°F$ ($330°C$), respectively. The pressurizer, eight steam generators, and the control rod mechanism are integrated into the pressure vessel with the reactor core. There is no huge pipe used to connect these components. This design avoids the large loss of coolant accident (LOCA). The entire control rod mechanism is mounted inside the pressure vessel to avoid the problem of penetrating the the pressure vessel head by the control rod. The test-bed is built using FORTRAN programming language. This FORTRAN model includes a reactor core model, a helical coil steam generator (HCSG) model.

The test-bed is implemented on a Quad Core 2.83 GHz CPU 8 GB RAM Workstation in the laboratory of Penn State. The IRIS simulator is operated in the *integrated control* mode through built-in PID controllers, which operates all three subsystems (i.e., turbine, feedwater flow and control rods) to combine the rapid response of a reactor-following system with the stability of a turbine-following system [23]. For a load increase in the turbine, the control system makes the control rods to be withdrawn and the feedwater flow rate to be increased. Simultaneously, the control system temporarily reduces the turbine header pressure set point. The turbine governor valves open to maintain the turbine speed and to reduce the main steam header pressure to the new set point. As reactor core power, primary to secondary heat transfer, and steam flow rate gradually increase, the turbine header pressure recovers and its set point is returned to the steady-state value. In this way, the control system achieves a rapid response and a smooth transition between turbine loads [23].

The IRIS test-bed is capable of simulating both normal conditions at different operational modes (e.g., normal power maneuvers, reactor start-up, and turbine loading) and a variety of anomalous scenarios that include:

- Actuator anomalies, e.g., feedwater pump trip, malfunctions of reactor coolant pump and control rod mechanism;
- Sensor failures, e.g., malfunctions of temperature, pressure, and flow-rate sensors;
- Internal anomalies, e.g., uncertainties in the fuel temperature coefficient of reactivity, coolant heat capacity, and feedwater heat capacity.

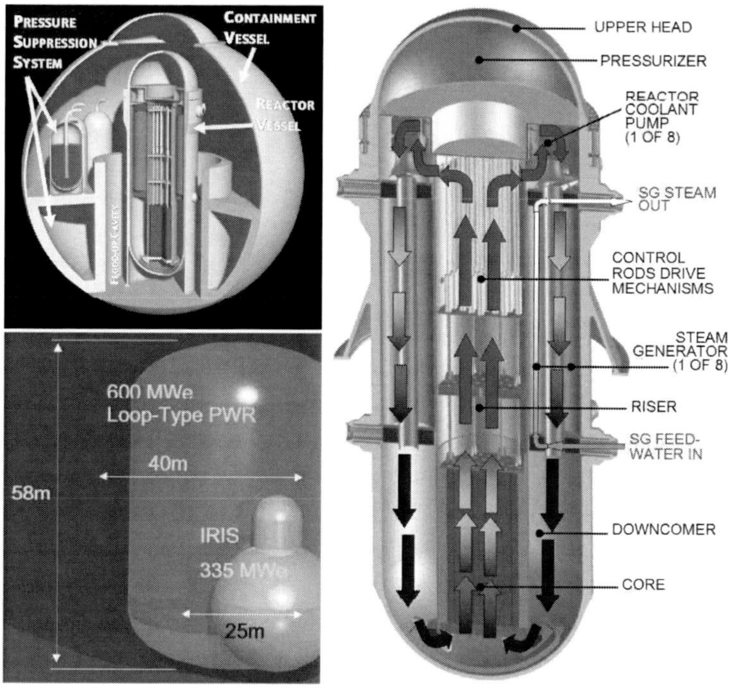

Figure 8. Layout of the primary side of the IRIS system [22].

In the IRIS test-bed, sensor degradations are realized as injected noise and disturbances. Depending on the location and modality of a sensor, there could be several different degradation levels. For example, the degradation levels in a pressure sensor have different characteristics from those of a temperature sensor. Furthermore, depending on the location and operating environment, even sensors of the same modality could have different degradation characteristics. In general, sensor degradation is categorized as the following [24, 25]:

- Constant bias and drift (i.e., slowly-varying bias);
- Change in sensor response time due to aging; and
- Change in the variance of sensor noise (e.g., due to large external electromagnetic radiation from electric motors).

Amongst the above sensor degradation types, only sensor degradation due to changes in the noise variance are investigated in this study. The rationale is that the sensors are assumed to be periodically tested and calibrated; hence, sensor degradation due to aging, bias, and drift is much less likely.

In a PWR plant, the primary coolant and feedwater temperatures are measured using resistance thermometer detectors (RTDs), and the temperature of the outlet water from the reactor core is measured using thermocouples (TCs). The TCs are mainly used for temperature monitoring, whereas the primary coolant RTDs typically feed the plant's control and safety system [25]. A case study has been presented in this work to validate the anomaly detection methodology, in which the reactor coolant pump (RCP) is chosen to be the location of a component-level degradation and the primary coolant temperature (T_{HL}) sensor (RTD) in the hot leg piping is chosen for anomaly detection in the RCP.

Since the plant controller receives feedback signals from the T_{HL} sensor, any degradation in this sensor could pervade through the plant, which will potentially affect the outputs of the remaining components due to the inherent electro-mechanical and control feedback. Component-level anomaly detection under different sensor noise variance is posed as a multi-class classification problem in the sequel.

2.4. Navy Ship-board Auxiliary Unit

A simulation test bed of shipboard auxiliary systems has been developed for testing the proposed algorithm of sensor fusion in distributed physical processes. The test bed is built upon the model of a notational hydraulic system that is coupled with a notional electrical system under an adjustable thermal loading.

The simulation model is implemented in MATLAB/Simulink as seen in Fig. 9. This distributed notional system is driven by an external speed command ω_{ref} that serves as a set point for the speed, ω_e, of the permanent magnet synchronous motor (PMSM). A mechanical shaft coupling connects the fixed displacement pump (FDP) to the PMSM. The torque load of the PMSM, T_m, is obtained from the shaft model of the hydraulic system. In turn, the speed of the PMSM, ω_e, is an input to determine the angular speed of the shaft, ω_s, which drives the FDP and the cooling fan in the thermal system. In turn, the FDP drives the hydraulic motor (HM) with a dynamic load, which consists of the thermal load, T_t, and a time-varying mechanical torque load. The proportional-integral (PI) controller regulates the

Figure 9. Notional coupled electrical, hydraulic and thermal systems.

PMSMs electrical frequency under dynamic loading conditions that arise due to fluctuations in hydraulic and thermal loading. There is a mechanical coupling between the PMSM and the FDP of the hydraulic system, which is modeled by a rotational spring-damper system applied to an inertial load. The mechanical system outputs the shaft velocity that drives the FDP and the cooling fan of the thermal system. The pump, in turn, drives the HM through the pipeline. The HM is subjected to a time-varying load with a profile defined by the user as well as the thermal load that varies with the fan efficiency of the cooling mechanism. The systems are further coupled with a feedback loop since the torque requirement of the HM is input to the PMSM of the electrical system. The model has multiple parameters that can simulate various fault conditions. There are multiple sensors in each system with different modalities such as Hall Effect sensors, torque, speed, current, temperature and hydraulic pressure sensors that are explained further in Section 5.2.

The governing equations of the electrical component model are as follows:

$$
\begin{aligned}
\frac{i_q}{dt} &= \frac{v_q - Ri_q - wL_d i_d - w\lambda_f}{L_q} \\
\frac{i_d}{dt} &= \frac{v_d - Ri_d + wL_q i_q}{L_d} \\
\frac{w_r}{dt} &= \frac{T_e - T_l - Bw_r}{J} \\
T_e &= \frac{3}{2}P[\lambda_f i_q + (L_d - L_q)i_d i_q] \\
w_r &= \frac{w}{P}
\end{aligned}
$$

where subscripts d and q have their usual significance of direct and quadrature axes in the equivalent 2-pole representation; v, i, and L are the corresponding axis voltages, stator currents and inductances; R and ω are the stator resistance and inverter frequency, respectively; λ_f is the flux linkage of the rotor magnets with the stator; P is the number of pole pairs; T_e is the generated electromagnetic torque; T_l is the load torque; B is the damping coefficient; ω_r is the rotor speed; and J is the moment of inertia.

The governing equations of the fixed displacement pump (FDP) model are as follows:

$$
\begin{aligned}
q_p &= D_p\omega_p - k_{leak}P_p \\
T_p &= \frac{D_p P_p}{\eta_m} \\
k_{leak} &= \frac{k_{HP}}{\nu\rho} \\
k_{HP} &= \frac{D_p\omega_{nom}(1-\eta_v)\nu_{nom}\rho}{P_{nom}}
\end{aligned}
$$

The governing equations of the hydraulic motor (HM) model are as follows:

$$
\begin{aligned}
\omega_m &= \frac{q_m - k_{leak}P_m}{D_m} \\
T_m &= D_m P_m \eta_m \\
k_{leak} &= \frac{k_{HP}}{\nu\rho} \\
k_{HP} &= \frac{D_m w_{nom}(1-\eta_v)\nu_{nom}\rho}{P_{nom}}
\end{aligned} \quad (12)
$$

where the subscripts p and m denote pump and motor parameters, respectively; the subscript nom denotes nominal values; q and P is the pressure differentials across delivery and terminal points; T is the shaft torque; D is the displacement; ω is the angular velocity; k_{leak} is the flow leakage coefficient; k_{HP} is the Hagen-Poiseuille coefficient; ν_v and ν_m are the volumetric and mechanical efficiencies, respectively; and ν and ρ are the fluid kinematic viscosity and density, respectively.

3. Data Space Abstraction

The first step of the SDF tool-chain involves abstraction of the data space that gives it a unique capability to fuse information from heterogeneous sensors. This also enables integration of mixed i.e., continuous, discrete and categorical data. While spatial partitioning is the major step of the abstraction process, often suitable domain transformation facilitates better feature extraction.

3.1. Data Preprocessing

Recently, it has been shown that Wavelet or Hilbert transformation of data before partitioning improves time-series feature extraction as both time and frequency domain information are considered. The wavelet space partitioning *WSP* is particulary effective for noisy data from high-dimensional dynamical systems. However, WSP has several other shortcomings such as identification of an appropriate basis function, selection of appropriate scales, and non-unique and lossy conversion of the two-dimensional scale-shift wavelet domain to a one-dimensional domain of scale-series sequences [12]. Subbu and Ray [13] have reported use of Hilbert transform to alleviate these problems. This method is called the analytic signal space partitioning (*ASSP*) as Hilbert transform converts data from time domain to analytic signal space. However, for noisy systems, it is expected that number of machine states typically increases largely, which in turn degrades computation efficiency (e.g., increased execution time and memory requirements) [6].

This section reviews a generalization of the classical Hilbert transform to modify *ASSP* for application to noisy systems. The objective here is to partition the transformed signal space such that D-Markov machines can be constructed with a small D without significant loss of information for noisy signals. The key idea is to provide a mathematical structure of the generalized Hilbert transform such that the low-frequency region is more heavily weighted than that in the classical Hilbert transform.

3.1.1. Generalization of Hilbert Transform

Hilbert transform and the associated concept of analytic signals, introduced by Gabor [26], have been widely adopted for time-frequency analysis in diverse applications of signal processing. Hilbert transform [27] of a real-valued signal $x(t)$ is defined as:

$$\widetilde{x}(t) \triangleq \mathcal{H}[x](t) = \frac{1}{\pi}\int_{\mathbb{R}} \frac{x(\tau)}{t-\tau} d\tau \tag{13}$$

That is, $\widetilde{x}(t)$ is the convolution of $x(t)$ with $\frac{1}{\pi t}$ over $\mathbb{R} \triangleq (-\infty, \infty)$, which is represented in the Fourier domain as:

$$\widehat{\widetilde{x}}(\omega) = -i\ sgn(\omega)\ \widehat{x}(\omega) \tag{14}$$

where $\widehat{x}(\omega) \triangleq \mathcal{F}[x](\omega)$ and $sgn(\omega) \triangleq \begin{cases} +1 & \text{if}\ \omega > 0 \\ -1 & \text{if}\ \omega < 0 \end{cases}$

Given the Hilbert transform of a real-valued signal $x(t)$, the complex-valued analytic signal [27] is defined as:

$$\mathcal{X}(t) \triangleq x(t) + i\,\widetilde{x}(t) \tag{15}$$

and the (real-valued) transfer function with input $\widehat{x}(\omega)$ and output $\widehat{\mathcal{X}}(\omega)$ is formulated as:

$$G(\omega) \triangleq \frac{\widehat{\mathcal{X}}(\omega)}{\widehat{x}(\omega)} = 1 + sgn(\omega) \tag{16}$$

Lohmann et al. [28] introduced the concept of a generalized Hilbert transform in the fractional Fourier space instead of the conventional Fourier space; a discrete version of this generalized Hilbert transform was developed later [29]. For geophysical applications, Luo et al. [30] proposed another type of generalized Hilbert transform that is essentially the windowed version of traditional Hilbert transform. However, the notion of generalization of Hilbert transform presented in the sequel is different from that in the previously published literature.

Let us define a generalized Hilbert transform as: $\mathcal{H}^\alpha$ of a real-valued signal $x(t)$ as the convolution:

$$\widetilde{x}^\alpha(t) \triangleq \mathcal{H}^\alpha[x](t) = x(t) * \left(\frac{sgn(t)}{\pi\ |t|^\alpha}\right) \text{ for } \alpha \in (0,1] \tag{17}$$

It is shown in the sequel that, as $\alpha \uparrow 1$ (i.e., the values of α form an increasing sequence of positive real numbers with the limit equal to 1), $\mathcal{H}^\alpha$ converges to $\mathcal{H}$, where $\mathcal{H}$ is the classical Hilbert transform defined in Eq. (13); that is, $\mathcal{H}^1 \equiv \mathcal{H}$.

Two lemmas are presented, which are necessary for derivation of the main results in the Fourier space (detailed proofs can be found in [14]).

Lemma 3.1.

$$\int_{-\infty}^{\infty} \frac{e^{-i\omega t}}{\pi|t|^\alpha} sgn(t)dt = -i\ sgn(\omega)\frac{2}{\pi}\frac{\Gamma(1-\alpha)}{|\omega|^{1-\alpha}} sin\left(\frac{\pi}{2}(1-\alpha)\right) \tag{18}$$

where $\alpha \in (0,1)$*; and* $\Gamma(1-\alpha) \triangleq \int_0^\infty \frac{e^{-y}}{y^\alpha}dy$.

Lemma 3.2. *As* $\alpha \uparrow 1$*, the integral* $\int_{-\infty}^{\infty} \frac{e^{-i\omega t}}{\pi|t|^\alpha} sgn(t)dt \rightarrow -i\ sgn(\omega)$*, i.e.,*

$$\lim_{\alpha\uparrow 1} \Gamma(1-\alpha)\left(\frac{2}{\pi} sin\frac{\pi}{2}(1-\alpha)\right) = 1. \tag{19}$$

Taking Fourier transform of the convolution in Eq. (17) and an application of Lemma 3.1 yield:

$$\begin{aligned}\widehat{\widetilde{x}^\alpha}(\omega) &= \mathcal{F}\left(x(t) * \frac{sgn(t)}{\pi|t|^\alpha}\right)\\ &= \mathcal{F}(x(t)) \cdot \mathcal{F}\left(\frac{sgn(t)}{\pi|t|^\alpha}\right)\\ &= -i\ sgn(\omega)\ \frac{2}{\pi}\ \frac{\hat{x}(\omega)\Gamma(1-\alpha)}{|\omega|^{1-\alpha}}\ sin\left(\frac{\pi}{2}(1-\alpha)\right)\end{aligned} \tag{20}$$

Since $\Gamma(1-\alpha) < \infty$ for $\alpha \in (0,1)$, the generalized Hilbert transform $\widetilde{x}^{\alpha}(t)$ can be evaluated by taking the inverse Fourier transform of $\widehat{\widetilde{x}^{\alpha}}(\omega)$.

The above formulation shows that reduced α puts more weight on the low frequency part of the signal $x(t)$ and hence more effectively attenuates the high-frequency noise than the classical Hilbert transform. Following Lemma 3.2, as $\alpha \uparrow 1$, Fourier transform of the signal $\frac{sgn(t)}{\pi|t|^{\alpha}}$ converges to $-i\ sgn(\omega)$. This leads to the fact, that as $\alpha \uparrow 1$, $\mathcal{H}^{\alpha}$ converges to $\mathcal{H}$, where $\mathcal{H}$ is the classical Hilbert transform defined in Eq. (13).

Analogous to the analytic signal in Eq. (15), the (complex-valued) generalized analytic signal of the real-valued signal $x(t)$ is defined as:

$$\mathcal{X}^{\alpha}(t) \triangleq x(t) + i\,\widetilde{x}^{\alpha}(t) \tag{21}$$

and the (real-valued) transfer function with input $\widehat{x}(\omega)$ and output $\widehat{\mathcal{X}}^{\alpha}(\omega)$ is formulated as:

$$G^{\alpha}(\omega) \triangleq \frac{\widehat{\mathcal{X}}^{\alpha}(\omega)}{\widehat{x}(\omega)} = 1 + sgn(\omega)\left(\frac{\omega_0(\alpha)}{|\omega|}\right)^{(1-\alpha)} \tag{22}$$

where $\omega_0(\alpha) \triangleq \left(\frac{2}{\pi\,\Gamma(1-\alpha)} sin\left(\frac{\pi}{2}(1-\alpha)\right)\right)^{\frac{1}{1-\alpha}}$.

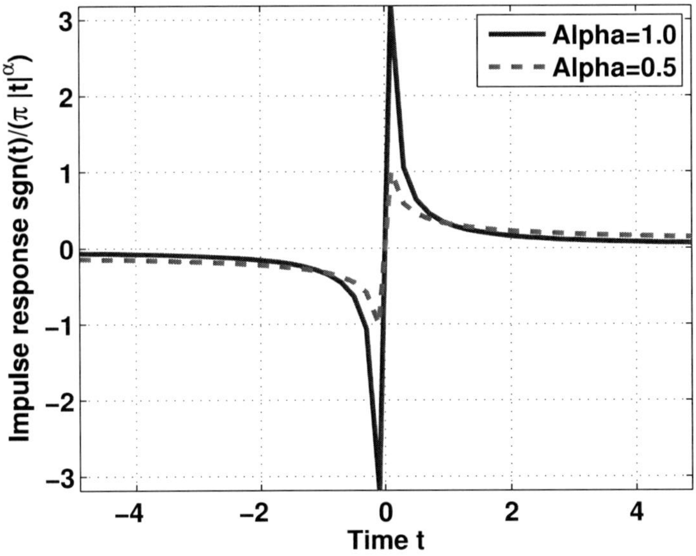

Figure 10. Impulse response $\left(\frac{sgn(t)}{\pi|t|^{\alpha}}\right)$ of the generalized Hilbert transform.

Remark 3.1. *For $\alpha = 1$, it follows from Eq. (17) that the real-valued signal $x(t)$ is convoluted with $\frac{1}{\pi t}$. The implication is that the effects of memory in the signal $x(t)$ reduce as fast as $\frac{1}{\pi|t|}$. As α is decreased, the tail of the impulse response of the generalized Hilbert transform $\widetilde{x}^{\alpha}(t)$ becomes increasingly fat as seen in Fig. 10. Hence, for $0 < \alpha < 1$, the generalized analytic signal $\mathcal{X}^{\alpha}(t)$ captures more (low-frequency) information from time series data than that for $\alpha = 1$.*

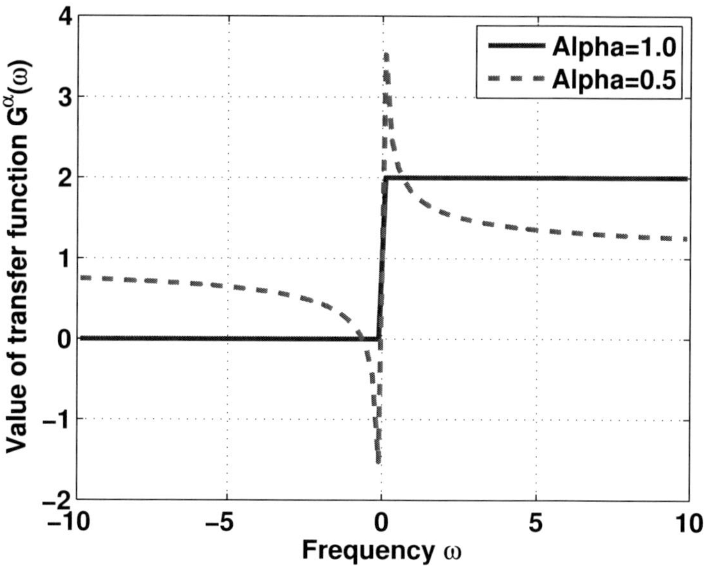

Figure 11. Transfer function $G^{\alpha}(\omega)$ of the generalized analytic signal.

Remark 3.2. *Fourier transform of a real-valued signal does not contain any additional information beyond what is provided by the positive frequency components, because of the symmetry of its spectrum. Therefore, in the construction of an analytic signal in Eq. (15) and its transfer function in Eq. (16), Hilbert transform removes the negative frequency components while doubling the positive frequency components. For* $\alpha < 1$, *it follows from Fig. 11 that the negative frequency components of the transfer function* $G^{\alpha}(\omega)$ *of a generalized analytic signal are no longer zero. Therefore, the generalized analytic signal in Eq. (21) is not an analytic signal in the sense of Gabor [26] for* $\alpha < 1$. *However, the transfer functions of both analytic and generalized analytic signals are real-valued almost everywhere in the range* $\omega \in \mathbb{R}$. *The phase of the (real-valued) transfer function* $G^{\alpha}(\omega)$ *is either* 0 *or* $-\pi$ *as explained below.*

- *The phase of* $G(\omega)$ *(i.e.,* $G^{\alpha}(\omega)$ *for* $\alpha = 1$*) is* 0 *radians in the frequency range* $(0, \infty)$. *Its magnitude in the negative frequency range* $(-\infty, 0)$ *is identically equal to* 0*; therefore, the phase in this range is inconsequential.*

- *For* $0 < \alpha < 1$, *the phase of* $G^{\alpha}(\omega)$ *is* $-\pi$ *radians in the frequency range* $(-\omega_0(\alpha), 0)$, *where* ω_0 *is defined in Eq. (22), and is* 0 *radians in the frequency range* $\big(-\infty, -\omega_0(\alpha)\big) \bigcup (0, \infty)$.

3.1.2. Test Results and Validation

The concept of generalized Hilbert transform is tested and validated by symbolic analysis of time series data, generated from the same apparatus of nonlinear electronic systems reported in the earlier publication [13].

The nonlinear active electronic system in the test apparatus emulates the forced Duffing equation:

$$\frac{d^2x}{dt^2} + \beta\frac{dx}{dt} + x(t) + x^3(t) = Acos(\omega t) \tag{23}$$

Having the system parameters set to $\beta = 0.24$, $A = 22.0$, and $\omega = 5.0$, time series data of the variable $x(t)$ were collected from the electronic system apparatus. These data sets do not contain any substantial noise because the laboratory apparatus is carefully designed to shield spurious signals and noise. Therefore, to emulate the effects of noise in the time series data, additive first-order colored Gaussian noise was injected to the collected time series data to investigate the effects of signal-to-noise ratio (SNR). The profile of a typical signal, contaminated with 10 *db* additive Gaussian noise (i.e., $SNR = 10$), is shown in Fig. 12.

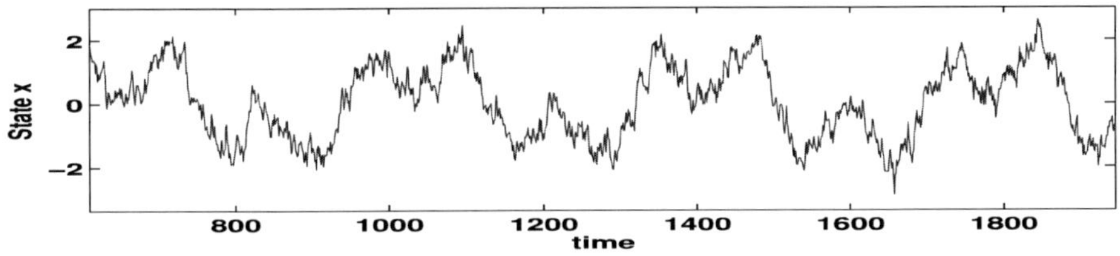

Figure 12. Signal contaminated with 10 *db* additive colored Gaussian noise.

Construction of the Transformed Phase Space: Let the real-valued noisy time-series data $x(t)$ contain *N* data points. Upon generalized Hilbert transformation of this data sequence, a complex-valued generalized analytic signal $\mathcal{X}^\alpha(t)$ is constructed. Similar to the procedure described in [13], $\mathcal{X}^\alpha(t)$ is represented as a one-dimensional trajectory in the two-dimensional pseudo-phase space. Let Ω be a compact region in the pseudo-phase space, which encloses the trajectory of *N* such data points.

Partitioning and Symbol Generation: The next task is to partition Ω into finitely many mutually exclusive and exhaustive segments, where each segment is labelled with a symbol or letter. The partitioning is based on the magnitude and phase of the complex-valued signal $\mathcal{X}^\alpha(t)$ as well as the density of data points in these segments, following the procedure described in [13]. Each point in the partitioned data set is represented by a pair of symbols; one belonging to the alphabet Σ_R based on the magnitude (i.e., in the radial direction) and the other belonging to the alphabet Σ_A based on the phase (i.e., in the angular direction). In this way, the complex-valued signal $\mathcal{X}^\alpha(t)$ is partitioned into a symbol sequence by associating each pair of symbols to a single symbol belonging to an alphabet Σ that is defined as:

$$\Sigma \triangleq \{(\sigma_i, \sigma_j) \in \Sigma_R \times \Sigma_A\} \text{ and } |\Sigma| = |\Sigma_R||\Sigma_A| \tag{24}$$

The results presented in this chapter are generated with $\Sigma_R = 8$ in the radial direction and $\Sigma_A = 5$ in the angular direction, i.e., $|\Sigma| = 40$.

State Transition Matrices: The symbol sequence is now used to construct *D*-Markov machine models [15]. The assumption of statistical stationarity of the symbol sequence is implicit in the construction of Markov models. In this study, Markov chain models of depth $D = 1$ and $D = 2$ have been constructed.

Modeling of the symbolic process as a $(D = 1)$ Markov chain involves evaluation of the Π^1 matrix, where the ij^{th} matrix element π^1_{ij} is defined as the probability that $(n+1)^{th}$ state is i given that the nth state was j, i.e.,

$$\pi^1_{ij} \triangleq P\left(q^1_{n+1} = i \mid q^1_n = j\right) \tag{25}$$

where q_k is the state at discrete time instant k. Evidently, the size of the Π matrix is $|\Sigma| \times |\Sigma|$, where $|\Sigma|$ is the number of symbols in the alphabet Σ.

Modeling the symbolic process as a $(D = 2)$ Markov chain involves evaluation of a 3-dimensional matrix, where the ijk^{th} matrix element π^2_{ijk} is defined as:

$$\pi^2_{ijk} \triangleq P\left(q^2_{n+2} = i | q^2_{n+1} = j, q^2_n = k\right) \tag{26}$$

and size of the (sparse) Π^2 matrix is $|\Sigma| \times |\Sigma| \times |\Sigma|$.

Remark 3.3. *Elements of both Π^1 and Π^2 matrices are estimated by conditional frequency count and their convergence requires a symbol sequence of sufficient length. This aspect has been discussed in [17][6] and is referred to as the stopping rule that assigns a bound on the length of the symbol sequence for parameter identification of the stochastic matrices Π^1 and Π^2.*

Computation of Mutual Information: Effectiveness of generalized Hilbert transform for Markov model construction has been examined from an information theoretic perspective [5]. The rationale is that, in a noise-corrupted system, higher values of mutual information imply less uncertainties in the symbol sequence. The mutual information $\mathcal{I}$ is expressed in terms of entropy $\mathcal{S}$ for both $(D = 1)$ and $(D = 2)$ Markov chains in the following set of equations:

$$\mathcal{I}\left(q_{n+3}; q_{n+2}\right) \triangleq \mathcal{S}\left(q_{n+3}\right) - \mathcal{S}\left(q_{n+3} | q_{n+2}\right) \tag{27}$$

$$\mathcal{S}\left(q_{n+3}\right) \triangleq -\sum_{\ell=1}^{|\Sigma|} P\left(q_{n+3} = \ell\right) \log_2 P\left(q_{n+3} = \ell\right) \tag{28}$$

Usage of maximum entropy partitioning [12] for symbol generation yields: $\mathcal{S}\left(q_{n+3}\right) = \log_2\left(|\Sigma|\right)$.

$$\mathcal{S}\left(q_{n+3} | q_{n+2}\right) \triangleq \sum_{\ell=1}^{|\Sigma|} P\left(q_{n+2} = \ell\right) \mathcal{S}\left(q_{n+3} | q_{n+2} = \ell\right) \tag{29}$$

where

$$\mathcal{S}\left(q_{n+3} | q_{n+2} = \ell\right) = -\sum_{j=1}^{|\Sigma|} P\left(q_{n+3} = j \mid q_{n+2} = \ell\right) \cdot \log_2 P\left(q_{n+3} = j | q_{n+2} = \ell\right) \tag{30}$$

$$\mathcal{I}\left(q_{n+3}; q_{n+2}, q_{n+1}\right) \triangleq \mathcal{S}\left(q_{n+3}\right) - \mathcal{S}\left(q_{n+3}|q_{n+2}, q_{n+1}\right) \tag{31}$$

$$\begin{aligned}&\mathcal{S}\left(q_{n+3}|q_{n+2}, q_{n+1}\right)\\ &\quad \triangleq -\sum_{i=1}^{|\Sigma|}\sum_{j=1}^{|\Sigma|} P\left(q_{n+2}=i, q_{n+2}=j\right)\cdot\\ &\quad \mathcal{S}\left(q_{n+3}|q_{n+2}=i, q_{n+1}=j\right)\end{aligned} \tag{32}$$

where

$$\begin{aligned}&\mathcal{S}\left(q_{n+3}|q_{n+2}=i, q_{n+1}=j\right)\\ &\quad = -\sum_{\ell=1}^{|\Sigma|} P\left(q_{n+3}=\ell|q_{n+2}=i, q_{n+1}=j\right)\cdot\\ &\quad \log_2 P\left(q_{n+3}=\ell|q_{n+2}=i, q_{n+j}=j\right)\end{aligned} \tag{33}$$

Based on Eq. (25) and Eqs. (27) to (30), the mutual information $\mathcal{I}\left(q_{n+3}; q_{n+2}\right)$ is calculated from the Π^1 matrix. Similarly, based on Eq. (26) and Eqs. (31) to (33), $\mathcal{I}\left(q_{n+3}; q_{n+2}, q_{n+1}\right)$ is calculated from the Π^2 matrix. Then, information gain (abbreviated as $\mathcal{I}_{\mathcal{G}}$) with $D = 2$ instead of $D = 1$ in the Markov chain construction is defined as:

$$\mathcal{I}_{\mathcal{G}} \triangleq \mathcal{I}\left(q_{n+3}; q_{n+2}, q_{n+1}\right) - \mathcal{I}\left(q_{n+3}; q_{n+2}\right) \tag{34}$$

Pertinent Results:

The pertinent results on mutual information and information gain are presented in Fig. 13 and Fig. 14, respectively. Although results are shown only for $SNR = \infty, 10, 4$ and 0, several other experiments with intermediate values of SNR between ∞ and 0 were performed, which shows the same trend.

The information gain is always a positive quantity as seen in Eq. (34). In other words, there is always a gain in information upon increasing the depth of the Markov chain model. Pertinent inferences, drawn from these results, are presented below.

1. Mutual information increases with decrease in α irrespective of D and SNR as seen in Fig. 13.

2. Information gain $\mathcal{I}_{\mathcal{G}}$ (see Eq. (34) and Fig. 14) is minimal for $SNR \to \infty$ (i.e., for the signal with no noise injection). Therefore, *D*=1 Markov chain should be adequate with $ASSP$ using conventional Hilbert transform (see Eq. 13) for low-noise signals.

3. As SNR is decreased (i.e., percentage of additive noise is increased), information gain $\mathcal{I}_{\mathcal{G}}$ increases for all values of α in the range of 1.0 down to about 0.2. As α is decreased, information gain decreases as seen in Fig. 14. Therefore, even for a considerable amount of noise, a smaller value of α should be able to achieve noise attenuation and thus allow usage of $D = 1$ in *D*-Markov machines.

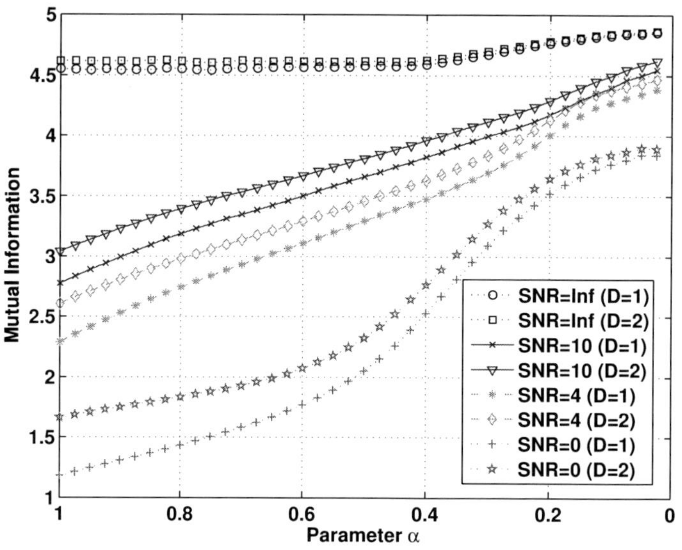

Figure 13. Profiles of mutual information.

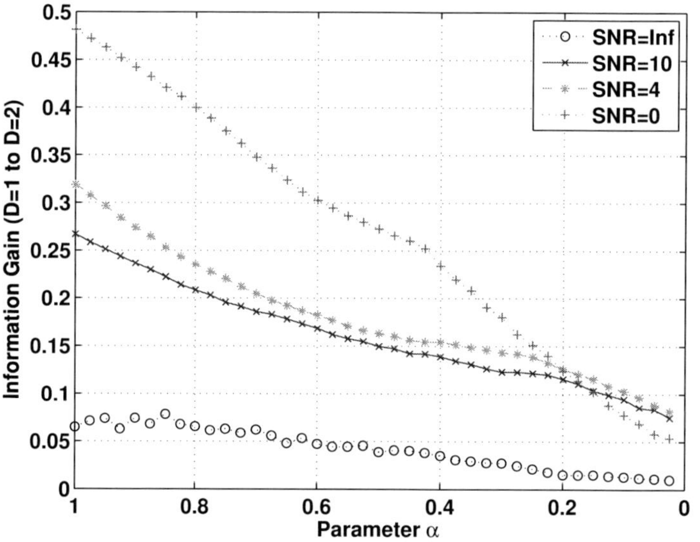

Figure 14. Profiles of information gain.

4. Results for a pathological case with $SNR \to 0$, (i.e., complete noise capture of the signal) in Fig. 13 and Fig. 14 show similar trends as above. The crossing of the information gain curves in Fig. 14 at low values of α (e.g., $\alpha \leq 0.2$) could possibly be attributed to the effects of coarse graining [7] due to symbol generation.

The rationale for the observed trends in Fig. 13 and Fig. 14 is reiterated as follows. An increase in depth D captures the effects of longer memory in the signal, and similarly reduced α puts more weight on the low-frequency components, which assists noise reduction.

The following conclusions are drawn from the validation results of the proposed generalization of Hilbert transform on a laboratory apparatus of nonlinear electronic systems as presented above.

- Generalized Hilbert transform with a smaller value of parameter α is capable of extracting more information from a data sequence irrespective of the depth of the D-Markov machine chosen for modeling.
- Information gain for a larger depth D reduces with smaller values of the parameter α.
- By selecting small values of the parameter α in the generalized Hilbert transform, it is possible to avoid using a computationally expensive larger depth D without loss of significant information.

3.2. Optimal Feature Extraction via Partitioning

Partitioning is the process of coarse-graining of the data-space under some notion of similarity, transforms continuous valued variables to a discrete symbolic space. This is an important pre-processing step essential for several applications such as *symbolic time-series analysis* of dynamical systems, data-mining and machine learning. Despite the necessity and importance of discretization, there is no standard way to approach it. This is because several factors such as the nature of the dynamical system or data set in question, choice of the similarity metric, and desired model simplicity affect the nature of a discretization scheme that is appropriate. Moreover, even if one can define the optimality criteria for discretization, the process will be NP-complete [31]. As pointed out by [32], while discretization is desirable pre-processing step, practical discretization schemes are necessarily heuristic in nature and a large number of such schemes have been proposed in the literature. The simplest methods include *equal interval width* (uniform) and *equal interval frequency* (maximum entropy) discretization or partitioning. One key distinction among different methods is whether the discretization process was *supervised* or *unsupervised*. In a supervised process a metric based on a discrete variable (e.g. class label) is used as a feedback to guide the discretization process, while a unsupervised method does not use such information. The literature is vast and there are several reviews available that summarize and compare different techniques [33, 34, 35, 36, 37]. This section reviews two recently developed multi-variate partitioning schemes with particular objective of symbolic dynamic analysis of time-series data from physical systems. While the first scheme (maximally bijective discretization) aims to optimize symbolic modeling of dynamical systems, the second scheme (optimal partitioning for classification) is particularly useful for time-series classification.

3.2.1. Maximally Bijective Partitioning

Consider a dynamical system with input/output signals as:

$$U(t) = 2\cos(0.25t) \tag{35}$$

$$Y(t) = \cos(0.5t) \tag{36}$$

To build a symbolic model of such a system, both input and output space need to be discretized. However, if the discretization of input and output spaces are completely independent, then important functional relationship in the continuous domain may not remain preserved in the discrete domain. Based on this key idea, this section reviews the recently developed methodology and algorithm for a discretization scheme that maximizes the degree of input-output symbol correspondence, hence named Maximally Bijective Discretization (MBD).

Although the required notations follow earlier discussion, it is repeated here for enhanced readability. Let the time series data of a variable generated from a physical complex system or its dynamical model be denoted as $\mathbf{q}$. A compact (i.e., closed and bounded) region $\Omega \in \mathbb{R}^n$, where $n \in \mathbb{N}$, within which the time series is circumscribed, is identified. Let the space of time series data sets be represented as $\mathbf{Q} \subseteq \mathbb{R}^{n \times N}$, where the $N \in \mathbb{N}$ is sufficiently large for convergence of statistical properties within a specified threshold. Note, n represents the dimensionality of the time-series and N is the number of data points in the time series. A discretization encodes Ω by introducing a partition $\mathbb{B} \equiv \{B_0, ..., B_{(p-1)}\}$ consisting of p mutually exclusive (i.e., $B_j \cap B_k = \emptyset \, \forall j \neq k$), and exhaustive (i.e., $\cup_{j=0}^{p-1} B_j = \Omega$) cells. For one-dimensional time series data, a discretization consisting of p cells is represented by $p - 1$ points that serve as cell boundaries. In the sequel, a p-cell discretization $\mathbb{B}$ is also expressed as $\Gamma_m \triangleq \{\gamma_1, \gamma_2, \cdots, \gamma_{m-1}\}$, where γ_i denotes a bin boundary.

Let the complex system under consideration has m real-valued output variables and n real-valued input variables that are denoted as $w_1, \cdots, w_m$ and $u_1, \cdots, u_n$ respectively. In the continuous domain, an input variable u_i for any i is represented as a function of all output variables for the purpose of discretization as follows:

$$u_i = h_i(w_1, \cdots, w_m) \tag{37}$$

Suppose a m-dimensional discretization is imposed on the space of output variables that divides the data for output variables into K discrete classes. The goal is to discretize each input variable based on the defined classes of output variables. This work develops a *Maximally Bijective* scheme of performing the discretization of each input variable separately. The problem is formulated in the sequel for any input variable u (omitting the subscript).

Let $\mathbb{B} \equiv \{B_0, ..., B_{(l-1)}\}$ be a discretization of one of the input variables, where $B_j, j \in \{0, ..., l-1\}$ is a bin of the discretization. With this setup, $correspondence$ between a class and a bin is defined as follows:

$$\begin{aligned} \text{If } i &= \arg\max_k P(C_k | x \in B_j) \\ \text{then, } C_i &\Rightarrow B_j \end{aligned} \tag{38}$$

where, $P(\cdot)$ denotes a probability function and $x \in \mathbb{R}$. In words, class C_i corresponds to bin B_j. Note, a class C_i may correspond to more than one disjoint bins. From this perspective, a reward function is defined as follows:

$$R(\mathbb{B}|x) = \{P(C_i|x \in B_j) : C_i \Rightarrow B_j\} \tag{39}$$

Note that this reward function signifies the notion of bijection, i.e., with higher reward, the probability of a class being corresponding to a bin increases. With this setup, the total expected reward is calculated as

$$TR(\mathbb{B}) = \int_X R(\mathbb{B}|x)P(x)dx \tag{40}$$

The goal here is to maximize this total reward function, hence the discretization scheme is called Maximally Bijective. It is clear from the formulation that maximizing the total reward function is equivalent to maximizing $R(\mathbb{B}|x)$ at each x.

$$\begin{aligned}\mathbb{B}^* &= \arg\max_{\mathbb{B}} \{R(\mathbb{B}|x)\ \forall x\} \\ &= \arg\max_{\mathbb{B}} \{P(C_i|x \in B_j) : C_i \Rightarrow B_j\ \forall x\}\end{aligned} \tag{41}$$

This observation leads to an algorithm that is developed and used to identify MBD in this work. Let $P_m(x)$ denotes $\{P(C_i|x \in B_j) : C_i \Rightarrow B_j\}$ at any x. Note, by this definition

$$P_m(x) = \max_i P(C_i|x)\ \forall x \tag{42}$$

With this setup an overview of the proposed algorithm is provided below:

Overview of the Algorithm

$x = \min(D)$
$k = 1$
while $x < \max(D)$ **do**
  Identify C_i such that $P(C_i|x) = P_m(x)$;
  Identify C_j such that $P(C_j|x + dx) = P_m(x + dx)$;
  if $i \neq j$ **then**
    $\gamma_k = x$ % Note, γ_k denotes the kth bin boundary
    $k \leftarrow k + 1$
  end if
  $x \leftarrow x + dx$
end while

However, the primary issue with the above process is reliable estimation of $P(C_i|x)$ from data [32]. This work adopts a basic frequency counting (over an interval) method to estimate this conditional probability. This means that the optimization process begins with considering a small window in the domain (around $\min(D)$) to identify the most probable class in that interval. Then the window is slided across the data range of the input variable and bin boundaries (γ_ks) are placed where the most probable class changes from one interval to the next. However, this approximation may result in sub-optimality of the solution. The rationale is as follows: Let $[a, b] \subset \mathbb{R}$ be an interval over which the frequencies of different classes are estimated to identify the most probable class. The frequency of class C_i in the interval $[a, b]$ is denoted as $n(C_i)|_{[a,b]}$ and it can be represented as

$$n(C_i)|_{[a,b]} = \int_a^b P(C_i|x)P(x)dx \tag{43}$$

To identify the most probable class in $[a, b]$, one needs to compare $n(C_i)|_{[a,b]}$ with

$n(C_j)|_{[a,b]}$. However, it is known that

$$\begin{aligned} \int_a^b P(C_i|x)P(x)dx &\gtrless \int_a^b P(C_j|x)P(x)dx \\ \nRightarrow P(C_i|x) &\gtrless P(C_j|x)\ \forall x \in [a,b] \end{aligned} \quad (44)$$

Consequently, the solution may become suboptimal due to the consideration of intervals of finite width. Also, in a realistic setting, the intervals need to be wide enough to avoid significant effects of noise in the data.

However, the following lemma can be stated in this scenario:

Lemma 3.3. *The total reward (TR) is a nondecreasing function of adding new bin boundaries.*

This ensures that the total reward at least does not reduce when a new bin boundary is introduced in the sequential algorithm proposed here. A rough proof sketch of this property is provided here:

Proof Sketch: First of all, it should be noted that introduction of a new bin boundary can be considered as splitting one of the current bins. Let B_j be the bin which is splitted into B_j^1 and B_j^2 with the introduction of a new bin boundary. Also, let class C_i be the most probable class of the bin B_j with frequency $n(C_i)|_{B_j}$. Let the total reward function before and after splitting be denoted as $TR(k)$ and $TR(k+1)$ respectively. Now, three cases are possible with this splitting.

- *Case I:* In both new bins B_j^1 and B_j^2, C_i is no longer the most probable class and the most probable classes in B_j^1 and B_j^2 are say C_p and C_q respectively. In this case,

$$\begin{aligned} TR(k+1) = n(C_p)|_{B_j^1} + n(C_q)|_{B_j^2} > \\ n(C_i)|_{B_j^1} + n(C_i)|_{B_j^2} = TR(k) \end{aligned} \quad (45)$$

- *Case II:* In one of the new bins, say in B_j^1, C_i is still the most probable class. However, in B_j^2, C_q is the most probable class. In this case,

$$\begin{aligned} TR(k+1) = n(C_i)|_{B_j^1} + n(C_q)|_{B_j^2} > \\ n(C_i)|_{B_j^1} + n(C_i)|_{B_j^2} = TR(k) \end{aligned} \quad (46)$$

- *Case III:* In both new bins, C_i is still the most probable class. In this case,

$$TR(k+1) = n(C_i)|_{B_j^1} + n(C_i)|_{B_j^2} = TR(k) \quad (47)$$

Therefore, $TR(k+1) \geq TR(k)$, i.e., total reward does not decrease when new bin boundaries are introduced sequentially in the algorithm described above. This is also compatible with the intuitive notion of increase in reward with increase in complexity.

Figure 15 shows the MBD for the illustrative example problem introduced in the beginning. It is clear from the result that in a numerically stable scenario, the present approach can identify the bin boundaries of an admissible discretization. Furthermore, Fig. 16

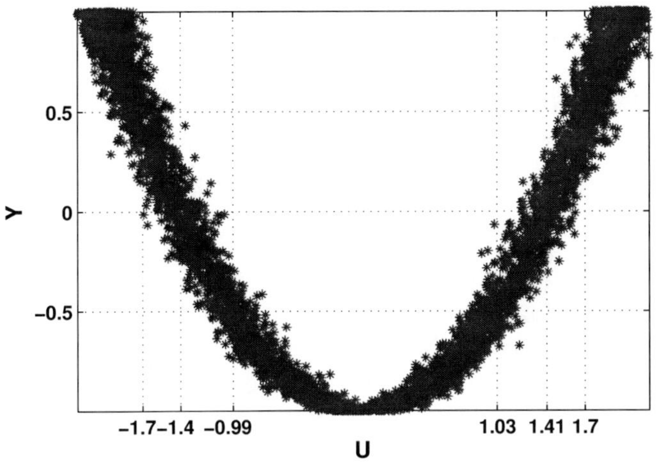

Figure 15. Illustrative example: MBD of noisy input U Given a Uniform discretization of output Y; Bin boundaries are marked on corresponding axes as grid lines.

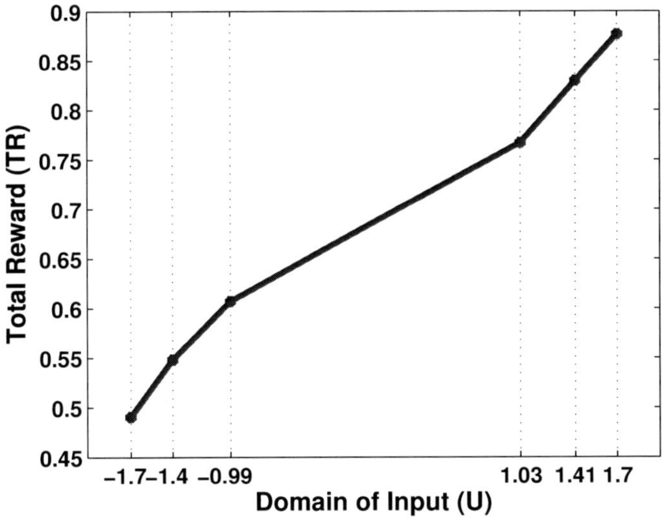

Figure 16. Monotonic increase in Total Reward with addition of new bin boundaries.

shows that the total reward monotonically increases with addition of new bin boundaries as stated in Lemma 3.3. The total reward of the MBD in this example is found to be 0.88. For comparison purposes, uniform discretization of same complexity, i.e., with same number of cells/bins (with bin boundaries $[-1.66, -1.00, -0.350.300.961.61]$) has total reward of 0.79 and maximum entropy discretization with same complexity (with bin boundaries $[-1.80 - 1.25 - 0.450.451.231.80]$) has total reward of 0.76. The following remark summarizes some of the key aspects the proposed discretization policy.

Remark 3.4. *The MBD scheme developed here avoids computationally expensive iterative search for bin boundaries in the data space. Instead, this algorithm scans data space once to sequentially place optimal bin boundaries and in turn identifies the optimal number of* *cells/bins in a stable numerical scenario. Furthermore, it should be noted that discretization of the output variables can be revised once MBD of the input variables are obtained. This process essentially involves removing certain bin boundaries of output variables that do not have effect in the MBD of the input variables. Finally, this scheme can be used for any complex dynamical system. However, it will be particularly useful for highly nonlinear systems where traditional uniform or maximum entropy discretizations may fail to preserve the functional relationships by a great deal.*

It should be noted that there are other popular cost functions (involving e.g., maximizing mutual information and minimizing Bayes risk) available in literature that may have similar effects on discretization as the current reward function proposed in this study. Therefore, comparison of the strategy presented here with the other similar cost/reward functions is an important topic of future investigation.

3.2.2. Optimal Partitioning for Classification

In the SDF framework, a symbol sequence $\{\mathbf{s}\}$ is compressed by a probabilistic finite state automaton ($PFSA$), where $j, k \in \{1, 2, ..., r\}$ are the states of the $PFSA$ with the $(r \times r)$ state transition matrix $\Pi = [\pi_{jk}]$ that is obtained at slow-scale epochs (different classes in this context). (Note: Π is a stochastic matrix, i.e., the transition probability $\pi_{jk} \geq 0$ and $\sum_k \pi_{jk} = 1$). To compress the information further, the state probability vector $\mathbf{p} = [p_1 \ \cdots \ p_r]$ that is the left eigenvector corresponding to the (unique) unity eigenvalue of the irreducible stochastic matrix Π is calculated. The vector $\mathbf{p}$ is the extracted feature vector and is a low-dimensional compression of the long time series data representing the dynamical system at the slow-scale epoch.

For classification using SDF, the reference time series, belonging to a class denoted as Cl_1, is symbolized by one of the standard partitioning schemes (e.g., Uniform Partitioning (UP) or Maximum Entropy partitioning (MEP)) [15, 12, 13]. Then, using the steps described earlier, a low-dimensional feature vector $\mathbf{p}^{Cl_1}$ is constructed for the reference slow-scale epoch. Similarly, from a time series belonging to a different class denoted as Cl_2, a feature vector $\mathbf{p}^{Cl_2}$ is constructed using the same partitioning as in Cl_1. The next step is to classify the data in the constructed low-dimensional feature space. In this respect, there are many options for selecting classifiers that could either be parametric or non-parametric. Among the parametric classifiers, one of the commonly used techniques relies on the second-order statistics in the feature space, where the mean feature is calculated for every class along with the variance of the feature space distribution in each class of the training set. Then, a test feature vector is classified by using the Mahalnobis distance [38] or the Bhattacharya distance [39] of the test vector from the mean feature vector of each class. However, these methods are not efficient for non-Gaussian distributions, where the feature space distributions may not be adequately described by the second order statistics. Consequently, a non-parametric classifier (e.g., k-NN classifier [40]) is potentially a better choice. In this study, the k-NN classifier has been used because the Gaussian probability distribution cannot be assured in the feature space. However, in general, any other suitable classifier, such as the Support Vector Machines (SVM) or the Gaussian Mixture Models (GMM) may also be used [40]. To classify the test data set, the time series sets are converted into feature vectors using the same partitioning that has been used to generate the training features. Then, using the labeled training features, the test features are classified by a k-NN classifier with suitable specifications (neighborhood size and distance metric).

Many optimization criteria have been reported in literature for feature extraction in multi-class classification problems. However, none can be more fundamental than minimization of classification error. Although, there have been attempts to approximate the classification error as the Bayes error using pairwise weighting functions for multi-class problems [41], the Bayes error itself cannot be expressed analytically except a few special cases. On the other hand, even if the Fisher criteria is not directly equivalent to classification error, it is widely used as a feature extraction optimization criteria in literature. This criteria involves maximization of the inter-class variance and minimization of the intra-class variance for the training data set. For a K-class ($K > 2$) classification problem, a general K-class Fisher criterion may be used; alternatively, a sum of $K(K-1)/2$ number of 2-class Fisher criteria may be used as well. This criterion is very useful for binary classification problems, especially when the samples are distributed in a Gaussian manner in the feature space. For multi-class problems [42], a criterion has been proposed based on the minimum classification error (MCE) that is calculated by the misclassification rate over the training samples. Formally, these two (fundamentally different) optimization criteria are known as the (i) Filter method and (ii) Wrapper method. While a filter method uses information content feedback (e.g., Fisher criterion, and statistical dependence) as optimization criteria for feature extraction, a wrapper method includes the classifier inside the optimization loop and attempts to maximize the predictive accuracy (e.g., classification rate) using statistical resampling or cross-validation [40]. In this methodology, the wrapper method is adopted (i.e., via minimization of the classification error on the training set) for optimization, primarily because of the non-binary nature of the problem at hand and the possible non-Gaussian distribution of training samples in the feature space.

In this context, ideally one should jointly minimize every element of the confusion matrix [43]. However, in that case, the dimension of the objective space may rapidly increase with an increase in the number of classes. To circumvent this situation in the present work, two costs are defined on the confusion matrix by using another weighting matrix, elements of which denote the relative penalty values for different confusions in the classification process. Formally, let there be $Cl_1, \cdots, Cl_n$ classes of labeled time-series data in the training set. A partitioning $\mathbb{B}$ is employed to extract features from each sample and a k-NN classifier $\mathbb{K}$ is used as a classifier. The confusion matrix $\mathbf{C}$ is obtained upon completion of the classification process, where the ij^{th} element c_{ij} of $\mathbf{C}$ denotes the frequency of data from class Cl_i being classified as data from Cl_j. Let $\mathbf{W}$ be the weighting matrix, where the ij^{th} element w_{ij} of $\mathbf{W}$ denotes the penalty incurred for classifying data from Cl_i as data from class Cl_j. (Note: Since there is no penalty for correct classification, the diagonal elements of $\mathbf{W}$ are identically equal to 0, i.e., $w_{ii} = 0 \; \forall \; i$.) With these specifications, two costs, $CostE$ and $CostW$, that are to be minimized are defined as follows.

The cost $CostE$ due to expected classification error is defined as:

$$CostE = \frac{1}{N_s}\left(\sum_i \sum_j w_{ij} c_{ij}\right) \tag{48}$$

where N_s is the total number of training samples including all classes. The above equation represents the total penalty for misclassification across all classes. Thus $CostE$ is related to

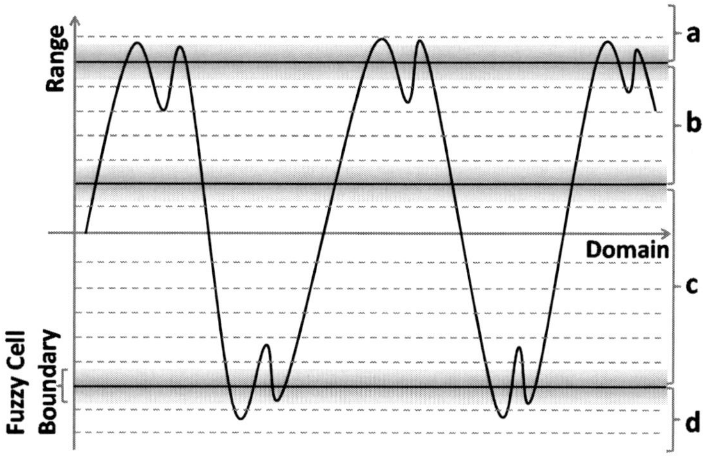

Figure 17. Fuzzy Cell Boundaries to obtain $CostE_{robust}$ and $CostW_{robust}$.

the expected classification error. The weights w_{ij} are selected based on the prior knowledge about the data and the user's requirements.

It is implicitly assumed in many supervised learning algorithms that the training data set is a statistically similar representation of the whole data set. However, this assumption may not be accurate in practice. A solution to this problem is to choose a feature extractor that minimizes the worst-case classification error [44]. In this setting, that cost $CostW$ due to worst-case classification error is defined as:

$$CostW = \max_i \left(\frac{1}{N_i} \sum_j w_{ij} c_{ij} \right) \tag{49}$$

where N_i is the number of training samples in the class Cl_i. (Note: In the present formulation, the objective space is two-dimensional for a multi-class classification problem and the dimension is not a function of the number of classes.)

Sensitivity and Robustness

Cost minimization requires a choice of partitioning that could be sensitive to class separation in the data space. However, the enhanced sensitivity of optimal partitioning may cause large classification errors due to noise and spurious disturbances in the data and statistical variations among training and testing data sets. Hence, it is important to invoke robustness properties into the optimal partitioning. In this study, robustness has been incorporated by modifying the costs $CostE$ and $CostW$, introduced above.

For one-dimensional time series data, a partitioning consisting of $|\Sigma|$ cells is represented by $(|\Sigma| - 1)$ points that serve as cell boundaries. In the sequel, a Σ-cell partitioning $\mathbb{B}$ is expressed as $\Lambda_{|\Sigma|} \triangleq \{\lambda_1, \lambda_2, \cdots, \lambda_{|\Sigma|-1}\}$, where λ_i denotes a partitioning boundary. Thus, a $Cost$ is dependent on the specific partitioning $\Lambda_{|\Sigma|}$ and is denoted by $Cost(\Lambda_{|\Sigma|})$. The key idea of a robust cost for a partitioning $\Lambda_{|\Sigma|}$ is that it is expected to remain invariant under small perturbations of the partitioning points, $\lambda_1, \lambda_2, \cdots, \lambda_{|\Sigma|-1}$ (i.e., fuzzy cell boundaries). To define a robust cost, a distribution of partitioning $f_{\Lambda_{|\Sigma|}}(\cdot)$ is considered, where a

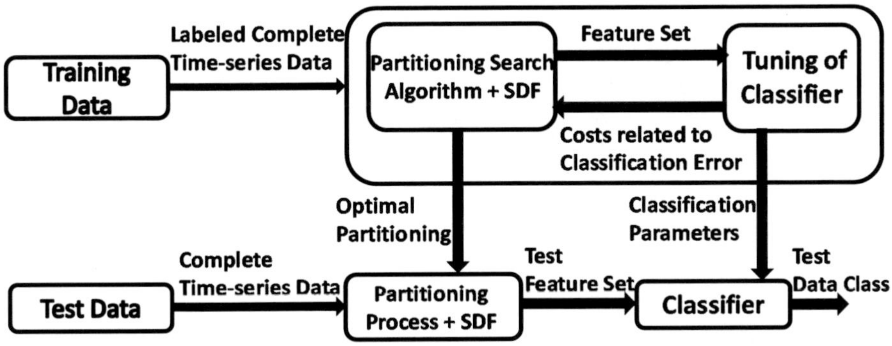

Figure 18. General Framework for Optimization of Feature Extraction.

sample of the distribution is denoted as $\tilde{\Lambda}_{|\Sigma|} = \{\tilde{\lambda}_1, \tilde{\lambda}_2, \cdots, \tilde{\lambda}_{|\Sigma|-1}\}$ and $\tilde{\lambda}_i$'s are chosen from independent Gaussian distributions with mean λ_i and a uniform standard deviation σ_λ; the choice of σ_λ is discussed later.

The resulting cost distribution is denoted as $f_{Cost(\Lambda_{|\Sigma|})}(\cdot)$, and $Cost_{robust}$ is defined as the cost value below which 95% of the population of distribution $f_{Cost(\Lambda_{|\Sigma|})}(\cdot)$ remains and is denoted as:

$$Cost_{robust} = P_{95}[f_{Cost(\Lambda_{|\Sigma|})}(\cdot)] \tag{50}$$

For the analysis to be presented here, it is assumed that sufficient samples are generated to obtain a good estimate of $Cost_{robust}$ as explained in Fig. 17, where the firm lines denote the boundaries that divide the range space into four cells, namely a, b, c and d. However, in general, the cell boundaries are fuzzy in nature, which leads to a probabilistic $Cost_{robust}$ instead of a deterministic cost based on the firm partitioning boundaries. Thus, a robust cost due to expected classification error, $CostE_{robust}$, and a robust cost due to worst-case classification error, $CostW_{robust}$, are defined for a given partitioning $\Lambda_{|\Sigma|}$.

Finally, a multi-objective overall cost $CostO$ is defined as a linear convex combination of $CostE_{robust}$ and $CostW_{robust}$ in terms a scalar parameter $\alpha \in [0, 1]$.

$$CostO \triangleq \alpha CostE_{robust} + (1 - \alpha) CostW_{robust} \tag{51}$$

Ideally, the optimization procedure involves construction of the Pareto front [45] by minimizing $CostO$ for different values of α that can be freely chosen as the operating point. However, for simplicity α is taken to be 0.5 in this work, i.e., equal weight for the costs $CostE_{robust}$ and $CostW_{robust}$. Thus, the optimal Σ-cell partitioning $\mathbb{B}^*$ is the solution to the following optimization problem:

$$\mathbb{B}^* = \arg\min_{\mathbb{B}} CostO(\mathbb{B}) \tag{52}$$

Figure 18 depicts a general outline of the classification process. Labeled time series data from the training set are partitioned. The low-dimensional feature vectors that are generated by symbolization and $PFSA$ construction are fed to the classifier. After classification, the training error costs, defined above, are computed and fed back to the feature extraction block. In the classification aspect, the classifier may be tuned to the obtain better classification rates. For example, for k-NN classifiers [40], the choice of neighborhood size or the distance metric can be tuned. Similarly, for support vector machines [40], an appropriate hyperplane should be selected to achieve good classification. The key idea is to update the partitioning to reduce the cost based on the feedback. The iteration is continued until the set of optimal partitioning in a multi-objective scenario and the correspondingly tuned classifier are obtained. Generally, choice of the optimal partitioning is made based on the choice of operating point α by the user. After the choice is made, the optimal partitioning and the tuned classifier are used to classify the test data set. Although this is the general framework that is being proposed for the optimization methodology, tuning of the classifier is not addressed in this chapter, because the main focus here is to choose the optimal partitioning to minimize the cost related to classification errors.

Optimization Procedure: A sequential search-based technique has been adopted in this study for optimization of the partitioning, which was previously proposed in [46]. As the continuity of the partitioning function with respect to the range space of classification error-related costs may not exist or at least are not adequately analyzed, gradient-based optimization methods are not explored here. To construct the search space, a suitably fine grid size depending on the data characteristics needs to be assumed. Each of the grid boundaries denotes a possible position of a partitioning cell boundary, as illustrated in Fig. 17. Here, the dotted lines denote the possible positions of a partitioning cell boundary and as discussed before, for a chosen partitioning (denoted by firm lines), the partitioning boundaries are perturbed to obtain a $Cost_{robust}$.

Let the data space region Ω be divided into G grid cells for search, i.e., there are $(G-1)$ grid boundaries excluding the boundaries. Thus, there are $|\Sigma|-1$ partitioning boundaries to choose among $(G-1)$ possibilities, i.e., the number of elements (i.e., $(|\Sigma|-1)$-dimensional partitioning vectors) in the space $\mathcal{P}$ of all possible partitioning is: ${}^{(G-1)}C_{(|\Sigma|-1)}$. It is clear from this analysis that the partitioning space $\mathcal{P}$ may become significantly large with an increase in values of G and $|\Sigma|$ (e.g., for $G >> |\Sigma|$, computational complexity increases approximately by a factor of $G/|\Sigma|$ with increase in the value of $|\Sigma|$ by one). Furthermore, for each element of $\mathcal{P}$, a sufficiently large number of perturbed samples need to be collected in order to obtain the $Cost_{robust}$. Therefore, usage of a direct search approach becomes infeasible for evaluation of all possible partitioning. Hence, a sub-optimal solution is developed in this chapter to reduce the computational complexity of the optimization problem.

The objective space consists of the scalar-valued cost $CostO$, while decisions are made in the space $\mathcal{P}$ of all possible partitionings. The overall cost is dependent on a specific partitioning Λ and is denoted by $CostO(\Lambda)$. This sub-optimal partitioning scheme involves sequential estimation of the elements of the partitioning Λ.

The partitioning process is initiated by searching the optimal cell boundary to divide the data set into two cells, i.e., $\Lambda_2 = \{\lambda_1\}$, where λ_1 is evaluated as

$$\lambda_1^* = \arg\min_{\lambda_1} CostO(\Lambda_2) \tag{53}$$

Now, the two-cell optimal partitioning is given by $\Lambda_2^* = \{\lambda_1^*\}$.

Note that to obtain $CostO$ (i.e., both $CostE_{robust}$ and $CostW_{robust}$) for a given partitioning, a suitable σ_λ needs to be chosen. Let the gap between two search grid boundaries (i.e., two consecutive dotted lines in Fig. 17) be l_λ, and σ_λ is chosen as $l_\lambda/3$ in this study.

The choice of such a standard deviation of the Gaussian perturbation is made for approximately complete coverage of the search space. Note that there could be an overlap of perturbation regions between two consecutive search grid boundaries, which leads to a smoother (that is essentially robust) cost variation across the domain space. In addition, depending on the gap l_λ and data characteristics, a suitable sample size is chosen to approximate the cost distribution under the fuzzy cell boundary condition.

The next step is to partition the data into three cells as Λ_3) by dividing either of the two existing cells of Λ_2^* with the placement of a new partition boundary at λ_2, where λ_2 is evaluated as

$$\lambda_2^* = \arg\min_{\lambda_2} CostO(\Lambda_3) \tag{54}$$

where $\Lambda_3 = \{\lambda_1^*, \lambda_2\}$. The optimal 3-cell partitioning is obtained as $\Lambda_3^* = \{\lambda_1^*, \lambda_2^*\}$. In this (local) optimization procedure, the cell that provides the largest decrement in $CostO$ upon further segmentation ends up being partitioned. Iteratively, this procedure is extended to obtain the $|\Sigma|$ cell partitioning as follows.

$$\lambda_{|\Sigma|-1}^* = \arg\min_{\lambda_{|\Sigma|-1}} CostO(\Lambda_{|\Sigma|}) \tag{55}$$

where $\Lambda_{|\Sigma|} = \Lambda_{|\Sigma|-1}^* \cup \{\lambda_{|\Sigma|-1}\}$ and the optimal $|\Sigma|$ cell partitioning is given by $\Lambda_{|\Sigma|}^* = \Lambda_{|\Sigma|-1}^* \cup \{\lambda_{|\Sigma|-1}^*\}$

In this optimization procedure, the cost function decreases monotonically with every additional sequential operation, under the assumption of correct estimation of $CostE_{robust}$, $CostW_{robust}$ and hence, $CostO$ under the fuzzy cell boundary condition. Formally, $CostO(\Lambda_{|\Sigma|-1}^*) \geq CostO(\Lambda_{|\Sigma|}^*)$. as explained below.

Let $\Lambda_{|\Sigma|-1}^*$ be the ($|\Sigma|$-1)-cell partitioning that minimizes $CostO$, based on the algorithm, $\Lambda_{|\Sigma|} = \Lambda_{|\Sigma|-1}^* \cup \{\lambda_{|\Sigma|-1}\}$). If $\lambda_{|\Sigma|-1}$ is chosen such that it already belongs to $\Lambda_{|\Sigma|-1}^*$, then there would be no change in the partitioning structure, i.e.,

$$CostO(\Lambda_{|\Sigma|}) = CostO(\Lambda_{|\Sigma|-1}^*) \text{ for } \lambda_{|\Sigma|-1} \in \Lambda_{|\Sigma|-1}^* \tag{56}$$

If $\lambda_{|\Sigma|-1} \in \Lambda_{|\Sigma|-1}^*$, then the partitioning $\Lambda_{|\Sigma|}$ is essentially treated as a $(|\Sigma| - 1)$-cell partitioning for the purpose of cost calculation. By definition,

$$CostO(\Lambda_{|\Sigma|}^*) \leq CostO(\Lambda_{|\Sigma|}) \ \forall \Lambda_{|\Sigma|} \tag{57}$$

Then, it follows that,

$$\min(CostO(\Lambda_{|\Sigma|-1})) \geq \min(CostO(\Lambda_{|\Sigma|})) \tag{58}$$

Or,

$$CostO(\Lambda_{|\Sigma|-1}^*) \geq CostO(\Lambda_{|\Sigma|}^*) \tag{59}$$

The monotonicity in the cost function allows formulation of a rule for termination of the sequential optimization algorithm. The process of creating additional partitioning cells is stopped if the cost decrease falls below a specified positive scalar threshold η_{stop} and the stopping rule is as follows.

$\Lambda^*_{|\Sigma|-1}$ is the optimal partitioning (and $|\Sigma|-1$ is the optimal Alphabet size) if

$$CostO(\Lambda^*_{|\Sigma|-1}) - CostO(\Lambda^*_{|\Sigma|}) \leq \eta_{stop}. \tag{60}$$

In contrast to the direct search of the entire space of partitioning, the computational complexity of this approach increases linearly with $|\Sigma|$. This approach also allows the user to have finer grid size for the partitioning search.

Validation Example 1: Fatigue Damage Classification- As described earlier, the process of fatigue damage is broadly classified into two phases: crack initiation and crack propagation. For validation purpose, the crack propagation stage is divided into four classes as presented below.

Class	Crack Length
1	0.5 to 1.75 mm
2	1.75 to 3.5 mm
3	3.5 to 5.5 mm
4	more than 5.5 mm

As described earlier, the weighing matrix W is chosen based on the user's requirement. In the present case, W is defined from the perspective of risk involved in miss-classification. For example, penalty is low when data in Class 1 (i.e., small crack length) is classified as in classes for larger crack length; however, penalty is high for the converse. For the results presented in this chapter, W matrix is defined as follows.

$$\mathbf{W} = \begin{pmatrix} 0 & 1 & 2 & 3 \\ 3 & 0 & 1 & 2 \\ 6 & 3 & 0 & 1 \\ 9 & 6 & 3 & 0 \end{pmatrix}$$

In the present work, a large volume of ultrasonic data has been collected for different crack lengths on different specimens to demonstrate the efficiency of the present classification scheme to account for the stochastic nature of material properties. The experimental procedure has been discussed earlier. Classification results for the crack propagation phase using the optimal partitioning along with results obtained by using Maximum Entropy and Uniform partitioning [12] are provided here. The classification process is started by wavelet transformation of the time series data with suitable scales and time shifts for a given basis function. Each transformed signal is normalized with the maximum amplitude of transformed signal obtained at the beginning of experiment, when there is no damage in the specimen. The data are normalized to mitigate the effects of variability in the placement of ultrasonic sensors during different experiments.

As described earlier, both training and test data sets are divided into four classes based on crack length as all the ultrasonic signals are labeled with the corresponding crack lengths. The sequential optimization of partitioning has been carried out on the training data set to find optimal partitioning. The specifications of SDF and the classifier remain same as in the first example. The optimal alphabet size is 6 with stopping threshold value $\eta_{stop} = 0.005$. The confusion matrices for Optimal, Uniform and Maximum Entropy Partitioning on the test data set are given by $\mathbf{C}^{OptP}_{test}$ $\mathbf{C}^{UP}_{test}$ and $\mathbf{C}^{MEP}_{test}$ respectively. Table 2 shows the comparison of classification performances using different partitioning processes.

$$\mathbf{C}_{test}^{OptP} = \begin{pmatrix} 93 & 7 & 0 & 0 \\ 5 & 89 & 6 & 0 \\ 0 & 4 & 92 & 4 \\ 0 & 3 & 7 & 90 \end{pmatrix}$$

$$\mathbf{C}_{test}^{UP} = \begin{pmatrix} 94 & 5 & 1 & 0 \\ 15 & 74 & 11 & 0 \\ 0 & 9 & 85 & 6 \\ 1 & 5 & 11 & 83 \end{pmatrix}$$

$$\mathbf{C}_{test}^{MEP} = \begin{pmatrix} 93 & 7 & 0 & 0 \\ 12 & 82 & 6 & 0 \\ 0 & 7 & 89 & 4 \\ 1 & 3 & 7 & 89 \end{pmatrix}$$

Table 2. Performance Comparison of Partitioning Schemes

Partitioning	$Cost_E$	$Cost_W$
OptP	0.255	0.5
UP	0.40333	0.68
MEP	0.31333	0.52

It is observed that both costs $Cost_E$ and $Cost_W$ are reduced for optimal partitioning as compared to maximum entropy and uniform partitioning. It is also evident from the confusion matrix that optimal partitioning has improved the classification results. A close observation of confusion matrix indicates that chances of higher damage level data samples being classified as a lower level damage is reduced.

Validation Example 2: Fault detection in Nuclear power plants - Component-level anomaly detection in nuclear power plants involves identification of the anomaly type and location & quantification of the anomaly level. Although the model configuration in the IRIS test-bed can be easily extended to simultaneous anomalies in multiple components, this work deals with a single component, namely, the reactor coolant pump (RCP), where the task is to detect an anomaly and identify its level for (possibly unknown) sensor noise variance. The RCP overspeed percentage (ψ_{RCP}) is chosen as a health parameter. Table 3 shows the approximate ranges of ψ_{RCP} under different anomaly levels. Here, the low anomaly level indicate very minimal overspeed in RCP and hence also includes the absolute nominal health condition ($\psi_{RCP} = 0$). Similarly, depending on the standard deviation of the noise in the hot-leg coolant temperature sensor T_{HL}, three sensor noise levels are selected. Table 4 shows the approximate ranges of the standard deviation of sensor noise under different levels. The primary coolant RTDs in PWR plants are typically calibrated to an accuracy of $0.5°F$ or better before installation [25]. In general, the level 1 noise is defined to be within the accuracy threshold of RTDs, while level 2 and 3 noise are beyond this accuracy threshold and hence are measurable by RTDs.

In the above context, $(3 \times 3) = 9$ classes of data sets are chosen to define each class by an RCP anomaly level and a noise level of the T_{HL} sensor. One hundred simulation

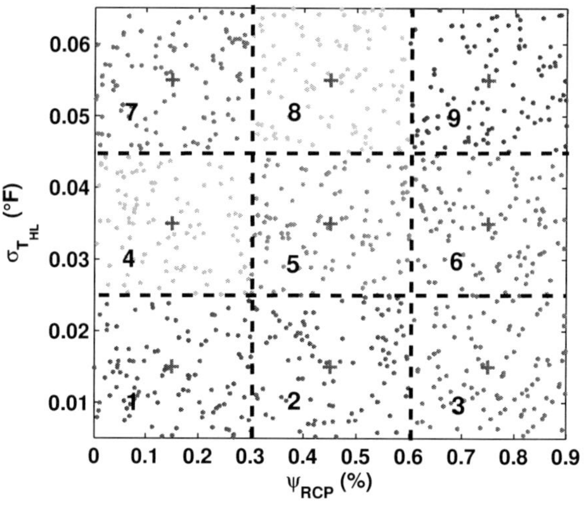

Figure 19. Original class labels for data collection.

runs are performed on the IRIS system for each class to generate time series data, among which 50 samples were chosen for training and the remaining samples for testing. The parameters, RCP overspeed percentage ψ_{RCP} and standard deviation of T_{HL} are chosen randomly from independent uniform distribution such that all of the parameter values are within the prescribed ranges given in Tables 3 and 4. Figure 19 plots the generated samples in the two-dimensional parameter space, where different classes of samples are shown in different colors and are marked within the class number. Note that the standard deviation ($\sigma_{T_{HL}}$) in Fig. 19 shows values of actual standard deviation, not as percents.

Often component degradations in nuclear plants occur on a slow-time scale, i.e., the anomaly gradually evolves over a relatively long time span. Sudden failures in plant components are disastrous and may occur only if a component accumulates sufficient amount of anomaly over a prolonged period and reaches the critical point, which justify early detection of an anomaly for enhancement of plant operational safety. However, component anomaly at the early stages of a malfunction is usually very small and is difficult to detect, especially under steady-state operations. One way to "amplify" the anomaly signature is to perturb the plant by (quasi-)periodically perturbing the turbine load such that the plant is in a transient state for a period that is sufficiently long for anomaly detection and insignificant from the perspectives of plant operation.

For each sample point in the parameter space, a time series is collected for T_{HL} sensor under persistent excitation of turbine load inputs that have truncated triangular profiles with the mean value of 99% of nominal output power, fluctuations within $\pm 1\%$ and frequency

Table 3. Anomaly Levels in Reactor Coolant Pump (RCP)

Anomaly Level	RCP Overspeed ψ_{RCP} (%)
Low	0.00 to 0.30
Medium	0.30 to 0.60
High	0.60 to 0.90

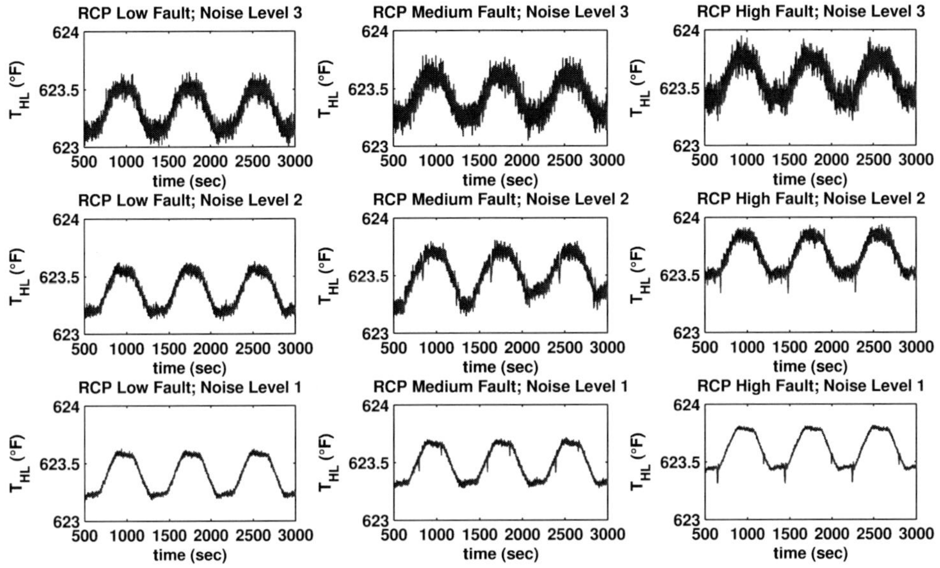

Figure 20. Representative time series data for RCP anomaly and T_{HL} degradation conditions.

of 0.00125Hz (period $T = 800$ sec). In other words, the turbine load is fluctuating between nominal power (i.e., 335 MW) and 98% of nominal power (i.e., 328.3 MW). Figure 21 shows the profile of turbine load and a typical example of turbine output power as a result of fluctuation in turbine load.

For each experiment, the sampling frequency for data collection is 1 Hz (i.e., the inter-sample time of 1 sec) and the length of the perturbation time window is 2,400 seconds, which generate 2,400 data points. The total perturbation period for each experiment is 3,000 seconds. In this study, to eliminate the possible transients, the data for the first 600 seconds have not been used so that quasi-stationarirty is achieved. Figure 20 shows representative examples of T_{HL} time series data from each of nine classes. Reduction of the perturbation period needs to be investigated for in-plant operations, which is a topic of future research.

With the objective of building a data-driven diagnostic algorithm that is robust to sensor noise (within an allowable range), a data class is defined to be only dependent on the RCP overspeed parameters. Thus, the 9 original classes are reduced to 3 classes as shown in Fig. 22. This is the final class assignment for the data set, where each class has

Table 4. Variable Noise Levels in the the Sensor T_{HL}

Noise Level	Standard Deviation Range $\sigma_{T_{HL}}$ ($^{\circ}F$)
Level 1	0.005 to 0.025
Level 2	0.025 to 0.045
Level 3	0.045 to 0.065

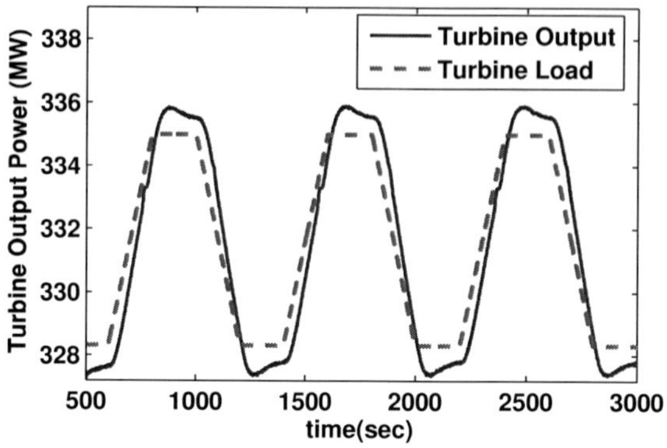

Figure 21. Profile of turbine load and turbine output response.

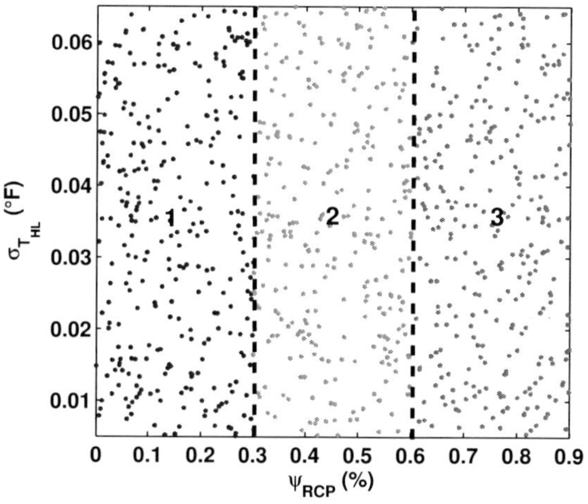

Figure 22. Revised class assignment for anomaly detection.

$(50 \times 3) = 150$ training samples and $(50 \times 3) = 150$ test samples. Thus, the problem of component level anomaly detection is formulated in presence of sensor noise as a multi-class classification problem; in the present scenario, number of classes is 3.

Pertinent results are presented here for the case study of anomaly detection in the reactor coolant pump (RCP) for comparative evaluation of the optimal partitioning-based SDF tool with those based on classical methods of partitioning as well as PCA.

At the beginning of the optimization procedure, a weighting matrix $\mathbf{W}$ needs to be defined to calculate the cost functionals $Cost_E$ and $Cost_W$ from the confusion matrix for the training data set. In this case study, $\mathbf{W}$ is defined according to the adjacency properties of classes in the parameter space, i.e., $w_{ii} = 0 \; \forall i \in \{1, 2, 3\}$, i.e. there is no penalty for correct classification. The weights are selected as: $w_{ij} = |i - j|$, $\forall i \in \{1, 2, 3\}$, i.e., given that a data sample originally from Cl_i is classified as a member of Cl_j, the penalty incurred by the classification process increases with increase in the separation of Cl_i and Cl_j in the parameter space. Then, it follows that:

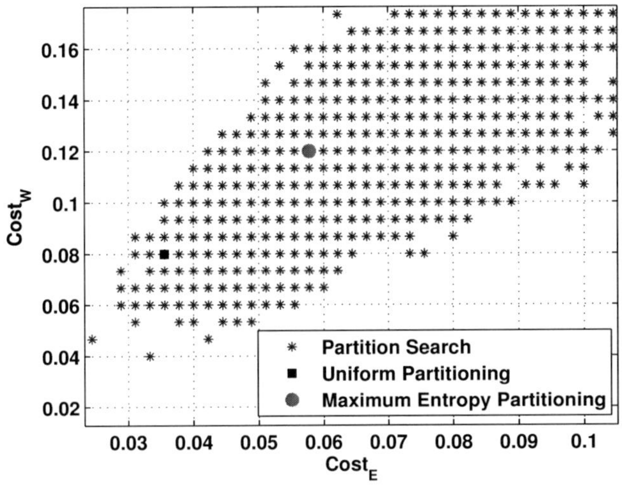

Figure 23. Two dimensional objective space for partitioning optimization.

$$\mathbf{W} = \begin{pmatrix} 0 & 1 & 2 \\ 1 & 0 & 1 \\ 2 & 1 & 0 \end{pmatrix}$$

The data space region Ω is divided into 40 grid cells, i.e., 39 grid boundaries excluding the boundaries of Ω. Each partitioning in the space $\mathcal{P}$ is evaluated by calculating $Cost_E$ and $Cost_W$ to identify the optimal partitioning. Figure 23 shows a relevant region of the (two-dimensional) $Cost_E$-$Cost_W$ objective space, where the elements of the space $\mathcal{P}$ are located. The Pareto front is also generated from this evaluation. The threshold ϵ, i.e., the maximum allowable $Cost_W$ is taken to be 0.041 in this case and the optimal partitioning (OptP) is chosen by the Neyman-Pearson criterion as discussed earlier. Thus, the user has the option of choosing a classification operating point with selected values of $Cost_E$ and $Cost_W$. The Pareto front [47] is generated after the threshold ϵ is chosen. Locations of the classical partitioning (i.e., uniform (UP) and maximum entropy (MEP)) are also plotted along with the elements of $\mathcal{P}$ in the figures for comparative evaluation.

For SDF analysis, the alphabet size is taken to be $|\Sigma| = 4$ and the depth for constructing PFSA is taken to be $D = 1$. Features are classified by a k-NN classifier using the Euclidean distance metric. Figure 24 shows locations of the training features in the three-dimensional plot using first three linearly independent elements of the feature vectors obtained by using the chosen optimal partitioning, OptP. Note, only $(|\Sigma|-1)$ out of its $|\Sigma|$ elements of a feature vector are linearly independent, because a training feature vector, $\mathbf{p}$ is also a probability vector, i.e., the sum of its elements is constrained to be equal to 1. The class separability is retained by the feature extraction (partitioning) process even after compressing a time series (with 2,400 data points) into 3 numbers.

For comparison purpose, classical partitioning schemes, such as, Uniform Partitioning (UP) and Maximum Entropy Partitioning (MEP) are also used with the same alphabet size, $|\Sigma| = 4$. Figures 25 and 26 show the location of each training time series in the three

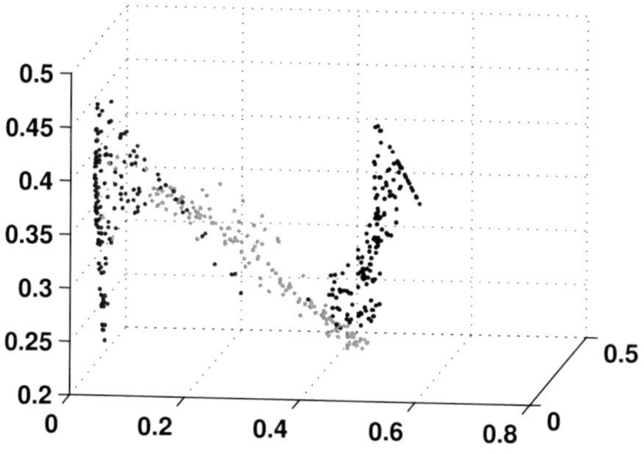

Figure 24. Feature space of the training set using optimal partitioning.

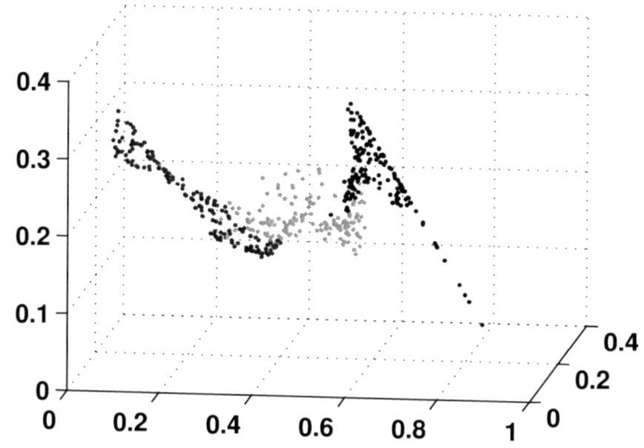

Figure 25. Feature space of training set – uniform partitioning (UP).

dimensional (using first three linearly independent elements of the feature vectors) feature space plot using UP and MEP, respectively.

Finally, the confusion matrices for the SDF-based methods (OptP, UP and MEP) with k-NN on the test data set are given by $\mathbf{C}_{test}^{OptP+kNN}$, $\mathbf{C}_{test}^{UP+kNN}$ and $\mathbf{C}_{test}^{MEP+kNN}$, respectively.

$$\mathbf{C}_{test}^{OptP+kNN} = \begin{pmatrix} 142 & 8 & 0 \\ 14 & 136 & 0 \\ 0 & 4 & 146 \end{pmatrix}$$

$$\mathbf{C}_{test}^{UP+kNN} = \begin{pmatrix} 145 & 5 & 0 \\ 17 & 132 & 1 \\ 0 & 6 & 144 \end{pmatrix}$$

$$\mathbf{C}_{test}^{MEP+kNN} = \begin{pmatrix} 143 & 7 & 0 \\ 18 & 132 & 0 \\ 0 & 5 & 145 \end{pmatrix}$$

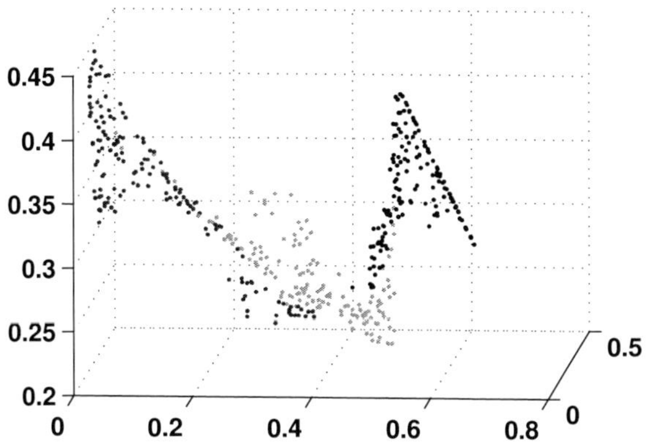

Figure 26. Feature space of training set – maximum entropy partitioning (MEP).

For comparative evaluation, the data sets are analyzed using other common pattern recognition tools [4][40]. In this case study, principal component analysis (PCA) is used as the feature extraction tool, while k-NN, linear discriminant analysis (LDA) and least squares algorithm (LS) are used as the pattern classifiers. For PCA anomaly detection, $M = 450$, $N = 2400$, and by choosing $\eta = 0.90$, the corresponding number of largest eigenvalues turns out to be $m = 20$. The confusion matrices for PCA with k-NN, LDA and LS are given by $\mathbf{C}_{test}^{PCA+kNN}$, $\mathbf{C}_{test}^{PCA+LDA}$ and $\mathbf{C}_{test}^{PCA+LS}$, respectively.

$$\mathbf{C}_{test}^{PCA+kNN} = \begin{pmatrix} 149 & 1 & 0 \\ 40 & 110 & 0 \\ 0 & 4 & 146 \end{pmatrix}$$

$$\mathbf{C}_{test}^{PCA+LDA} = \begin{pmatrix} 136 & 14 & 0 \\ 18 & 126 & 6 \\ 0 & 3 & 147 \end{pmatrix}$$

$$\mathbf{C}_{test}^{PCA+LS} = \begin{pmatrix} 145 & 5 & 0 \\ 27 & 115 & 8 \\ 0 & 0 & 150 \end{pmatrix}$$

Figure 27 shows how the neighbor size in the k-NN classifiers affect the classification performance, and the performances of Opt, MEP, UP and PCA with different neighbor size k are compared. Only odd values of neighbor size are shown because when using an even value for k, it might be necessary to break a tie in the number of nearest neighbors by selecting a random tiebreaker or using the nearest neighbor among the tied groups. It is seen in Fig. 27 the classification performance of SDF-based methods is consistently better than that of PCA-based method and almost independent of the neighbor size, whereas the classification error of PCA increases with the number of neighbor size. This results show that the feature vectors extracted by SDF retain better separability among classes than those extracted by PCA.

Table 5 presents the comparison of the classification error related costs for SDF-based methods and PCA-based methods on the test data set. The observation made from these

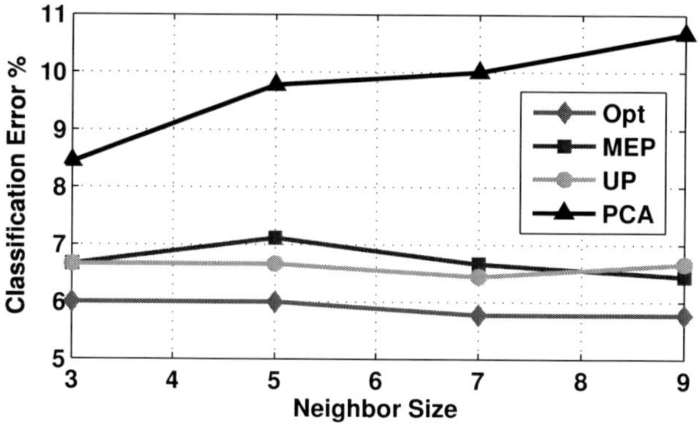

Figure 27. Classification error vs. neighbor size in k-NN classifiers.

Table 5. Comparison of Classification Performances of Different Methods on Test-data set (50×9 samples)

Methods	$CostE$	$CostW$	Classification Error %
OptP+kNN	0.0578	0.0933	5.78
UP+kNN	0.0644	0.1200	6.44
MEP+kNN	0.0667	0.1200	6.67
PCA+kNN	0.1000	0.2667	10.00
PCA+LDA	0.0911	0.1200	9.11
PCA+LS	0.0889	0.1800	8.89

results indicate that the classification performance of SDF-based methods are superior to that of the PCA-based methods. The classification performance of MEP and UP are close, and UP performs slightly better in this case study. The costs due to worst-case classification error and expected classification error are reduced compared to that of the uniform partitioning scheme by optimizing the partitioning process over a representative training set.

It is noted that, for some problems, the classical partitioning schemes may perform similar to the optimal one. Therefore, the optimization procedure may also be used to evaluate the capability of any partitioning scheme towards achieving a better classification rate. The evaluation can be performed by using a part of the labeled training data set as the validation set. Although the construction of the cost functions allow solutions of problems with a large number of classes, its upper limit could be constrained by the alphabet size used for data partitioning which determines the dimension of the feature space.

4. Symbolic Dynamic Analysis of Short-Length Transient Time-Series Motifs

Many classification algorithms are not well-suited for applications to nonlinear dynamical systems [48][49][50]. This phenomenon has led to departure from the conventional continuous-domain modeling towards a formal language-theoretic paradigm in the symbolic domain [51]. The theory of symbolic dynamic filtering (SDF) has been developed as a tool for anomaly detection [15][12] and also for feature extraction [52] in pattern classification problems [1].

A symbol string is obtained from the output of the dynamical system by partitioning (also called quantization) of the time-series data. Thereafter, a probabilistic finite state automaton (PFSA) is constructed from the (finite-length) symbol sequence via one of the construction algorithms (e.g., [15], [53], and [54]). Due to the quasi-stationarity assumption in SDF, it may not be feasible to obtain sufficiently long strings of symbols in both training and testing phases of classification. Therefore, the estimated parameters of the resulting PFSA model may not be precise.

The goal of this section is to identify the hidden structure of the symbol string and to construct a Bayesian classifier for identification of the probability morph matrices of PFSA based on the data length in both training and testing phases of transient time-series motifs. In this context, Dirichlet and multinomial distributions have been used to construct the *a priori* and *a posteriori* models of uncertainties, respectively. The algorithms are formulated by quantitatively incorporating the effects of finite-length symbol strings in both training and testing phases of pattern classification. In this setting, pertinent information (e.g., probability morph matrix and state transition function) derived from a PFSA model serves as a feature vector even though the PFSA representations may have dissimilar structures. The Proposed algorithm is validated on an industrial application which is as follows: Transient Fault Detection in Aircraft Gas Turbine Engine.

4.1. Theory

A symbol string is obtained from the output of system under consideration by partitioning, also called quantization, of the time-series data. Thereafter, a probabilistic finite state automaton (PFSA) model is developed from the (finite-length) symbol sequence [15] as explained in the subsection 1.1. This section constructs a Bayesian classifier for identification of the probability morph matrices of PFSA based on the transient data in both training and testing phases. The Dirichlet and multinomial distributions have been used to construct the *a priori* and *a posteriori* models of uncertainties, respectively. This formulation quantitatively incorporates the effects of finite-length symbol strings in both training and testing phases of pattern classification.

4.1.1. Partitioning of Time Series Data

The sensor time series is encoded by data partitioning in the range of the signal [55][12], where the conversion to symbol strings is achieved by substituting each (real-valued) data point in the time series by a symbol corresponding to the region (i.e., interval) within which the data point lies. This step enables transformation of the sensory information from the continuous domain to the symbolic domain; in other words, the sensor data at each sampling instant is replaced by a symbol. Sarkar et al. [55][56] have reported details of data partitioning and its usage for fault detection in gas turbine engines.

4.1.2. Modeling via Probabilistic Finite State Automaton

The symbolic sequence is modeled as a probabilistic finite state automaton (PFSA) that is constructed as a tuple $G \triangleq (Q, \Sigma, \delta, \Pi)$, where the alphabet Σ is a nonempty finite set of symbols and the set of states Q is constrained to be nonempty and finite. To simplify the PFSA model construction, this section considers only a class of PFSA, known as D-Markov machines [15], where the states are strings of the D past symbols; the positive integer D is called the depth of the machine and the number of states $|Q| \leq |\Sigma|^D$. Given the previous state and an observed symbol, the state transition function $\delta : Q \times \Sigma \rightarrow Q$ yields the new state. In addition, the morph function $\pi : Q \times \Sigma \rightarrow [0, 1]$ is an output mapping that satisfies the condition: $\sum_{\sigma \in \Sigma} \pi(q, \sigma) = 1$ for all $q \in Q$. The morph function π has a matrix representation Π, called the (probability) morph matrix of dimension $(|Q| \times |\Sigma|)$. Each row sum of Π is equal to 1 and each matrix element Π_{ij} is strictly positive due to the finite length constraint of time series from which PFSA models are constructed [57].

Remark 4.1. *The PFSA $G \triangleq (Q, \Sigma, \delta, \Pi)$ is not dependent on the initial state $q_0 \in Q$ due to the assumption of quasi-stationarity of the observed sensor data.*

4.1.3. The Online Classification Problem

Let there be K symbolic systems (i.e., classes) of interest, denoted by $C_1, C_2, \ldots, C_K$, over the same alphabet Σ. Each class C_i is modeled by an ergodic (equivalently, irreducible) PFSA $G^i = (Q^i, \Sigma, \delta^i, \Pi^i)$, where $i = 1, 2 \ldots, K$.

During the training phase, a symbol string $S^i \triangleq s_1^i s_2^i \ldots s_{N_i}^i$ is generated from each class C_i. The state transition function δ and the set of states Q of the D-Markov machine are fixed by choosing an appropriate depth D. Thus, Π^i's become the only unknowns and could be selected as the feature vectors for the purpose of classification. The distribution of the morph matrix Π^i is computed in the training phase from the finite length symbol sequences for each class.

In the testing phase, let another symbol string $\widetilde{S}$ be obtained from a sensor time series data. Then, the task is to determine which class this observed symbol string $\widetilde{S}$ belongs to. While the previous work [15][12] has aimed at identification of a PFSA from a given symbol string, the objective of this section is to imbed the uncertainties due to the finite length of the symbol string in the identification algorithm that would influence the final classification decision.

In the training phase, each row of Π^i is treated as a random vector. Let the mth row of Π^i be denoted as Π_m^i and the nth element of the mth row as $\Pi_{mn}^i > 0$ and $\sum_{n=1}^{|\Sigma|} \Pi_{mn}^i = 1$. The *a priori* probability density function $f_{\Pi_m^i | S^i}$ of the random row-vector Π_m^i, conditioned on a symbol string S^i, follows the Dirichlet distribution [58] [59] as described below.

$$f_{\Pi_m^i | S^i}(\boldsymbol{\theta}_m^i | S^i) = \frac{1}{B(\boldsymbol{\alpha}_m^i)} \prod_{n=1}^{|\Sigma|} (\theta_{mn}^i)^{\alpha_{mn}^i - 1} \tag{61}$$

where $\boldsymbol{\theta}_m^i$ is a realization of the random vector Π_m^i, namely,

$$\boldsymbol{\theta}_m^i = \begin{bmatrix} \theta_{m1}^i & \theta_{m2}^i & \cdots & \theta_{m|\Sigma|}^i \end{bmatrix}$$

and the normalizing constant is

$$B(\boldsymbol{\alpha}_m^i) \triangleq \frac{\prod_{n=1}^{|\Sigma|} \Gamma(\alpha_{mn}^i)}{\Gamma(\sum_{n=1}^{|\Sigma|} \alpha_{mn}^i)} \tag{62}$$

where $\boldsymbol{\alpha}_m^i \triangleq \left[\alpha_{m1}^i \alpha_{m2}^i \cdots \alpha_{m|\Sigma|}^i\right]$ with $\alpha_{mn}^i = N_{mn}^i + 1$ and N_{mn}^i is the number of times the symbol σ_n in S^i is emanated from the state q_m, i.e.,

$$N_{mn}^i \triangleq \left|\{(s_k^i, v_k^i) : s_k^i = \sigma_n, v_k^i = q_m\}\right| \tag{63}$$

where s_k^i is the kth symbol in S^i and v_k^i is the kth state as derived from the symbolic sequence S^i. Recall that a state is defined as a string of D past symbols. Then, the number of occurrence of the state q_m in the state sequence is given by $N_m^i \triangleq \sum_{n=1}^{|\Sigma|} N_{mn}^i$. It follows from Eq. (62) that

$$B(\boldsymbol{\alpha}_m^i) = \frac{\prod_{n=1}^{|\Sigma|} \Gamma(N_{mn}^i + 1)}{\Gamma(\sum_{n=1}^{|\Sigma|} N_{mn}^i + |\Sigma|)} = \frac{\prod_{n=1}^{|\Sigma|} (N_{mn}^i)!}{(N_m^i + |\Sigma| - 1)!} \tag{64}$$

by use of the relation $\Gamma(n) = (n-1)!\ \forall n \in \mathbb{N}_1$.

By the Markov property of the PFSA G^i, the $(1 \times |\Sigma|)$ row-vectors, $\{\Pi_m^i\}, m = 1, \ldots |Q|$, are statistically independent of each other. Therefore, it follows from Eqs. (61) and (64) that the *a priori* joint density $f_{\Pi^i|S^i}$ of the probability morph matrix Π^i, conditioned on the symbol string S^i, is given as

$$\begin{aligned} f_{\Pi^i|S^i}(\boldsymbol{\theta}^i|S^i) &= \prod_{m=1}^{|Q|} f_{\Pi_m^i|S^i}(\boldsymbol{\theta}_m^i|S^i) \\ &= \prod_{m=1}^{|Q|} (N_m^i + |\Sigma| - 1)! \prod_{n=1}^{|\Sigma|} \frac{(\boldsymbol{\theta}_m^i)^{N_{mn}^i}}{(N_{mn}^i)!} \end{aligned} \tag{65}$$

where $\boldsymbol{\theta}^i = \left[(\boldsymbol{\theta}_1^i)^T\ (\boldsymbol{\theta}_2^i)^T\ \cdots\ (\boldsymbol{\theta}_{|q|}^i)^T\right] \in [0,1]^{|Q| \times |\Sigma|}$

In the testing phase, the probability of observing a symbol string $\widetilde{S}$ belonging to a particular class of PFSA $(Q, \Sigma, \delta, \Pi^i)$ is a product of independent multinomial distribution [60] given that the exact morph matrix Π^i is known.

$$\begin{aligned} &\Pr\left(\widetilde{S}|Q, \delta, \Pi^i\right) \\ &= \prod_{m=1}^{|Q|} (\widetilde{N}_m)! \prod_{n=1}^{|\Sigma|} \frac{\left(\Pi_{mn}^i\right)^{\widetilde{N}_{mn}}}{(\widetilde{N}_{mn})!} && (66) \\ &\triangleq \Pr\left(\widetilde{S}|\Pi^i\right) \quad \text{as } Q \text{ and } \delta \text{ are kept invariant} && (67) \end{aligned}$$

Similar to N_{mn}^i defined earlier for S^i, $\widetilde{N}_{mn}$ is the number of times the symbol σ_n is emanated from the state $q_m \in Q$ in the symbol string $\widetilde{S}$ in the testing phase, i.e.,

$$\widetilde{N}_{mn} \triangleq \left|\{(\tilde{s}_k, \tilde{v}_k) : \tilde{s}_k = \sigma_n, \tilde{v}_k = q_m\}\right| \tag{68}$$

where $\tilde{s}_k$ is the kth symbol in the string $\widetilde{S}$ and $\tilde{v}_k$ is the kth state derived from $\widetilde{S}$. It is noted that $\widetilde{N}_m \triangleq \sum_{n=1}^{|\Sigma|} \widetilde{N}_{mn}$.

The results, derived in the training and testing phases, are now combined. Given a symbol string S^i in the training phase, the probability of observing a symbol string $\widetilde{S}$ in the testing phase is obtained as follows.

$$\begin{aligned}
\Pr(\widetilde{S}|S^i) &= \int \cdots \int \Pr\left(\widetilde{S}|\Pi^i = \boldsymbol{\theta}^i\right) f^i_{\Pi^i|S^i}(\boldsymbol{\theta}^i|S^i) d\boldsymbol{\theta}^i \\
&= \int \cdots \int \left[\prod_{m=1}^{|Q|} (\widetilde{N}_m)! \prod_{n=1}^{|\Sigma|} \frac{\left(\theta^i_{mn}\right)^{\widetilde{N}_{mn}}}{(\widetilde{N}_{mn})!}\right] \\
&\times \prod_{m=1}^{|Q|} \left[(N^i_m + |\Sigma| - 1)! \prod_{n=1}^{|\Sigma|} \frac{\left(\theta^i_{mn}\right)^{N^i_{mn}}}{(N^i_{mn})!} d\theta^i_{mn}\right] \\
&= \prod_{m=1}^{|Q|} (\widetilde{N}_m)!\,(N^i_m + |\Sigma| - 1)! \\
&\times \frac{\int \cdots \int \prod_{n=1}^{|\Sigma|} \left(\theta^i_{mn}\right)^{\widetilde{N}_{mn} + N^i_{mn}} d\theta^i_{mn}}{\prod_{n=1}^{|\Sigma|} (\widetilde{N}_{mn})!(N^i_{mn})!}
\end{aligned} \tag{69}$$

The integrand in Eq. (69) is the density function for the Dirichlet distribution up to the multiplication of a constant. Hence, it follows from Eq. (64) that

$$\begin{aligned}
&\int \cdots \int \prod_{n=1}^{|\Sigma|} \left(\theta^i_{mn}\right)^{\widetilde{N}_{mn} + N^i_{mn}} d(\theta^i_{mn}) \\
&= \frac{\prod_{n=1}^{|\Sigma|} (\widetilde{N}_{mn} + N^i_{mn})!}{\left(\widetilde{N}_m + N^i_m + |\Sigma| - 1\right)!}
\end{aligned}$$

Then, it follows from Eq. (69) that

$$\begin{aligned}
\Pr(\widetilde{S}|S^i) &= \prod_{m=1}^{|Q|} \frac{(\widetilde{N}_m)!\,(N^i_m + |\Sigma| - 1)!}{\left(\widetilde{N}_m + N^i_m + |\Sigma| - 1\right)!} \\
&\times \prod_{n=1}^{|\Sigma|} \frac{(\widetilde{N}_{mn} + N^i_{mn})!}{(\widetilde{N}_{mn})!(N^i_{mn})!}
\end{aligned} \tag{70}$$

It might be easier to compute the logarithm of $\Pr(\widetilde{S}|S^i)$ by using Stirling's approximation formula $\log(n!) \approx n\log(n) - n$ [2] because both N^i and $\widetilde{N}$ would be large numbers. The posterior probability of a symbol string S belonging to the class C_i is denoted as $\Pr(C_i|\widetilde{S})$ and is given as

$$\Pr(C_i|\widetilde{S}) = \frac{\Pr(\widetilde{S}|S^i)\Pr(C_i)}{\sum_{j=1}^{K} \Pr(\widetilde{S}|S^j)\Pr(C_j)}, \quad i = 1, 2, \cdots, K \tag{71}$$

where $\Pr(C_i)$ is the known prior distribution of the class C_i. Then, the classification decision is made as follows.

$$\begin{aligned} D_{class} &= \arg\max_i \Pr(C_i|\widetilde{S}) \\ &= \arg\max_i \Big(\Pr(\widetilde{S}|S^i) \Pr(C_i) \Big) \end{aligned} \tag{72}$$

Algorithms 1 and 2 respectively summarize the training and testing phases for classification of time-series data.

Algorithm 1 Training

Input: Time-series data, one for each of the K classes
User defined Input: Symbol alphabet Σ, depth D of PFSA
the symbols are defined $\sigma_1, \sigma_2, ..., \sigma_{|\Sigma|}$ and states are
defined as $q_1, q_2, ..., q_{|\Sigma|^D}$
Output: N^i_{mn}, where $i = 1, ..., K, n = 1, ..., |\Sigma|$
and $m = 1, ..., |\Sigma|^D$
Initialize: All $N^i_{mn} = 0$.
for all $i \in \{1, 2, ..., K\}$ **do**
Symbolize time-series data for class i to obtain a symbol
sequence S^i
Obtain N^i_{mn}, the number of times the symbol σ_n in S^i
is observed immediately after the state q_m
end for
return N^i_{mn}, where $i = 1, ..., K, n = 1, ..., |\Sigma|$
and $m = 1, ..., |\Sigma|^D$

The concept of the online classifier can be clarified using a simple example [61] as follows. Figure 28 shows two single-state PFSA with the alphabet $\Sigma = \{1, 2, 3\}$, which have identical algebraic structures, but they differ in their morph probabilities; these PFSA belong to the respective classes, C_1 and C_2. Since both PSFA have only one state, each symbol in a string is independent and identically distributed. Given an observed symbol string $\widetilde{S}$, the task is to identify one of the two classes to which $\widetilde{S}$ belongs, i.e., to select one of the two PFSA that would more likely generate the symbol string $\widetilde{S}$. In this example, two training symbol strings S^1 and S^2 are chosen, one from each class. In the testing phase, the quantities $\widetilde{N}^1_{01}$, $\widetilde{N}^1_{02}$, $\widetilde{N}^1_{03}$ are obtained following Eq. (68), which are essentially the frequency counts of the symbols 1, 2, and 3, respectively. Let η_1, η_2, and η_3 be the normalized frequency counts obtained by $\eta_k \triangleq \frac{\widetilde{N}^1_{0k}}{\widetilde{N}^1_{01}+\widetilde{N}^1_{02}+\widetilde{N}^1_{03}}$, $k = 1, 2, 3$. Two classes are generated on the simplex plane in Fig. 29 that shows how the classification boundary changes as the length of a symbol string in the testing phase is increased.

4.2. Transient Fault Detection in Aircraft Gas Turbine Engine

Performance monitoring of aircraft gas turbine engines is typically conducted on quasi-stationary steady-state data collected during cruise conditions. This issue has been ad-

Algorithm 2 Testing

Input: Test time-series data, and output of training N^i_{mn}
User defined Input: Prior probability of each class $Pr(C_i)$,
symbol alphabet Σ, and depth D of PFSA,
the symbols are defined as $\sigma_1, \sigma_2, ..., \sigma_{|\Sigma|}$, and states are defined as $q_1, q_2, ..., q_{|\Sigma|^D}$
Output: Posterior probability of each class.
Initialize: All $\tilde{N}^i_{mn} = 0$.
Symbolize test time-series data to obtain a symbol sequence $\tilde{S}^i$
Obtain $\tilde{N}^i_{mn}$, the number of times the symbol σ_n in $\tilde{S}$ is observed immediately after the state q_m is reached
for all $i \in \{1, 2, ..., K\}$ **do**
  Evaluate Pr($\tilde{S}|S^i$) using Eq. (70)
end for
for all $i \in 1, 2, ..., K$ **do**
  Evaluate Pr($C_i|\tilde{S}$) using Eq. (71)
end for
return Pr($C_i|\tilde{S}$), $i \in \{1, 2, ..., K\}$

dressed by several researchers by using a variety of analytical tools. For example, Lipowsky et al. [62] made use of Bayesian forecasting for change detection and performance analysis of gas turbine engines. Guruprakash and Ganguli [63] reported optimally weighted recursive median filtering for gas turbine diagnostics from noisy signals. However, transient operational data (e.g., from takeoff, climb, or landing) can be gainfully utilized for early detection of incipient faults. Usually engines operate under much higher stress and temperature conditions under transient operations compared to those under steady-state conditions. Signatures of certain incipient engine faults (e.g., bearing faults, controller miss-scheduling, and starter system faults) tend to magnify during the transient conditions [64], and appropriate usage of transient data could enhance the probability of fault detection [65]. Along

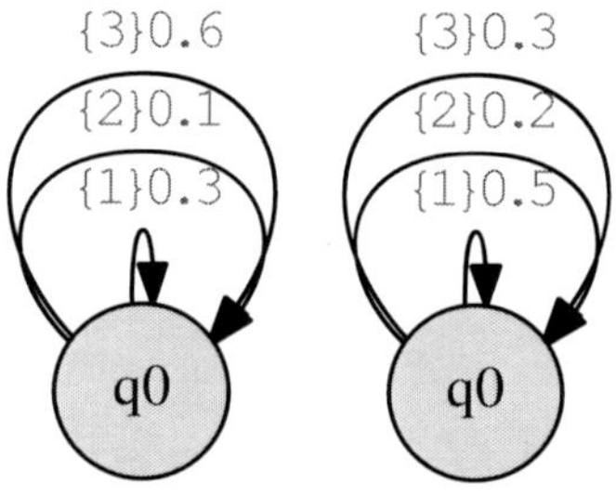

(a) Class C_1 (b) Class C_2

Figure 28. Two one-state PFSA with different morph probabilities.

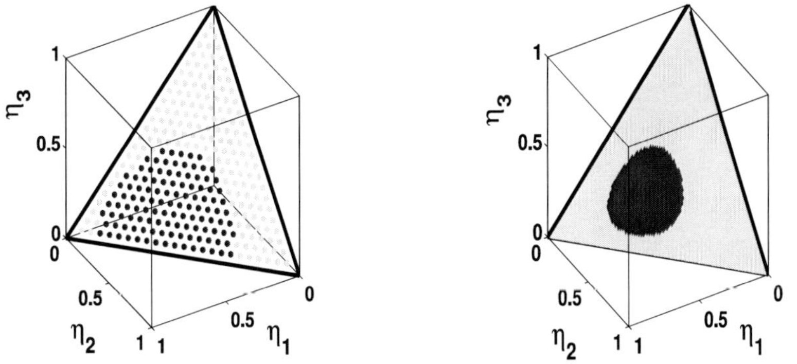

(a) Testing data length = 25 (b) Testing data length = 80

Figure 29. Both subfigures (a) and (b) show the classification boundaries for the two classes C_1 (in light shade) and C_2 (in dark shade) for different lengths of testing string $\widetilde{S}$. The training strings S^1 and S^2 for classes C_1 and C_2 are obtained from the two machines and have lengths 30 and 80, respectively.

this line, several model-based and data-driven fault diagnosis methods have been proposed by making use of transient data. For example, Wang et al. [66] developed a neural network-based fault diagnosis method for automotive transient operations, and Surendar and Ganguli [67] used an adaptive Myriad filter to improve the quality of transient operations for gas turbine engines. However, model-based diagnostics may suffer from inaccuracy and loss of robustness due to low reliability of the transient models. This problem is partially circumvented through usage of data-driven diagnostics; for example, Menon et al. [68] used hidden Markov models (HMMs) for transient analysis of gas turbine engines.

A recently developed data-driven technique, called the symbolic dynamic filtering (SDF) [15], has been shown to yield superior performance in terms of early detection of anomalies and robustness to measurement noise in comparison to other techniques such as Principal Component Analysis (PCA), Neural Networks (NN) and Bayesian techniques [6]. Recently, in a two-part paper [69][70], an SDF-based algorithm for detection and isolation of engine subsystem faults (specifically, faults that cause efficiency degradation in engine components) has been reported and an extension of that work to estimate simultaneously occurring multiple component-level faults has been presented in [56]. Furthermore, an optimized feature extraction technique has been developed under the same semantic framework in [55]. However, all of the above studies were conducted on steady-state cruise flight data that conform with the assumption of quasi-stationarity made in SDF. Due to this assumption, SDF may not be able to adequately handle transient data that are usually of limited length. To handle this limitation the above mentioned algorithm is applied using the *Transient Test-case Generator* [71] of the Commercial Modular Aero Propulsion System Simulation (C-MAPSS) test bed, developed by NASA to detect transient fault.

4.2.1. Fault Injection in C-MAPSS Test Bed

A stochastic damage model has been developed and incorporated in the C-MAPSS *Transient Test-case Generator* [71] (developed by NASA), based on the experimental data for trending the natural deterioration of the engine components as explained in section 2. Although injection of faults is described in the transient test-case generator code [71], it is explained in this section for completeness. For all five rotating components, faults exhibit random magnitudes (F_m), and a random health parameter ratio (HPR). While F_m and HPR directly determines the change in efficiency health parameter $\psi(C)$ of a component C, a change in the flow health parameter ζ_C is determined by HPR for a given perturbation in ψ_C. Formally, the following two relations are used.

$$\delta_{\psi_C} = -\frac{F_m}{\sqrt{1+HPR^2}} \quad \text{and} \quad \delta_{\zeta_C} = \delta_{\psi_C} \cdot HPR \tag{73}$$

where δ_{ψ_C} and δ_{ζ_C} denote the changes in ψ_C and ζ_C respectively.

In the case study, fault magnitude (F_m) follows a random uniform distribution ranging from 1 to 7. Health parameter ratios (HPR) for Fan, LPC, and HPC are uniformly distributed between 1.0 and 2.0, whereas HPR s for HPT and LPT are uniformly distributed between 0.5 to 1.0. The changes in health parameters occur from certain base values of ψ_C and ζ_C.

4.3. Performance of Transient Fault Classifier

This section presents the simulation results of symbolic analysis of transient time series and incipient fault detection on the C-MAPSS test bed. The goal here is to demonstrate how the binary faults (i.e., $K = 2$ in Algorithms 1 and 2) are detected and identified at an early stage in time-critical operations like engine health management. To establish the relationship between detection time and detection accuracy, numerous simulation experiments have been performed during the "takeoff" operation, where the Mach number is varied from 0 to 0.24 in 60 seconds keeping the altitude at zero (i.e., sea level) and TRA at 80%. Faulty operations in three different components, Fan, LPC and HPT, are simulated along with the ideal engine condition. The HPT exit temperature (T_{48}) sensor (that captures the above failure signatures in the gas path) is chosen to provide the transient response. The rationale behind choosing the T_{48} sensor is attributed to the following physical facts: (i) T_{48} is placed between HPT and LPT, and (ii) LPT is mechanically connected to the Fan and LPC via the outer shaft. The sampling frequency of the T_{48} sensor data is 66.7 Hz. For the purpose of training, the duration for each simulation run is chosen to be 60 sec (i.e., a time series of length ~ 4000) and the transient data set for each fault class is constructed by concatenating 50 such blocks of time series data that were individually generated on the simulation test bed under similar transient operating conditions. It is noted that these blocks of data are statistically similar but they are not identical.

The next step is to partition the data sets to generate respective symbol strings. The range of the time series is partitioned into 5 intervals (i.e., the alphabet size $|\Sigma| = 5$), where each segment of the partitioning corresponds to a distinct symbol in the alphabet Σ. The training phase commences after the symbol strings are obtained for each of the four classes, i.e., one nominal and three faulty conditions. In this application, it suffices to choose the depth $D = 1$ [15], which implies that the probability of generation of a future symbol depends only on the last symbol, and hence the set of states is isomorphic to the symbol alphabet (i.e., $Q \equiv \Sigma$). For each class C_i, the parameters N^i_{mn} are obtained by counting the number of times the symbol σ_n is emitted from state q_m.

The testing phase starts with a new time series from one of the classes, where the symbol sequence is generated by using the same alphabet and partitioning as in the training phase. Following Eq. (71), the posterior probability of each class is calculated as a function of the length of the testing data set. Figure 30 shows the posterior probability of each class as a function of the length of the observed test data. It is seen that the observed sequence is correctly identified to belong to the class of LPC faults as the posterior probability of the corresponding class approaches one, while it approaches zero for each of the remaining classes (i.e., the other two component faults and nominal condition). By repeating the same classification technique on 50 new test runs of LPC fault, it is observed that a data length of 500 is sufficient to detect the fault with a reasonable confidence. For other two component faults, the posterior probability for the correct fault class approaches unity within the data length of 50, as seen in Fig. 31. The rationale is that the fault signatures for fan and HPT in T_{48} response are dominant, even if they are incipient.

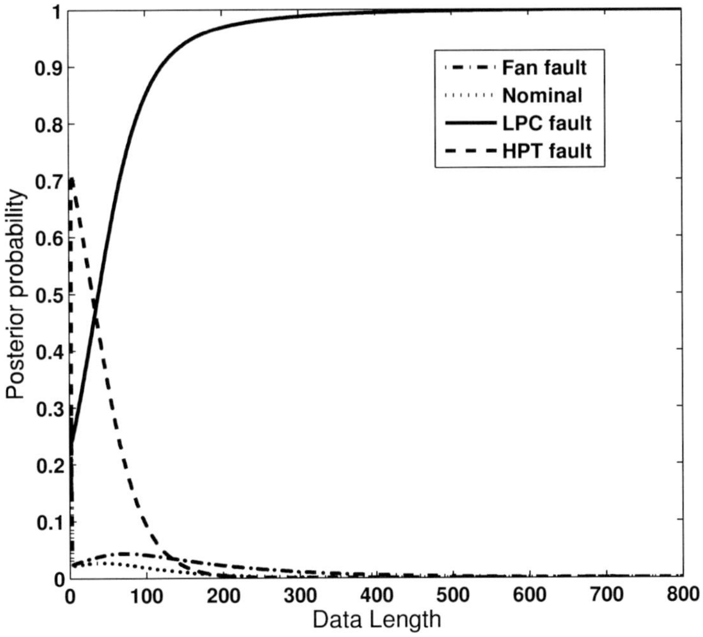

Figure 30. Detection in a multi-fault framework (Ground truth: a faulty LPC).

In a hierarchical fault detection and isolation strategy, at a low resolution level, the goal is to identify the faulty component (e.g., Fan, LPC or HPT). Once the fault is located, the next step is to quantify the severity of the located fault. The symbolic transient fault detection technique can be extended to classify different levels of fault in single components of a gas turbine engine. For verification, samples of nominal data are injected with a low-level fan fault, where the fault magnitude (F_m) follows a random uniform distribution ranging

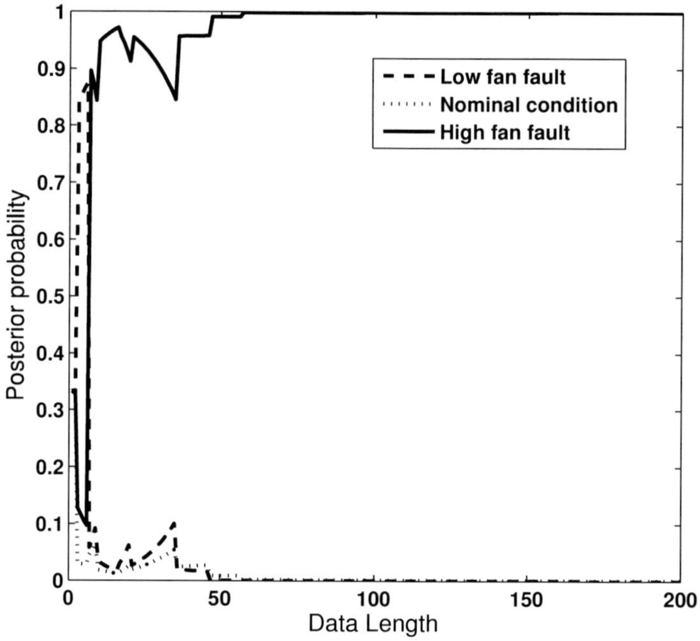

Figure 31. High-level fan fault detection (Ground truth: A high-level fan fault).

from 1 to 3. The remaining samples of nominal data are injected with a high-level fan fault of magnitude (F_m) within the range of 5 to 7. In this case, the alphabet size is chosen to be $|\Sigma| = 6$ to obtain better class separability. Figure 31 shows that the posterior probability for a high fan fault saturates to the value of 1 within a data length of 70, which is equivalent to ~ 1 sec. (Note: The oscillations in posterior probabilities in Fig. 31 are due to randomness of the estimated parameters for data lengths smaller than 50. But when the test case is a low-fan fault (see Fig. 32), the posterior probability of a high-level fan fault remains dominant until the data length reaches ~ 400. This would result in a false classification as the posterior probability for the true class reaches 1 monotonically at around the data length of 1000 at, about 15 seconds, as seen in Fig. 32. This result agrees well with the intuition that the detection time increases with smaller the fault levels.

To examine the performance of symbolic transient fault detection, a family of receiver operating characteristics (ROC) is constructed for different lengths of test data. A binary classification scenario is constructed, which consists of two classes: the nominal engine condition belonging to the class C_1, and a faulty fan condition belonging to the class C_2. The training data length is the same as described in the previous simulation runs. The general classification rule [72] in a symbol string $\widetilde{S}$ is given by

$$\frac{\Pr(\widetilde{S}|C_1)}{\Pr(\widetilde{S}|C_2)} \underset{C_2}{\overset{C_1}{\gtrless}} \lambda \tag{74}$$

where the threshold parameter λ is varied to generate the ROC curves. For the binary classification problem at hand, an ROC curve provides the trade-off between the probability

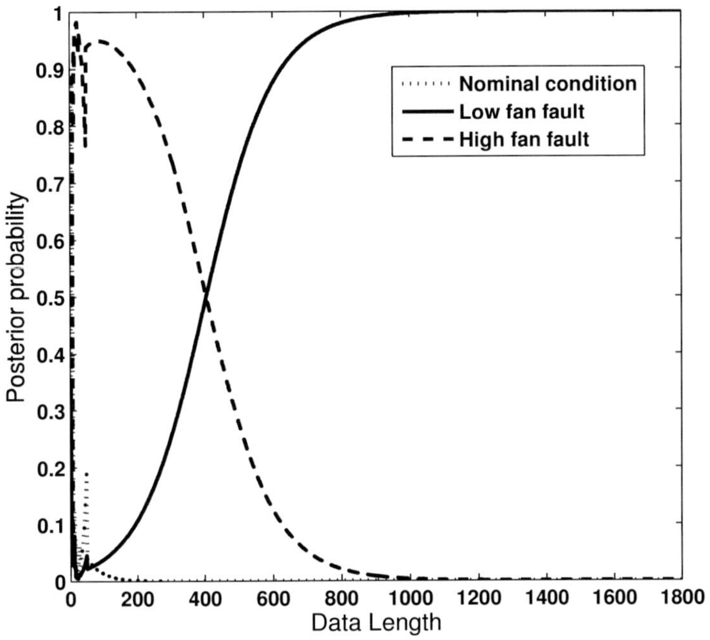

Figure 32. Low-level fan fault detection (Ground truth: A low-level fan fault).

of detection $P_D = \Pr\{\text{decide } C_2|C_2 \text{ is true}\}$ and the probability of false alarms $P_F = \Pr\{\text{decide } C_2|C_1 \text{ is true}\}$. Figure 33 exhibits a family of ROC curves for the proposed fault detection technique with varying lengths of test data. For each ROC curve, λ is varied between 0.01 and 2 with steps of 0.01. It is observed that the ROC curves yield improved performance (i.e., movement toward the top left corner) progressively as the test data length is increased from $N_{test} = 20$ to $N_{test} = 200$. Thus, based on a family of such ROC curves, a good combination of P_D and N_{test} is obtained for a given P_F.

5. Semantic Framework for Multi-sensor Data Interpretation and Fusion

Sensor fusion has been one of the focused topics in the data analysis of distributed physical processes, where the individual sensory information is often used to reveal the underlying process dynamics and to identify potential changes therein. Distributed physical processes are usually equipped with multiple sensors having (possibly) different modalities over a sensor network to accommodate both model-based and data-driven diagnostics & control. The ensemble of distributed and heterogeneous information needs to be fused to generate accurate inferences about the states of critical systems in real time. Various sensor fusion methods have been reported in the literature to address the fault detection & classification problems; examples are linear and nonlinear filtering, adaptive model reference methodologies and neural network-based estimation schemes.

Researchers have used multi-layer perception [73] and radial basis function [74] con-

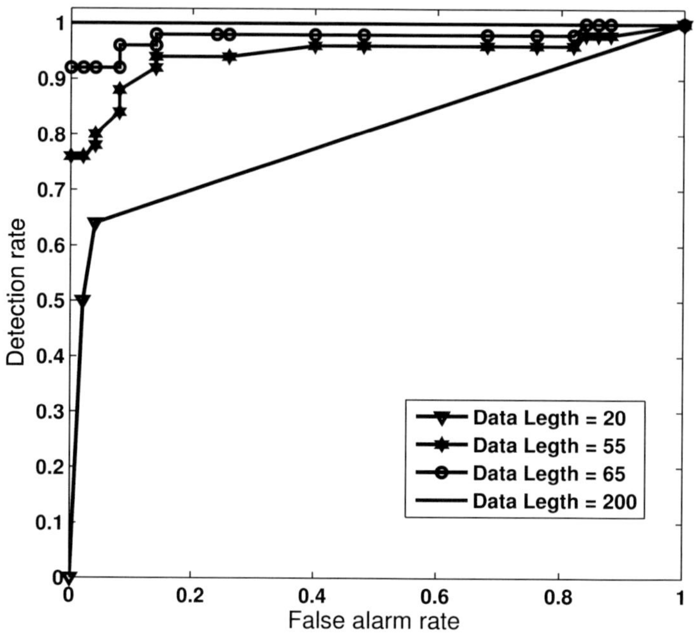

Figure 33. ROC curves for fan fault identification with different test data lengths (λ varying between 0.01 and 2 with steps of 0.01).

figurations of neural networks for detection & classification of plant component, sensor and actuator faults [75]. Similarly, principal component analysis [76] and kernel regression [77] techniques have been proposed for data-driven pattern classification. These approaches address nonlinear dynamics as well as scaling and data alignment issues. However, the effectiveness of data-driven techniques may often degrade rapidly for extrapolation of non-stationary data in the presence of multiplicative noise. Some of the above difficulties can be alleviated to a certain extent by simplifying approximations along with a combination of model-based and data-driven analysis as discussed below.

Robust filtering techniques have been developed to generate reliable estimations from sensor signals, because sensor time series data are always noise-contaminated to some extent [78][79]. Recent literature has also reported Monte Carlo Markov chain (MCMC) techniques (e.g., particle filtering [80] and sigma point techniques [81]) that yield numerical solutions to Bayesian state estimation problems and have been applied to diverse non-linear dynamical systems [82]. The performance and quality of estimation largely depend on the modeling accuracy which is the central problem in the filtering approach; either the dynamics must be linear or linearized, or the data must be strictly periodic or stationary for the linear models to be good estimators. It is noted that the estimation error could be considerably decreased with the availability of high-fidelity models and usage of nonlinear filters, which require numerical solutions; such numerical methods are usually computationally expensive and hence may not be suitable for real-time estimation. Many techniques of reliable state estimation have been reported in literature; examples are multiple model schemes [83], techniques based on analytical redundancy and residuals [84], and nonlinear observer theory [85]. In essence, the information from multiple sources must be synergistically aggregated for diagnosis and control of distributed physical processes (e.g., shipboard auxiliary systems). In this context, information fusion requires processing of non-stationary sensor information to detect parametric or nonparametric changes in the underlying system.

This section presents the development of a sensor data fusion method for fault detection & classification in distributed physical processes with an application to shipboard auxiliary systems, where the process dynamics are interactive. For example, the electrical system is coupled with hydraulic system with time-dependent thermal load. The challenge here is to mitigate several inherent difficulties that include: (i) non-stationary behavior of signals, (ii) diverse nonlinearities of the process dynamics, (iii) uncertain input-output & feedback interactions and scaling, and (iv) alignment of multi-modal information and multiplicative process noise.

The sensor fusion concept, proposed in this section, is built upon the algorithmic structure of symbolic dynamic filtering (SDF) [15]. A spatiotemporal pattern network is constructed from disparate sensors and the fully connected network is then pruned by applying an information-theoretic (e.g., mutual information-based) approach to reduce computational complexity. The developed algorithms are demonstrated on a test bed that is constructed based on a notional MATLAB/Simulink model of Shipboard Auxiliary Systems, where a notional electrical system is coupled with a notional hydraulic system under a thermal load. A benchmark problem is created and the results under different performance metrics are presented.

The section is organized in three sections including the present one. Section 5.1 presents the semantic framework for multi-sensor data modeling and explains how the proposed technique is used to prune the heterogenous sensor network for information fusion. Section 5.2 presents a fault injection scheme in ship-board auxiliary unit (section 2.4) for conducting simulation exercises and validates the fault detection accuracy for different scenarios in the proposed method of information fusion.

5.1. Multi-sensor Data Modeling and Fusion

This section presents a semantic information fusion framework that aims to capture temporal characteristics of individual sensor observations along with co-dependence among spatially distributed sensors. The concept of spatiotemporal pattern networks (STPNs) represent temporal dynamics of each sensor and their relational dependencies as probabilistic finite state automata (PFSA). Patterns emerging from individual sensors and their relational dependencies are called atomic patterns (AP) and relational patterns (RP), respectively. Sensors, APs, and RPs are represented as nodes, self-loop links, and links between pairs of nodes, respectively, in the STPN framework.

5.1.1. Modeling of Temporal Dynamics of Individual Sensor Data

This subsection briefly describes the concept of symbolic dynamic filtering (SDF) [15] for extracting atomic patterns from single-sensor data. The key concepts of SDF are succinctly presented below for completeness of the section. For more detail subsection 1.1

should be consulted.

Symbolic feature extraction from time series data is posed as a two-time-scale problem as explained before. Symbolic dynamic filtering (SDF) encodes the behavior of (possibly nonlinear) dynamical systems from the time series by symbolization and subsequent state machine construction. This is followed by computation of the state probability vectors (or symbol generation matrices) that are representatives of the evolving statistical characteristics of the dynamical system. To this end, the time series data are partitioned into a mutually exclusive and exhaustive set of finitely many cells; maximum-entropy partitioning (MEP) [12] has been adopted to construct the symbol alphabet Σ for generating symbol sequences, where the information-rich regions of the data set are partitioned finer and those with sparse information are partitioned coarser in order that the Shannon entropy [5] of the generated symbol sequence is maximized.

Consequently a D-Markov machine is constructed from the symbol sequence based on: (i) state splitting that generates symbol blocks of different lengths according to their relative importance; and (ii) state merging that assimilates histories from symbol blocks leading to the same symbolic behavior [86]. Words of length D on a symbol sequence are treated as the states of the D-Markov machine before any state-merging operation is executed. Thus, on an alphabet Σ, the total number of possible states becomes less than or equal to $|\Sigma|^D$; and operations of state merging may significantly reduce the number of states.

Let the state of a sensor A at the kth instant be denoted as q_k^A. With this notation, the ij^{th} matrix element π_{ij}^A of the (stationary) state transition matrix Π^A (atomic pattern for sensor A) is the probability that q_{k+1}^A state is i given that the q_k^A state was j, i.e.,

$$\pi_{ij}^A \triangleq P\left(q_{k+1}^A = i \mid q_k^A = j\right) \text{ for an arbitrary instant } k$$

5.1.2. Pattern Analysis of Multi-sensor Information

Relational patterns (that are necessary for construction of STPN) are essentially extracted from the relational probabilistic finite state automata (PFSA). These PFSA are obtained as xD-Markov machines to determine cross-dependence as defined below; the underlying algorithm is described in Subsection 5.1.2.

Construction of Relational PFSA: xD-Markov Machine

This section describes the construction of xD-Markov machines from two symbol sequences $\{s_1\}$ and $\{s_2\}$ obtained from two different sensors (possibly of different modalities) to capture the symbol level cross-dependence. A formal definition is as follows:

Definition 5.1 (xD-Markov). *Let $\mathcal{M}_1$ and $\mathcal{M}_2$ be the PFSAs corresponding to symbol streams $\{s_1\}$ and $\{s_2\}$ respectively. Then a xD-Markov machine is defined as a 5-tuple $\mathcal{M}_{1\to 2} \triangleq (\mathcal{Q}_1, \Sigma_1, \Sigma_2, \delta_1, \tilde{\Pi}_{12})$ such that:*

- $\Sigma_1 = \{\sigma_0, ..., \sigma_{|\Sigma_1|-1}\}$ *is the alphabet set of symbol sequence* $\{s_1\}$
- $\mathcal{Q}_1 = \{q_1, q_2, \ldots, q_{|\mathcal{Q}_1|}\}$ *is the state set corresponding to symbol sequence* $\{s_1\}$
- $\Sigma_2 = \{\sigma_0, ..., \sigma_{|\Sigma_2|-1}\}$ *is the alphabet set of symbol sequence* $\{s_2\}$

- $\delta_1 : \mathcal{Q}_1 \times \Sigma_1 \rightarrow \mathcal{Q}_1$ *is the state transition mapping that maps the transition in symbol sequence* $\{s_1\}$ *from one state to another upon arrival of a symbol in* $\{s_1\}$

- $\tilde{\Pi}_{12}$ *is the symbol generation matrix of size* $|\mathcal{Q}_1| \times |\Sigma_2|$*; the* ij^{th} *element of* $\tilde{\Pi}_{12}$ *denotes the probability of finding the symbol* σ_j *in the symbol string* $\{s_2\}$ *while making a transition from the state* q_i *in the symbol sequence* $\{s_1\}$

In practice, $\tilde{\Pi}_{12}$ is reshaped into a vector of length $|\mathcal{Q}_1| \times |\Sigma_2|$ and is treated as the extracted feature vector that is a low-dimensional representation of the relational dependence between $\{\mathbf{s}_1\}$ and $\{\mathbf{s}_2\}$. This feature vector is called a relational pattern (RP). When both symbol sequences are the same, RPs are essentially the atomic pattern (AP) corresponding to the symbol sequence; in that case, the xD-Markov machine reduces to a simple D-Markov machine. It is noted that an RP between two symbol sequences is not necessarily symmetric; therefore, RPs need to be identified for both directions. It is also useful to quantify cross state transition matrices Π^{AB} and Π^{BA} to quantify the state level cross-dependence between sensors A and B. As illustrated in Fig. 34, elements of the state transition matrices Π^{AB} and Π^{BA} corresponding to the cross machines are expressed as:

$$\pi_{k\ell}^{AB} \triangleq P\left(q_{n+1}^{B} = \ell \mid q_{n}^{A} = k\right) \; \forall n$$

$$\pi_{ij}^{BA} \triangleq P\left(q_{n+1}^{A} = j \mid q_{n}^{B} = i\right) \; \forall n$$

where $j, k \in Q^A$ and $i, \ell \in Q^B$. For a xD-Markov machine, the cross state transition matrix is constructed from symbol sequences generated from two sensors by identifying the probability of occurrence of a state in one sensor from another state in the second sensor. For depth $D = 1$ in a a xD-Markov machine, a cross state transition matrix and the corresponding cross symbol generation matrix are identical.

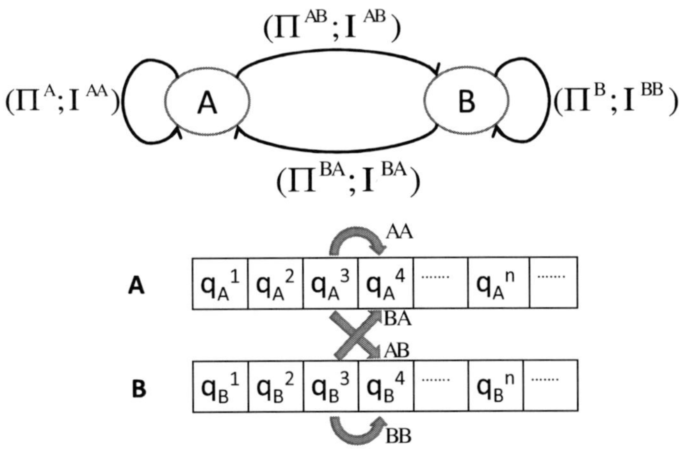

Figure 34. Illustration of Spatiotemporal Pattern Network (STPN).

Pruning of STPN

From the system perspectives, all APs and RPs need to be considered in order to model the nominal behaviors and to detect anomalies. However, it is obvious that there is a scalability issue if there is a significant number of sensors because the number of relational patterns increases quadratically with the number of sensors; for example, the number of RPs could be $S(S-1)$ where S is the total number of sensors and total number of patterns become S^2. The explosion of the pattern space dimension may prohibit the use of a *complete* STPN approach for monitoring of large systems under computational and memory constraints. However, for many real systems, a large fraction of relational patterns may have a very low information content due to the lack of their physical (e.g., electro-mechanical or via feedback control loop) dependencies. Therefore, a pruning process needs to be established to identify a *sufficient* STPN for a system. This section adopts an information-theoretic measure based on *Mutual Information* to identify the importance of an AP or an RP. Mutual information-based criteria have been very popular and useful in general graph pruning strategies [87, 88] including structure learning of Bayesian Networks [89]. In the present context, mutual information quantified on the corresponding state transition matrix essentially provides the information contents of APs and RPs. The concept of network pruning strategy is briefly described below.

Mutual information for the atomic pattern of sensor A is expressed as:

$$I^{AA} = I(q_{n+1}^A; q_n^A) = H(q_{n+1}^A) - H(q_{n+1}^A | q_n^A)$$

where

$$H(q_{n+1}^A) = -\sum_{i=1}^{Q_A} P(q_{n+1}^A = i) \log_2 P(q_{n+1}^A = i)$$

$$H(q_{n+1}^A | q_n^A) = -\sum_{i=1}^{Q_A} P(q_n^A = i) H(q_{n+1}^A | q_n^A = i)$$

$$H(q_{n+1}^A | q_n^A = i) = -\sum_{l=1}^{Q_A} P(q_{n+1}^A = l | q_n^A = i) \cdot \log_2 P(q_{n+1}^A = l | q_n^A = i)$$

The quantity I^{AA} essentially captures the temporal self-prediction capability (self-loop) of the sensor A. However, as an extreme example, the AP for a random sensor data may not be very informative and its self mutual information becomes zero under ideal estimation.

Similarly, mutual information for the relational pattern R^{AB} is expressed as:

$$I^{AB} = I(q_{n+1}^B; q_n^A) = H(q_{n+1}^B) - H(q_{n+1}^B | q_n^A)$$

where

$$H(q_{n+1}^B | q_n^A) = -\sum_{i=1}^{Q_A} P(q_n^A = i) H(q_{n+1}^B | q_n^A = i)$$

$$H(q_{n+1}^B|q_n^A = i) = -\sum_{l=1}^{Q_B} P(q_{n+1}^B = l|q_n^A = i) \cdot \\ \log_2 P(q_{n+1}^B = l|q_n^A = i)$$

The quantity I^{AB} essentially captures sensor A's capability of predicting sensor B's outputs and vice versa for I^{BA}. Similar to atomic patterns, an extreme example would be the scenario where sensors A and B are not co-dependent (i.e., sensor A completely fails to predict temporal evolution of sensor B). In this case, R^{AB} is not very informative and I^{AB} will also be zero under ideal estimation.

Therefore, mutual information is able to assign weights on the patterns based on their relative importance (i.e., information content). The next step is to select certain patterns from the entire library of patterns based on a threshold on the metric. In this section, patterns are selected based on a measure of information gain due to atomic and relational patterns. Formally, let the total information gain I_G^{tot} is defined as the sum of mutual information for all patterns, i.e.,

$$I_G^{tot} = \sum_{(A,B)\in\mathcal{S}\times\mathcal{S}} I^{AB} \tag{75}$$

where $\mathcal{S}$ is set of all sensors. Now, the goal is to eliminate insignificant patterns from the set $\mathcal{S}\times\mathcal{S}$ of all patterns. Let the set of rejected patterns be denoted as $\mathcal{P}^{rej} \subset \mathcal{S}\times\mathcal{S}$ and the corresponding information gain be denoted as I_G^{rej}. The set of rejected patterns is chosen such that, for a specified $\eta \in (0, 1)$,

$$\frac{I_G^{rej}}{I_G^{tot}} < \eta \tag{76}$$

where mutual information for any pattern in the reject set should be smaller than the mutual information for any pattern in the accepted set $\mathcal{P}^{acc} \triangleq (\mathcal{S}\times\mathcal{S}) \setminus \mathcal{P}^{rej}$, which is expressed as $I^{AB}|_{(A,B)\in\mathcal{P}^{rej}} \ll I^{CD}|_{(C,D)\in\mathcal{P}^{acc}}$ for a sufficiently small η. In this section, η is chosen as 0.1 for the validation experiments. In the pruning strategy, it is possible that all patterns related to a certain sensor might be rejected, implying that the sensor is not useful for the purpose at hand. However, the user may choose an additional constraint to keep at least one (atomic or relational) pattern for each sensor in the accepted set of patterns.

Remark 5.1. *In order to use the STPN for fault detection, a network of PFSA can be identified following the above process under the nominal condition. Faulty conditions can then be detected by identifying the changes in parameters related to the accepted patterns. However, under some severe faults, the causal dependency characteristics (i.e., the structure itself) among the sensor nodes may not remain invariant. In such cases, new structures of the STPN can signify severely faulty conditions.*

5.2. Fault Detection in Ship-Board Auxiliary Unit via Sensor Fusion

This section presents and discusses the results of validation of the the sensor fusion algorithm on the simulation test bed of shipboard auxiliary systems described in subsection 2.4.

Table 6. Sensors of the system

System	Sensor	Physical Quantity
Electrical	T_e	Torque output of PMSM
	w_e	Rotor speed of PMSM
Hydraulic	w_{hp}	Angular Velocity of Hydraulic Pump
	P_{hm}	Pressure across Hydraulic Motor (HM)
	T_{hm}	Torque output of HM
	w_{hm}	Angular Velocity of output shaft of HM
Thermal	T_f	Temperature

Table 7. Fault parameters of the system

System	Fault parameter	Symbol	Range
Electrical	Flux linkage of PMSM	W_b	Nominal:0.05 ± 0.005
			Fault:0.03 ± 0.005
Hydraulic	Volumetric Efficiency of HM	ν_{vm}	Nominal:0.9 ± 0.02
			Fault:0.8 ± 0.02
	Total Efficiency of HM	ν_{tm}	Nominal:0.8 ± 0.02
			Fault:0.65 ± 0.02
Thermal	Thermal efficiency	ν_{th}	Nominal:0.9 ± 0.02
			Fault:0.8 ± 0.02

5.2.1. Fault Injection, Sensors, Data Partitioning

Each of the electrical, hydraulic and thermal subsystems of the shipboard auxiliary system is provided with a set of sensors as listed in Table 6. In the simulation test bed, selected parameters in each subsystem can be perturbed to induce faults as listed in Table 7.

The pertinent assumptions in the execution of fault detection algorithms are delineated below.

- At any instant of time, the system is subjected to at most one of the faults mentioned in Table 7, because the occurrence of two simultaneous faults is rather unlikely in real scenarios.
- The mechanical efficiency of a hydraulic motor or a pump is assumed to stay constant over the period of observation as the degradation of machines due to wear and tear occurs at a much slower rate with respect to the drop in efficiency.
- The dynamical models in the simulation test bed are equipped with standard commercially available sensors. Exploration of other feasible sensors (e.g., ultra-high temperature sensors) to improve fault detection capabilities is not the focus of this study.

Figure 35 depicts a typical electromagnetic torque output for nominal and PMSM fault cases. As it is seen that the data itself is very noisy and, due to feedback control actions, there is no significant observable difference in the two cases in Fig. 35. Information integration from disparate sensors has been performed to enhance the detection & classification accuracy in such critical fault scenarios.

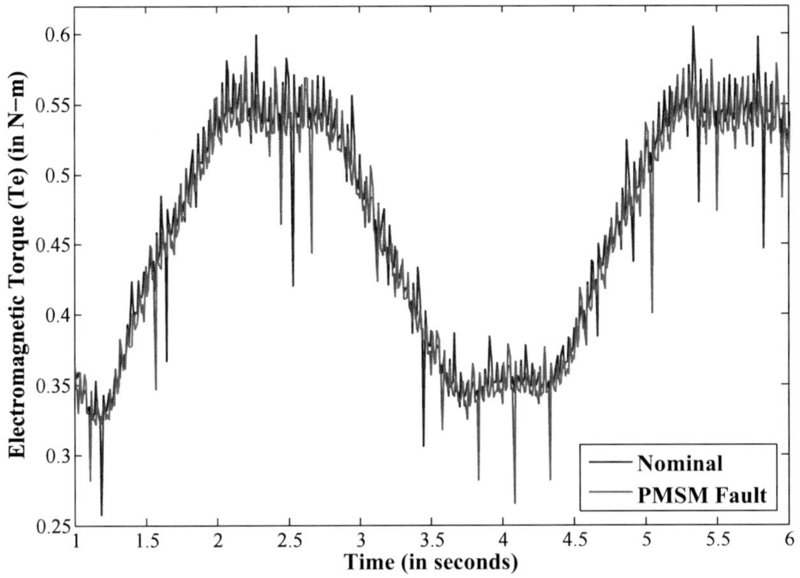

Figure 35. Electromagnetic torque under nominal and PMSM fault conditions.

For all the fault scenarios 100 samples from each sensor are equally divided into two parts for training and testing purposes. For symbolization, maximum entropy partitioning is used with alphabet size, $|\Sigma| = 6$ for all sensors although $|\Sigma|$ does not need to be same for individual sensors. The depth for constructing $PFSA$ states is taken to be $D = 1$ for construction of both atomic pattern and relational pattern. A reduced set of these patterns are aggregated to form the composite pattern, which serves as the feature classified by a k-NN classifier (with $k = 5$) using the Euclidean distance metric for fault detection [40].

5.2.2. Fusion with Complete STPN

One sensor from each of the subsystems (i.e., T_f from thermal subsystem, T_e from electrical subsystem and T_{hm} from hydraulic subsystem) is selected for sensor fusion to identify component faults in the system. The composite patterns (CPs) are formed by concatenating atomic and relational patterns. Therefore, while patterns with high information content (based on the formulation above) help distinguishing between classes, patterns with low information content dilutes the ability of separating classes. Therefore, removing non-informative patterns may lead to reduction of both false alarm, missed detection rates and computational complexity. In this study, CPs consist of all possible APs and RPs of T_f, T_e and T_{hm}. It is seen in Table 8 that CPs perform better for detection of the nominal condition than individual sensors, but the false alarm rate is still high.

Table 8. Fault classification accuracy by exhaustive fusion

class	T_f	T_e	T_{hm}	CP
Nominal	32%	42%	32%	68%
PMSM fault	30%	100%	40%	84%
HM fault	40%	100%	100%	100%
Thermal fault	100%	58%	44%	100%

5.2.3. Pruning of STPN

Pruning of large sensor networks of the given system is attempted here to reduce the complexity of fusion and improve the detection accuracy by capturing the essential spatiotemporal dynamics of the system. Left half of the Fig. 36 shows a fully connected graph of seven sensors of the system where each node is a sensor; bi-directional arcs among them depict the RPs in both directions and self-loops are the APs corresponding to sensors.

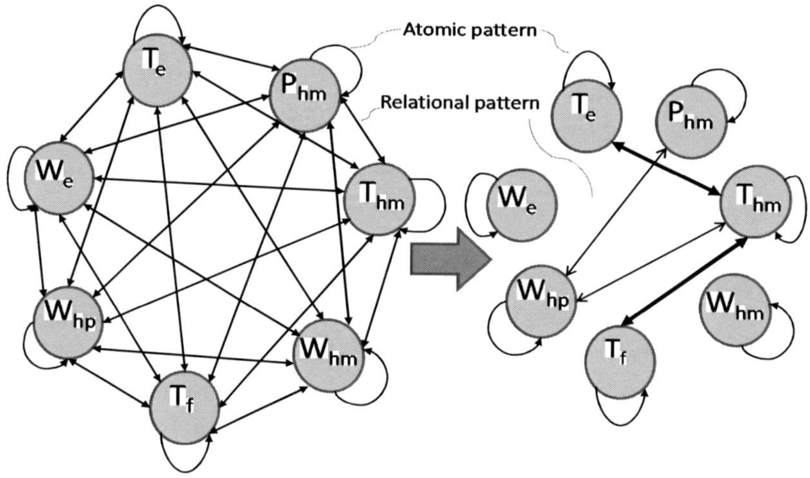

Figure 36. Pruning of STPN.

The right half of Fig. 36 demonstrates the pruned STPN, where the thickness of arcs represents the intensity of mutual information of the RPs among sensors. Both directions of arrows are preserved as the mutual information of the two oppositely directed RPs for a pair of sensors are comparable. In this example, all the self loops are kept intact and arcs with negligible mutual information are omitted from the graph. In this simulation study, the structure of the reduced STPN is observed to remain stable for all the fault classes. The reduction in complexity of network graph is more significant in larger STPNs. The following two scenarios are chosen to justify the credibility of the pruned STPN in the light of fault detection accuracy.

Table 9. Comparison of false alarm rate generated by exhaustive sensor fusion and pruned STPN

Fusion type	False alarm rate
complete STPN	32%
Pruned STPN	8%

Reduction of False Alarm Rates

The same set of sensors, namely, T_f, T_e and T_{hm}, are selected as the STPN and it is subjected to the proposed pruning technique, which results in a composite pattern of AP of T_e (Π^{T_e}) and two RPs ($\Pi^{T_{hm}T_e}$, $\Pi^{T_{hm}T_f}$ as shown by two thick arcs in Fig. 36). This action significantly reduces the false alarm rate as seen in Table 9, where APs of T_f and T_{hm} are dropped from the CP because these patterns do not facilitate better detection. Also PMSM fault detection accuracy does not degrade from 100% unlike fusion with complete STPN. Hence, this pruning technique reduces a CP containing 9 patterns (i.e., 3 APs, 6 RPs) to a CP of three APs and two RPs along with providing better class separability.

Adaptability to Malfunctioning Sensors

In a distributed physical process, such as the shipboard auxiliary system under consideration, malfunctioning of primary sensors in a subsystem is a plausible event. One of the current challenges in the fault detection area is to identify a fault in the subsystem with malfunctioning sensors from the sensor responses of the subsystems that are electromechanically connected. To simulate that situation, three prime heterogenous sensors from the hydraulic subsystem, namely, w_{hp}, P_{hm} and T_{hm}, are selected and a fault is injected to the thermal subsystem by degrading the thermal efficiency (see Table 7). The T_f sensor of thermal subsystem is chosen to be the malfunctioning sensor and hence it is not incorporated in the detection process of thermal fault.

Table 10. Thermal fault detection by sensors of hydraulic subsystem

	w_{hp}	P_{hm}	T_{hm}	CP
Detection accuracy	58%	18%	18%	70%

As the individual sensors of the hydraulic subsystem performs rather poorly in detecting thermal faults as seen in Table. 10, the information from these three sensors are fused by applying the proposed pruning technique on the hydraulic subsystem. The pruned STPN yields a CP consisting of an AP of w_{hp} ($\Pi^{w_{hp}}$) and two RPs, namely, $\Pi^{P_{hm}w_{hp}}$ and $\Pi^{T_{hm}w_{hp}}$, depicted by two thin arcs in Fig. 36; it results in a decent detection accuracy of 70% as seen in Table 10.

Summary, Conclusions and Future Research

The chapter presents some of the key recent developments of the stochastic time-series analysis tool called the symbolic dynamic filtering (SDF) with a special emphasis on its application in the field of fault detection and diagnostics of complex industrial systems. Three major aspects of the tool are highlighted: (i) data-space abstraction via partitioning, (ii) encoding and pattern classification of time-series motifs and (iii) spatio-temporal information fusion. Validation results are presented on high fidelity industrial testbeds of various domains, such as aircraft gas turbine engines, fatigue damage problems, nuclear power plants and ship-board auxiliary systems. The major advantages of SDF for detecting both small and large changes in a dynamical system are listed below:

- Robustness to measurement noise and spurious operational disturbances [12]
- Adaptability to low-resolution sensing due to the coarse graining in space partitions [15]
- Capability for early detection of anomalies because of sensitivity to signal distortion [90]
- real-time execution on commercially available inexpensive platforms [6][90].

While the above mentioned advantages have been demonstrated in multiple industrial system testbeds and small-scale laboratory setups, a key next step is to validate the concepts on real-life systems in order to reduce possible risks and increase robustness. Apart from that, some of the broader research directions that are currently being pursued are:

- *From early detection to prognostics*: SDF has been particularly useful for early detection of anomalies in complex systems. An important current work is concentrating on how to translate this capability to a prognostics capability which is absolutely crucial especially for safety-critical systems.
- *Contextual adaptation*: To become a sustainable health monitoring tool over the life-cycles of industrial systems, SDF will be required to adapt based on contexts regarding internal operations or external environment. Future research will be dedicated to build in and automate different contextual adaptation features in SDF.
- *Soft data integration and HMI*: It is increasingly becoming obvious in the machine learning community that exploiting human knowledge and intent can be significantly more rewarding compared to designing a completely automated systems. To this end, a unified mathematical framework is being investigated that can accommodate both human and machine information.
- *Mathematics of PFSA*: A recent development on Hilbert space formulation of Probabilistic Finite State Automata (PFSA) [91] demonstrated a path of formalizing the SDF developments that can be particularly useful for information fusion purposes. Further studies are being pursued to investigate this area in a greater detail.

Acknowledgments

The authors acknowledge the assistance and technical contributions of their colleagues Dr. Kushal Mukherjee, Dr. Yicheng Wen, Dr. Abhishek Srivastav and Dr. Shashi Phoha in the reported work.

References

[1] S. Gupta and A. Ray, "Statitical mechanics of complex systems for pattern identification," *Journal of Statistical Physics*, vol. 134, no. 2, pp. 337–364, 2009.

[2] R. Pathria, *Statistical Mechanics*. Oxford, UK: Butterworth-Heinemann, 2nd ed., 1996.

[3] D. Lind and M. Marcus, *An Introduction to Symbolic Dynamics and Coding*. Cambridge University Press, Cambridge, U.K., 1995.

[4] R. O. Duda, P. E. Hart, and D. G. Stork, *Pattern Classification*. New York, NY: Wiley Interscience, 2nd edition ed., 2001.

[5] T. Cover and J. Thomas, *Elements of Information Theory, 2nd ed.* New York, NY, USA: Wiley, 2006.

[6] C. Rao, A. Ray, S. Sarkar, and M. Yasar, "Review and comparative evaluation of symbolic dynamic filtering for detection of anomaly patterns," *Signal, Image and Video Processing*, vol. 3, no. 2, pp. 101–114, 2009.

[7] C. Beck and F. Schlögl, *Thermodynamics of chaotic systems: an introduction*. Cambridge University Press, United Kingdom, 1993.

[8] R. Badii and A. Politi, *Complexity hierarchical structures and scaling in physics*. Cambridge University Press, United Kingdom, 1997.

[9] C. S. Daw, C. E. A. Finney, and E. R. Tracy, "A review of symbolic analysis of experimental data," *Review of Scientific Instruments*, vol. 74, no. 2, pp. 915–930, 2003.

[10] H. Kantz and T. Schreiber, *Nonlinear Time Series Analysis, 2nd ed.* Cambridge, U.K.: Cambridge University Press, 2004.

[11] M. B. Kennel and M. Buhl, "Estimating good discrete partitions form observed data: Symbolic false nearest neighbors," *Physical Review E*, vol. 91, no. 8, pp. 084–102, 2003.

[12] V. Rajagopalan and A. Ray, "Symbolic time series analysis via wavelet-based partitioning," *Signal Processing*, vol. 86, no. 11, pp. 3309–3320, 2006.

[13] A. Subbu and A. Ray, "Space partitioning via hilbert transform for symbolic time series analysis," *Applied Physics Letters*, vol. 92, no. 8, pp. 084107–1 to 084107–3, February 2008.

[14] S. Sarkar, K. Mukherjee, and A. Ray, "Generalization of hilbert transform for symbolic analysis of noisy signals," *Signal Processing*, vol. 89, no. 6, pp. 1245–1251, 2009.

[15] A. Ray, "Symbolic dynamic analysis of complex systems for anomaly detection," *Sig. Process.*, vol. 84, no. 7, pp. 1115–1130, 2004.

[16] C. Rao, K. Mukherjee, S. Sarkar, and A. Ray, "Statistical estimation of multiple parameters via symbolic dynamic filtering," *Signal Processing*, vol. 89, no. 6, pp. 981–988, June 2009.

[17] A. Ray, "Signed real measure of regular languages for discrete-event supervisory control," *Int. Journal of Control*, vol. 78, no. 12, pp. 949–967, 2005.

[18] R. Bapat and T. Raghavan, *Nonnegative Matrices and Applications*. Cambridge, U.K: Cambridge University Press, 1997.

[19] Y. Wen and A. Ray, "A stopping rule for symbolic dynamic filtering,," *Signal, Image, and Video Processing*, vol. 7, no. 1, pp. 189–195, 2013.

[20] Y. Wen, A. Ray, and Q. Du, "A variance-estimation-based stopping rule for symbolic dynamic filtering," *Signal, Image, and Video Processing*, vol. 7, no. 1, pp. 189–195, 2013.

[21] D. K. Frederick, J. A. DeCastro, and J. S. Litt, "Users guide for the commercial modular aero-propulsion system simulation (C-MAPSS)," October 2007. NASA/TM2007-215026.

[22] M. D. Carelli, "The design and safety features of the IRIS reactor," *Nuclear Engineering and Design*, vol. 230, pp. 151–167, 2004.

[23] S. C. Stultz and J. B. Kitto, *Steam: its generation and use*. Babcock & Wilcox Company, 40 ed., 1992.

[24] F. Koushanfar, M. Potkonjak, and A. Sangiovanni-Vincentelli, "On-line fault detection of sensor measurements," *IEEE Sensors*, pp. 974–980, 2003.

[25] H. M. Hashemian, *Maintenance of Process Instrumentation in Nuclear Power Plants*. Springer, Berlin, Germany, 2006.

[26] D. Gabor, "Theory of communications," *J. Inst. Electrical Engineering*, vol. 93, p. 429457, 1946.

[27] L. Cohen, *Time-Frequency Analysis*. Prentice Hall PTR, 1995.

[28] A. Lohmann, D. Mendlovic, and Z. Zalevsky, "Fractional hilbert transform," *Opt. Lett.*, vol. 21, no. 4, p. 281283, 1996.

[29] S.-C. Pei and M.-H. Yeh, "Discrete fractional hilbert transform," *IEEE Transactions on Circuits and SystemsII: Analog and Digital Signal Processing*, vol. 47, no. 11, 2000.

[30] Y. Luo, S. Al-dossary, M. Mahroon, and M. Alfaraj, "Generalized hilbert transform and its applications in geophysics," *The Leading Edge*, March 2003.

[31] B. S. Chlebus and S. H. Nguyen, "On finding optimal discretizations for two attributes," in *Proceedings of the First International Conference on Rough Sets and Current Trends in Computing*, RSCTC '98, (London, UK, UK), pp. 537–544, Springer-Verlag, 1998.

[32] Y. Yang and G. I. Webb, "Discretization for naive-bayes learning: managing discretization bias and variance," *Machine Learning*, vol. 74, pp. 39–74, 2009.

[33] S. Garcia, J. Luengo, J. A. Saez, V. Lopez, and F. Herrera, "A survey of discretization techniques: Taxonomy and empirical analysis in supervised learning," *IEEE Transactions on Knowledge and Data Engineering*, vol. 99, no. PrePrints, 2012.

[34] H. Liu, F. Hussain, C. Tan, and M. Dash, "Discretization: An enabling technique," *Data Mining and Knowledge Discovery*, vol. 6, pp. 393–423, 2002.

[35] J. Dougherty, R. Kohavi, and M. Sahami, "Supervised and Unsupervised Discretization of Continuous Features," in *Proceedings of the Twelfth International Conference on Machine Learning* (A. Prieditis and S. Russell, eds.), pp. 194–202, 1995.

[36] Y. Yang, G. I. Webb, and X. Wu, "Discretization methods," in *Data Mining and Knowledge Discovery Handbook, 2nd ed.*, pp. 101–116, Springer, New York, NY, USA, 2010.

[37] A. Bakar, Z. Othman, and N. Shuib, "Building a new taxonomy for data discretization techniques," in *Data Mining and Optimization, 2009. DMO '09. 2nd Conference on*, pp. 132 –140, Oct. 2009.

[38] G. J. Mclachlan, *Discriminant Analysis and Statistical Pattern Recognition.* (Wiley Series in Probability and Statistics) Wiley-Interscience, 2004.

[39] E. Choi and C. Lee, "Feature extraction based on the bhattacharyya distance," *Pattern Recognition*, vol. 36, pp. 1703 – 1709, August 2003.

[40] C. M. Bishop, *Pattern Recognition and Machine Learning (Information Science and Statistics).* Secaucus, NJ, USA: Springer-Verlag New York, Inc., 2006.

[41] M. Loog, R. Duin, and R. Haeb-Umbach, "Multiclass linear dimension reduction by weighted pairwise fisher criteria," *IEEE Trans. Pattern Anal. Mach. Intell.*, vol. 23, no. 7, p. 762766, 2001.

[42] A. Biem, S. Katagiri, and B. Juang, "Pattern recognition using discriminative feature extraction," *IEEE Trans. Signal Process.*, vol. 45, no. 2, p. 500504, 1997.

[43] R. Kohavi and F. Provost, "Glossary of terms," *Machine Learning*, vol. 30, no. 2/3, pp. 271–274, 1998.

[44] R. Alaiz-Rodrguez, A. Guerrero-Curieses, and J. Cid-Sueiro, "Minimax classifiers based on neural networks," *Pattern Recognition*, vol. 38, no. 1, pp. 29 – 39, 2005.

[45] K. Miettinen, *Nonlinear Multiobjective Optimization.* Kluwer Academic, Boston, MA, USA, 1998.

[46] X. Jin, S. Sarkar, K. Mukherjee, and A. Ray, "Suboptimal partitioning of time-series data for anomaly detection," in *Proceedings of 48th IEEE Conference on Decision and Control*, (Shanghai, China), pp. 1020–1025, December 2009.

[47] R. Steuer, *Multiple Criteria Optimization: Theory, Computations, and Application.* John Wiley & Sons, Inc., New York, USA, 1986.

[48] L. Ljung, *System identification: Theory for the User, 2nd ed.* Prentice Hall PTR, Upper Saddle River, NJ, USA, 1999.

[49] O. Nelles, *Nonlinear system identification from classical to neural networks and fuzzy models.* Springer, Berlin, Germany, 2001.

[50] H. Garnier and L. Wang, *Identification of continuous-time models from sampled data from classical to neural networks and fuzzy models.* Springer, London, U.K., 2008.

[51] I. Chattopadhyay and A. Ray, "Structural transformation of probabilistic finite state machines," *Int. J. Control*, vol. 81, no. 5, pp. 820–835, 2008.

[52] X. Jin, S. Gupta, K. Mukherjee, and A. Ray, "Wavelet-based feature extraction using probabilistic finite state automata for pattern classification," *Pattern Recognition*, vol. 44, no. 7, pp. 1343–1356, 2011.

[53] C. R. Shalizi and K. L. Shalizi, "Blind construction of optimal nonlinear recursive predictors for discrete sequences," in *AUAI '04: Proceedings of the 20th conference on Uncertainty in artificial intelligence*, (Arlington, Virginia, United States), pp. 504–511, AUAI Press, 2004.

[54] I. Chattopadhyay, Y. Wen, A. Ray, and S. Phoha, "Unsupervised inductive learning in symbolic sequences via recursive identification of self-similar semantics," in *Preprints Proceedings of American Control Conference, San Francisco, CA, USA*, June-July 2011.

[55] S. Sarkar, X. Jin, and A. Ray, "Data-driven fault detection in aircraft engines with noisy sensor measurements," *Journal of Engineering for Gas Turbines and Power-Transactions of the ASME*, vol. 133, no. 8, p. 081602 (10 pages), 2011.

[56] S. Sarkar, C. Rao, and A. Ray, "Statistical estimation of multiple faults in aircraft gas turbine engines," *Proceedings of the I Mech E Part G: Journal of Aerospace Engineering*, vol. 223, no. 4, pp. 415–424, 2009.

[57] P. Adenis, Y. Wen, and A. Ray, "An inner product space on irreducible and synchronizable probabilistic finite state automata," *Math. Control Signals Syst.*, vol. 23, no. 4, pp. 281–310, 2012.

[58] T. Ferguson, "Ea bayesian analysis of some nonparametric problems," *The Annals of Statistics*, vol. 1, no. 2, 1973.

[59] A. Stentz, "Optimal and efficient path planning for partially-known environments," in *ICRA'94*, pp. 3310–3317, 1994.

[60] S. Wilks, *Mathematical Statistics.* John Wiley, New York, NY, USA, 1963.

[61] Y. Wen, K. Mukherjee, and A. Ray, "Adaptation in symbolic dynamic systems for pattern classification," in *2012 American Control Conference*, (Monreal, Canada), 2012.

[62] H. Lipowsky, S. Staudacher, M. Bauer, and K.-J. Schmidt, "Application of Bayesian forecasting to change detection and prognosis of gas turbine performance," *Journal of Propulsion and Power*, vol. 132, no. 3, p. 031602, 2010.

[63] V. Guruprakash and R. Ganguli, "Three-and seven-point optimally weighted recursive median filters for gas turbine diagnostics," *Proceedings of the I Mech E Part G: Journal of Aerospace Engineering*, vol. 222, no. 3, pp. 307–318, May 2008.

[64] Y. G. Li, "A gas turbine diagnostic approach with transient measurements," *Proceedings of IMechE, Part A: Journal of Power and Energy*, vol. 217, no. 2, pp. 169–177, 2003.

[65] G. Merrington, O. Kwon, G. Goodwin, and B. Carlsson, "Fault detection and diagnosis in gas turbines," *J. Eng. Gas Turbines Power*, vol. 113, pp. 276–282, 1991.

[66] X. Wang, N. McDowell, U. Kruger, G. McCullough, and G. W. Irwin, "Semi-physical neural network model in detecting engine transient faults using the local approach," in *Proceedings of the 17th World Congress The International Federation of Automatic Control*, July 6-11, 2008.

[67] V. P. Surender and R. Ganguli, "Adaptive myriad filter for improved gas turbine condition monitoring using transient data," *J. Eng. Gas Turbines Power*, vol. 127, pp. 329–339, 2005.

[68] S. Menon, O. Uluyol, K. Kim, and E. O. Nwadiogbu, "Incipient fault detection and diagnosis in turbine engines using hidden markov models," *ASME Conference Proceedings*, vol. 2003, no. 36843, pp. 493–500, 2003.

[69] S. Gupta, A. Ray, S. Sarkar, and M. Yasar, "Fault detection and isolation in aircraft gas turbine engines: Part i - underlying concept," *Proceedings of the I Mech E Part G: Journal of Aerospace Engineering*, vol. 222, no. 3, pp. 307–318, May 2008.

[70] S. Sarkar, M. Yasar, S. Gupta, A. Ray, and K. Mukherjee, "Fault detection and isolation in aircraft gas turbine engines: Part ii - validation on a simulation test bed," *Proceedings of the I Mech E Part G: Journal of Aerospace Engineering*, vol. 222, no. 3, pp. 319–330, May 2008.

[71] J. Armstrong, "Users guide for the transient test case generator," September 2009. NASA GRC Internal Report.

[72] J. Pearl, *Probabilistic Reasoning in Intelligent Systems: Networks of Plausible Inference.* San Francisco, CA, USA: Morgan Kaufmann Publishers Inc., 1988.

[73] J. Liu and J. M. A. Scherpen, "Fault detection method for nonlinear systems based on probabilistic neural network filtering," *International Journal of Systems Science*, vol. 33, pp. 1039–1050, 2002.

[74] S. Haykin, *Neural Networks : A Comprehensive Foundation*. Upper Saddle River, NJ: Prentice Hall, 1999.

[75] M. R. Napolitano, Y. An, and B. A. Seanor, "A fault tolerant flight control system for sensor and actuator failures using neural networks," *Aircraft Design*, vol. 3, pp. 103–128, 2000.

[76] K. Fukunaga, *Statistical Pattern Recognition*. Boston, MA, USA: Academic Press, 2 ed., 1990.

[77] J. Shawe-Taylor, *Kernel Methods for Pattern Analysis*. Cambridge, U.K.: Cambridge University Press, 2004.

[78] M. S. Grewal and A. P. Andrews, *Kalman Filtering: Theory and Practice Using MATLAB*. John Wiley & Sons, 2 ed., 2001.

[79] A. Gelb, ed., *Applied Optimal Estimation*. The MIT Press, 1974.

[80] C. Andrieu, A. Doucet, S. Singh, and V. B. Tadic, "Particle methods for change detection, system identification, and control," *Proceedings of IEEE*, vol. 92, pp. 423–438, 2004.

[81] S. Julier, J. Uhlmann, and H. F. Durrant-Whyte, "A new method for the nonlinear transformation of means andcovariances in filters and estimators," *IEEE Transactions on Automatic Control*, vol. 45, pp. 477–482, 2000.

[82] P. Li and V. Kadirkamanathan, "Particle filtering based likelihood ratio approach to fault diagnosis in nonlinear stochastic systems," *IEEE Transactions on Systems, Man and Cybernetics*, vol. 31, pp. 337–343, 2001.

[83] M. Gopinathan, J. D. Boskovic, R. K. Mehra, and C. Rago, "A multiple model predictive scheme for fault-tolerant flight control design," in *Conference on Decision and Control*, pp. 1376–1381, 2001.

[84] J. J. Gertler, "Survey of model-based failure detection and isolation in complex plants," *IEEE Control Systems Magazine*, vol. 8, pp. 3–11, 1988.

[85] E. A. Garcia and P. M. Frank, "Deterministic nonlinear observer-based approaches to fault diagnosis: A survey," *Control Engineering Practice*, vol. 5, pp. 663–670, 1997.

[86] P. Adenis, K. Mukherjee, and A. Ray, "State splitting and state merging in probabilistic finite state automata," in *American Control Conference, San Francisco, CA, USA*, pp. 5145–5150, 2011.

[87] H. Kretzschmar, C. Stachniss, and G. Grisetti, "Efficient information-theoretic graph pruning for graph-based slam with laser range finders," in *International Conference on Intelligent Robots and Systems (IROS)*, pp. 865 – 871, 2011.

[88] A. J. Butte and I. S. Kohane, "Mutual information relevance networks:functional genomic clustering using pairwise entropy measurements," in *Pacific Symposium on Biocomputing*, pp. 418–429, 2000.

[89] L. M. de Campos, "A scoring function for learning bayesian networks based on mutual information and conditional independence tests," *Journal of Machine Learning Research*, vol. 7, pp. 2149–2187, 2006.

[90] S. Gupta, A. Ray, and E. Keller, "Symbolic time series analysis of ultrasonic data for early detection of fatigue damage," *Mechanical Systems and Signal Processing*, vol. 21, no. 2, pp. 866–884, 2007.

[91] Y. Wen, A. Ray, and S. Phoha, "Hilbert space formulation of symbolic systems for signal representation and analysis," *Signal Processing*, vol. 93, no. 9, pp. 2594–26, September 2013.

In: Fault Detection
Editors: F. P. G. Márquez and M. Papaelias
ISBN: 978-1-62808-999-8

Chapter 4

EXPERT SYSTEM FOR INDUCTION MOTOR FAULT DETECTION BASED ON VIBRATION ANALYSIS AND SUPPORT VECTOR MACHINES

Željko S. Kanović
Faculty of Technical Sciences, Novi Sad, Serbia

ABSTRACT

Machine condition monitoring with fault diagnosis plays an important role in maintenance system. It reduces maintenance costs, improves productivity and efficiency and increases machine availability. This chapter presents an expert system for online and offline induction motor fault detection based on vibration analysis and support vector machines (SVM). Vibration analysis is used because of its easy measurability, high accuracy and reliability. Vibration signals of more types of healthy and faulty induction motors are collected and characteristic features, as indicators of fault presence, are calculated, in both time and frequency domain. The most frequent faults present at induction motors are considered, such as bearing wear, static eccentricity and broken rotor bar. Obtained feature sets were then used for training of support vector machines classifiers, a type of artificial intelligence classification technique which can determine whether some of considered faults is present or not. The performance of SVM classifiers is improved using General Particle Swarm Optimization (GPSO) algorithm for SVM parameter tuning. This expert system can function online, when vibration sensors are connected to the motor and fault detection is carried out simultaneously, or offline, when collected vibration signal is recorded and processed afterwards. The system was tested and validated on a number of healthy and faulty motors in industrial facility for sunflower oil processing during the maintenance period. Obtained results show that this system can detect faults in early stages with high accuracy and reliability, which could provide malfunction and failure prevention and improve overall performance and efficiency of industrial systems.

Keywords: Fault Detection, Induction Motor, Vibration Analysis, Support Vector Machines

1. INTRODUCTION

Induction motors play an important role as prime movers in manufacturing, process industry and transportation due to their reliability and simplicity in construction. Although induction motors are reliable, the possibility of unexpected faults is unavoidable. The issue of robustness and reliability is very important to guarantee the good operational condition. Therefore, condition monitoring of induction motors has received considerable attention in recent years. Early fault diagnosis and condition monitoring can reduce the consequential damage, breakdown maintenance and reduce the spare parts of inventories (Matić et al., 2010). Moreover it can prolong the machine life and increase the performance and the availability of the machine.

Many researchers have proposed techniques and systems for the diagnosis process. Various techniques have been used, such as motor current signature analysis (Kulić et al., 2010), electromagnetic torque measurement (Thollon et al., 1993), acoustic analysis (Lee et al., 1994) and partial discharge (Stone et al., 1996). One of the most popular techniques, used with induction motors fault detection in particular and also in condition monitoring in general is vibration analysis, due to its easy measurability, high accuracy and reliability (Marquez et al., 2012; Yang et al., 2003.)

Support vector machines (SVMs) have been extensively employed to solve classification problems. In machine condition monitoring and fault diagnosis, some researchers have used SVMs as a tool for classification of different kind of faults, such as ball bearing faults (Jack and Nandi, 2002), gear faults (Samanta, 2004), condition classification of small reciprocating compressor (Yang et al., 2005a), cavitation detection of butterfly valve (Yang et al., 2005b) and so on.

To perform good classification using SVMs, the preparation of data inputs for classifier needs special treatment to guarantee the good performance. Recently, the use of feature extraction and feature selection for data preparation to avoid the redundancy before inserting into classifier has received considerable attention (Cao et al., 2003, Marquez, 2009).

There are numerous papers and studies that describe laboratory experiments and applied techniques conducted in purpose of different kinds of induction motor faults detection. However, application of these techniques and their results in real industrial systems is not so common. In this chapter, an expert system for induction motor fault detection is presented, which represents an attempt to implement very well known fault detection techniques in real industrial system. This system consists of several modules, which perform all necessary tasks in fault detection and classification process (vibration signal acquisition, data processing, SVM fault detection and visualization and archiving of obtained results). Three kinds of faults are considered, bearing wear, static eccentricity and broken rotor bar. The system is developed and tested using real data from laboratory and sunflower oil processing industry. It can be used as stand-alone tool for condition monitoring of induction motors. Use of this system in industrial facilities could assist in early fault detection and prevent malfunctions and failures of production systems, improving their overall performance and efficiency.

The chapter is organized as follows. In section 2 fault detection process using vibration analysis and support vector machines is described, presenting shortly all included phases and applied techniques. Section 3 illustrates the structure of expert system and its embedded modules. In section 4, results of system testing performed in laboratory and oil industry facility are presented.

2. Vibration Analysis and Support Vector Machines in Fault Detection

Vibration signals are frequently used for fault diagnosis of mechanical systems since they carry information on the dynamic state of the mechanical elements themselves. Fault diagnosis is conducted typically in the following phases (as shown in Figure 1):

- vibration data acquisition
- feature extraction and
- fault detection and classification

Vibration Data Acquisition

Vibration signals are initially obtained as a series of digital values representing acceleration in the time domain. These signals are then analyzed using many different techniques proposed in the literature, in time and/or frequency domain. In the present case, both time and frequency domain features of acquired signals were used for fault detection.

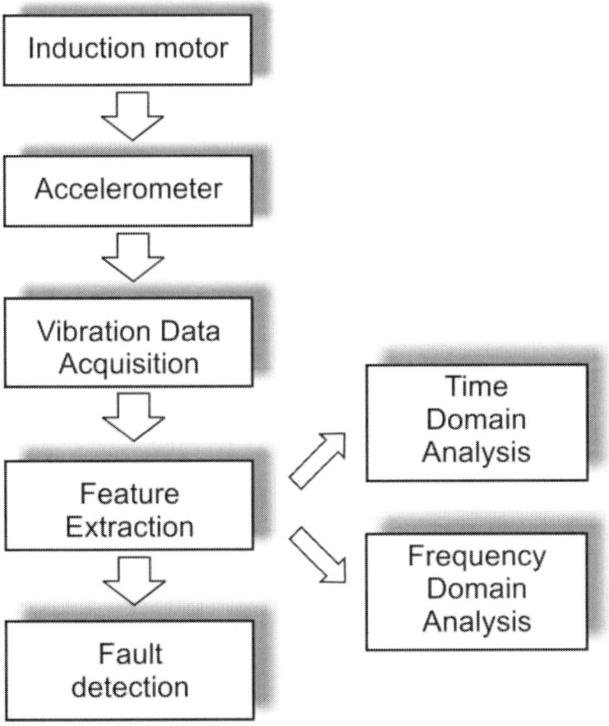

Figure 1. Diagram of fault detection process using vibration signal analysis.

In time domain, we chose to observe eight statistical features: mean, root mean square, skewness, kurtosis, C factor, L factor, S factor and I factor, often used in the literature (Stepanić et al., 2009). In frequency domain, we chose to observe power spectrum, whose amplitude is the square of the amplitude of the spectrum, which is an effective method to diagnose machinery faults (Yang et al., 2003). Ten characteristic frequency-domain features were used. They represent the sum of amplitudes of power spectrum in the region around characteristic frequencies. The summation is performed in the band $f_c \pm 3Hz$, where f_c is characteristic frequency.

First feature represents the sum around twice supply frequency ($f_s = 50Hz$), whose increased amplitude is the indicator of static eccentricity in induction motor (Dorrell and Thomson, 1997; Kanović et al., 2011).

The following five features are related to condition of bearings. If a ball bearing is damaged, the increased amplitude of spectrum in bearing characteristic frequencies will occur. These frequencies are determined by geometry of the bearing. According to that, as features in the frequency domain the sum of amplitudes of power spectrum around these characteristic frequencies are used, which are:

outer race fault frequency: $f_{rpfo} = f_r \times \frac{N}{2}\left(1 - \frac{d}{D}\cos\phi\right)$ (1)

inner race fault frequency: $f_{bpfi} = f_r \times \frac{N}{2}\left(1 + \frac{d}{D}\cos\phi\right)$ (2)

rotation frequency of the rolling element: $f_{bsf} = f_r \times \frac{D}{2d}\left[1 - \left(\frac{d}{D}\right)^2 \cos^2\phi\right]$ (3)

rolling element fault frequency: $f_{bff} = 2 \times f_{bsf}$ (4)

cage fault frequency: $f_{ftf} = f_r \times \frac{1}{2}\left(1 - \frac{d}{D}\cos\phi\right)$ (5)

where N is the number of rolling elements in the bearing, ϕ is the contact angle of the rolling element, d is the rolling element diameter and D is the diameter of the bearing shell (Tandon and Choudhury, 1999; Stepanić et al., 2009).

The next four features are related to broken rotor bar. If a broken rotor bar exists, no current will flow in that bar. As a result, the field in the rotor around that particular bar will not exist.

Therefore the force applied to that side of the rotor would be different from that on the other side of the rotor, creating an unbalanced magnetic force that rotates at one times rotational speed and modulates at a frequency equal to slip frequency times the number of poles, which is known as pole pass frequency. It means that in vibration spectrum increased amplitudes will occur at the rotation frequency f_r and its sidebands

$$f_{brb1} = f_r \pm f_p \tag{6}$$

where f_p is pole pass frequency defined as

$$f_p = \left(f_{sync} - f_r\right) \cdot P \tag{7}$$

with P being number of poles and f_{sync} being synchronous speed [4]. These sidebands occur also in higher harmonics of rotation speed ($2f_r$, $3f_r$,). Detection of broken bar based on vibration analysis and observation of features (6) is common in literature and widely used in practical applications [6]. In the present system, we used the sum of amplitudes of power spectrum around two characteristic sideband frequencies in the first and the second harmonic of rotation speed, which makes total four broken bar features.

Based on both time and frequency domain features, after proper dimensional reduction, which is in this study conducted using principal component analysis, artificial intelligence classifier can be applied to detect faults in operation of monitored object.

Feature Extraction

Obtained feature set has totally eighteen features for every recorded signal. Using all of them for fault classification might be inefficient, since this feature set contains redundancy. Its dimensionality is thus reduced using principal component analysis.

Principal component analysis (PCA) involves a mathematical procedure that transforms a number of possibly correlated variables into a smaller number of uncorrelated variables called principal components. The first principal component accounts for as much of the variability in the data as possible, and each succeeding component accounts for as much of the remaining variability as possible.

PCA is mathematically defined as an orthogonal linear transformation that transforms the data to a new coordinate system such that the greatest variance by any projection of the data comes to lie on the first coordinate (called the first principal component), the second greatest variance on the second coordinate, and so on. PCA is theoretically the optimum transform for given data in least square terms. Given a set of centered input vectors $\mathbf{x}_t$ ($t = 1, \ldots, l$ and $\sum_{t=1}^{l} \mathbf{x}_t = 0$), each of which is of m dimension $\mathbf{x}_t = (x_t(1), x_t(2), \ldots, x_t(m))^{\mathrm{T}}$, usually $m<l$, PCA linearly transforms each vector $\mathbf{x}_t$ into a new one $\mathbf{s}_t$ by

$$\mathbf{s}_t = \mathbf{U}^{\mathrm{T}} \mathbf{x}_t \tag{8}$$

where $\mathbf{U}$ is the $m \times m$ orthogonal matrix whose ith column, $\mathbf{u}_i$ is the eigenvector of the sample covariance matrix

$$\mathbf{C} = \frac{1}{l}\sum_{t=1}^{l} \mathbf{x}_t \mathbf{x}_t^{\mathbf{T}} \tag{9}$$

The new components $\mathbf{s}_t$ are called principal components. By using only the first several eigenvectors sorted in descending order of the eigenvalues, the number of principal components in $\mathbf{s}_t$ can be reduced, so PCA has the dimensional reduction characteristic (Jollife, 1986). In the present expert system, after thorough experimental research, we determined that first six principal components carry sufficient amount of information to perform reliable fault classification, so we reduced the dimensionality of feature set from original eighteen to only six dimensions.

Support Vector Machines in Fault Detection

SVMs are a kind of learning machine based on statistical learning theory (Vapnik, 1995). The basic idea of applying SVM to pattern classification can be stated as follows: first, map the input vectors into one features space, either linearly or nonlinearly, which is relevant with the kernel function. The mapped data belong to one of the two classes, marked as circles, representing negative class with labels $y_i = -1$, and squares, representing positive class with labels $y_i = 1$, as shown in Figure 2.

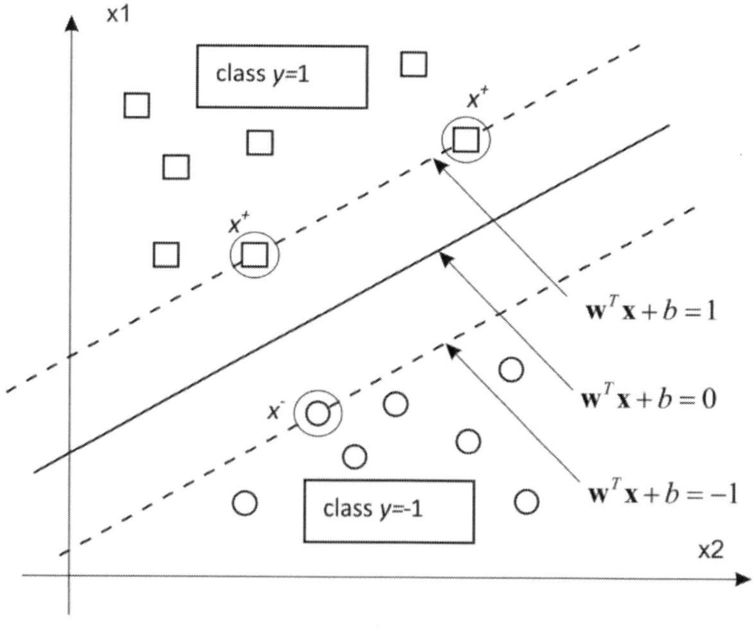

Figure 2. Classification using SVM

The SVM attempts to place a linear separation hyperplane (a line, in two-dimensional case shown in Figure 2) called boundary between the two different classes, and orient it in such a way that the margin (represented by dotted lines) is maximized. The boundary is then placed in the middle of this margin between the two points. The nearest data points are used to define the margins and are known as support vectors (SV), denoted as x^+ and x^- and encircled in Figure 2. Once the support vectors are selected, the rest of the feature set can be discarded, since the SVs contain all the necessary information for the classifier.

The separating hyperplane can be expressed as

$$f(\mathbf{x}) = \mathbf{w}^T \cdot \mathbf{x} + b = \sum_{j=1}^{N} w_j x_j + b = 0, \quad \mathbf{w} \in R^N, \quad b \in R \tag{10}$$

where $\mathbf{w}$ defines the boundary, $\mathbf{x}$ is the input vector of dimension N and b is scalar representing threshold.

Using proper scaling of parameters $\mathbf{w}$ and b, the equations for positive and negative class at the margins are, respectively:

$$f(\mathbf{x}) = \mathbf{w}^T \cdot \mathbf{x} + b = 1 \tag{11}$$

and

$$f(\mathbf{x}) = \mathbf{w}^T \cdot \mathbf{x} + b = -1. \tag{12}$$

For all points, we have

$$y_i\left(\mathbf{w}^T \mathbf{x}_i + b\right) \geq 1,\ i = 1, 2, \ldots N.$$

From the geometry, the margin is found to be $\|w\|^{-2}$, so the optimal separating hyperplane can be found as a solution to the optimization problem defined as:

$$\text{minimize } \frac{1}{2}\|\mathbf{w}\|^2$$

subject to constraints

$$y_i(\mathbf{w} \cdot \mathbf{x} + b = 0) \geq 1, \quad i = 1, \ldots, N. \tag{13}$$

By introducing Lagrange multipliers α_i, the problem can be described as

$$\text{minimize } L(\mathbf{w},b,\alpha_i)=\frac{1}{2}\mathbf{w}^T\mathbf{w}+\sum_{i=1}^{M}\alpha_i\left[y_i\left(\mathbf{w}^T\mathbf{x}_i+b\right)-1\right]$$

where L denotes the Lagrangian function of the original problem. The solution is a unique globally optimized result, with parameters **w** and b defined as

$$\mathbf{w}=\sum_{i}^{N}\alpha_i y_i \mathbf{x}_i \tag{14}$$

and

$$b_O=\frac{1}{N_{SV}}\left[\sum_{s=1}^{N_{SV}}\left(\frac{1}{y_s}-\mathbf{x}_s{}^T\mathbf{w}_O\right)\right] \tag{15}$$

where N_{SV} denotes the number of support vectors, having corresponding values of $\alpha_i > 0$.

The process of determining optimal parameters **w** and b for the separating hyperplane is referred as SVM training. One can now define the decision function

$$g(\mathbf{x})=sign\left(\sum_{i=1}^{N}\alpha_i y_i\left(\mathbf{x}\cdot\mathbf{x}_i\right)+b\right) \tag{16}$$

and the obtained sign of this decision function result determines the class label of considered point (in our case, the set of features).

For a linear non-separable case, SVMs perform a nonlinear mapping of the input vector **x** from the input space $\Re^d$ into a higher dimensional Hilbert space, where the mapping is determined by kernel function (Gaussian RBF, polynomial,..). According to the different classification problems, the different kernel function can be selected to obtain the optimal classification results. Some typical kernel functions are presented in Table 1.

Table 1. Typical kernel function for nonlinear SVMs

Kernel function	Classifier type
$K(\mathbf{x}_i,\mathbf{x}_j)=(1+\mathbf{x}_i^T\mathbf{x}_j)^p$	Polynomial
$K(\mathbf{x}_i,\mathbf{x}_j)=\exp(-\frac{\left\|\mathbf{x}_i-\mathbf{x}_j\right\|^2}{2\sigma^2})$	Gauss RBF (*Radial Basis Function*)
$K(\mathbf{x}_i,\mathbf{x}_j)=\tanh(\beta_0\mathbf{x}_i^T\mathbf{x}_j+\beta_1)$	Sigmoidal or MLP (*Multilayer perceptron*)

In the present application, RBF type kernel function is used, which is characterized by the value of the parameter σ. This parameter is usually calculated iteratively. We chose to optimize the value of σ in order to improve the performance of SVM classifier. In that purpose, we used Generalized Particle Swarm Optimization algorithm (GPSO), global optimization algorithm based on social behavior of animals moving in large groups, which is robust, easy to implement and computationally inexpensive. More details on SVM optimization procedure can be seen in (Kanović et al., 2011).

3. Expert System Structure

Expert system for induction motor fault detection presented in this chapter implements all previously stated phases of detection process. Accordingly, it consists of following modules:

- Vibration signal acquisition module, which collects signals from two accelerometers and transforms it in textual data file. The module is implemented in LabView software, compatible with hardware used for signal acquisition. This module also calculates signal spectrum and maps obtained data in textual file.
- Data processing module; this module is implemented in Matlab. It imports textual data files outputted by previous module and performs characteristic features calculation, described in section 2. This module also applies PCA to reduce dimensionality of obtained feature set, decreasing number of features from eighteen in original set to six in reduced feature set, which is empirically proven to be satisfactory for successful fault detection process.
- SVM fault detection module, also implemented in Matlab, which performs classification of input feature set and detects the presence of the fault. SVMs are trained on training data set, formed using feature sets of vibration signals of different induction motors. Using these trained SVMs, obtained feature set of the observed motor is classified as faulty or healthy, concerning all considered faults. The result of the classification is then forwarded to next module.
- Module for graphical representation, which illustrates classification results, creates diagrams for visual fault detection and performs data archiving.

The expert system is implemented in Microsoft Windows Visual Basic 6, with support of Microsoft Access for data archiving, and runs on Microsoft Windows operating system. This way, the executable file is rather small, since it uses only basic elements of Windows operating system, and thus has minor demands concerning operating memory. On the other hand, this implementation approach provides good interconnection with other software packages and is therefore suitable for application purposes.

The expert system can run in two different modes, manual and automatic. Manual mode implies that the application is used occasionally, on demand, for example in maintenance period or if there is any reason to suspect that some motor is faulty. Fault detection process is then performed once, providing information on motor condition. When running in automatic mode, application is executed continuously, monitoring condition of the motor during the entire production process. In both modes, obtained results are archived in data base, containing information on event's date and time, operation mode (manual / automatic), presence of the fault(s), diagrams of recorded signal and calculated signal spectrum. Also, the analysis can be carried out online, when all fault detection stages are conducted in continuous sequence, or offline, when detection stages are separated (e. g. data collection is conducted, text file with signal data is saved and the analysis is performed afterwards).

UML Use Case diagram of expert system is depicted in Figure 3, showing all provided cases of use. One can see that operator can access and edit Matlab scripts for signal processing, feature calculation and SVM classification. Data stored in database are also available for review, control and monitoring. The operator can also adjust parameters concerning program execution and characteristic motor data, such as power, nominal speed, and bearing geometry data, which are needed for feature calculation. All these activities are processed using operator interface, which also shows sequences and phases of detection process, classification results and diagrams of vibration signals on both channels (vertical and horizontal accelerometers) and their spectrums.

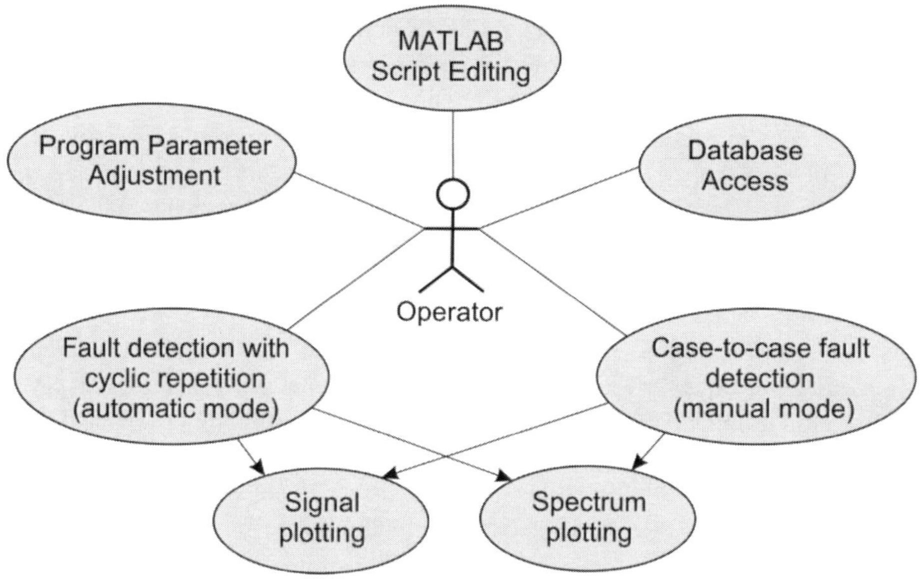

Figure 3. UML Use Case diagram of expert system.

4. System Testing and Results

The system was tested and validated on real data, obtained in a sunflower oil production facility, a heating plant facility and in the laboratory environment. Totally 320 vibration signals from induction motors were collected, using different speed and load. These signals are used for SVM classifier training. The structure of obtained signal set is presented in Table 1. Static eccentricity is denoted as Fault_1, bearing wear as Fault_2 and broken rotor bar as Fault_3.

More different types of motors were considered. First type is 5.5kW motor with one pair of poles, nominal speed of 2925 rpm and nominal current of 10.4 A. These motors are driving the screw conveyors in the process plant. Two motors of this type were healthy, one motor had wear on both inner and outer race of the bearing, one motor had static eccentricity level of 30% and one motor had both faults present at the same time (inner and outer race wear and 50% static eccentricity).

Table 1. Overview of vibration signals of faulty and fault-free motors

Number of signals					Total number of signals
Fault-free	Fault_1 only	Fault_2 only	Fault_1 and Fault_2	Fault_3	
140	80	40	20	40	320

Second type is 15 kW motor driving the crushed oilseed conditioners, with one pair of poles, nominal speed of 2940 rpm and nominal current of 26.5 A. Two of these motors were healthy, two had static eccentricity level of 30% and 50% and one had defects on bearing ball, inner and outer race (Kanović, 2012).

Third type is a 3.2 MW high-voltage induction motor driving high- and low-pressure pumps in city heating plant, with nominal speed of 2994 rpm. Two motors of this type were considered, one healthy and one with two broken rotor bars. The layout of the motor and mounted equipment is shown in Figure 4.

Figure 4. Instrumentation for vibration signal acquisition and fault detection.

Fourth and fifth motor types were observed in the laboratory. Fourth type is a 1.5 kW motor with nominal speed of 2860 rpm and nominal current of 5.65 A. Two motors of this type were used, one healthy and one with 50% static eccentricity. Fifth type is a 1.1 kW motor with nominal speed of 1415 rpm and 4.4 A nominal current. This motor was used with two rotors, one healthy and one with rotor bar damaged by drilling.

Multiple vibration measurements were conducted on each motor. Two high-sensitivity IEPE accelerometers (100 mV/g) were used for collecting signals of horizontal and vertical vibrations. Data acquisition card NI-9234 (4 Channel, ±5V, 52,2 kS/s, with 2mA IEPE accelerometer excitation) was used for signal collection. Figure 4 shows vibration measurement instrumentation layout.

Signals were acquired for ten seconds with sampling frequency of 10 kHz. The recorded signals were then used to calculate time and frequency domain features. Using obtained feature sets, three SVM classifiers were trained, one for each considered fault. Parameters of these trained classifiers were then imported in module for SVM classification providing a base for following fault detection.

It should be noted that characteristic features are common for different types of motors and that they are a function of construction parameters, such as nominal speed, slip and bearing geometry. The trained SVM classifier can be applied to any similar motor, if proper parameters for feature calculation are known. However, to obtain universal applicability of this system, other types of motors should also be considered in SVM training to improve the classifier performance.

Testing of the expert system was performed using a part of collected data set. In order to obtain objective and reliable results, we used different motors in training and test phase. Testing was conducted in both manual and automatic mode. Test data set structure and obtained classification results are shown in Table 2.

Table 2. Fault classification testing results

		Static eccentricity classifier	Bearing wear classifier	Broken bar classifier
Test data set	Faulty	20	20	20
	Fault-free	40	40	20
False fault classifications		1	1	0
Classification error [%]		1.66	1.66	0

These results demonstrate the efficiency of presented expert system for induction motor fault detection. For static eccentricity and bearing wear, there was only one false classification per fault, which results in classification error less than 2 percent. The broken bar classifier, however, had no false classifications. These experimental results proved that presented expert system can be successfully applied in fault detection, providing accurate, efficient and reliable fault classifiers applicable in real industrial systems.

Conclusion

In this chapter an expert system for induction motor fault detection based on vibration signal analysis and support vector machine classifiers is presented. Using programming software tools and packages combined with appropriate hardware and implementing well known fault detection and diagnosis procedures and techniques, also presented within the chapter, this system is intended to be used as practical diagnostic tool for condition monitoring and fault, malfunction and failure prevention in industrial systems. The system is tested in real industrial environment and obtained results proved its accuracy and efficiency.

Further research could be conducted in several directions. First of them would be to extend the number of faults covered by the classification, such as dynamic eccentricity, misalignment, looseness of motor parts and so on. Also, the number of motor types used in SVM training should be extended, to cover wider range of motors. The other direction would be improvement of feature selection method, considering some other linear or nonlinear transformations or application of optimization procedures in feature selection. This would enable more efficient fault detection based on more comprehensive feature sets, which would improve maintenance procedure and decrease total cost in production processes.

REFERENCES

Cao, L. J., Chua, K. S., Chong, W. K., Lee, H. P., Gu, Q. M. (2003) A comparison of PCA, KPCA and ICA for dimensional reduction in support vector machine. *Neurocomputing,* 55, 321–336.

Dorrell, D., Thomson, W. (1997) Analysis of Airgap Flux, Current and Vibration Signals as a Function of the Combination of Static and Dynamic Airgap Eccentricity in 3-Phase Induction Motors. *IEEE Transactions on Industry Applications,* 33 (1), 24-34.

Jack, L. B., Nandi, A. K. (2002). Fault detection using support vector machines and artificial neural network, augmented by genetic algorithms. *Mechanical System and Signal Processing*, 16, 373–390.

Jolliffe, I. J. (1986). *Principal component analysis*. New York: Springer.

Kanović, Ž., Rapaić, M., Jeličić, Z. (2011). Generalized particle swarm optimization algorithm – Theoretical and empirical analysis with application in fault detection. *Applied Mathematics and Computation*, 217, 10175-10186.

Kanović, Ž. (2012) Modifikacije algoritma optimizacije rojem čestica sa primenom u detekciji kvarova na objektima upravljanja sa kontinualnom dinamikom. PhD Thesis, Faculty of Technical Sciences, Novi Sad, Serbia. *In Serbian.*

Kulić, F., Kanović, Ž., Petković, Milena, Matić, D. (2010). Expert system for induction motor fault detection. *Journal on Processing and Energy in Agriculture,* vol. 14, No 4, 173-177.

Lee, Y. S., Nelson, J. K., Scarton, H. A., Teng, D., Azizi-Ghannad, S. (1994). An acoustic diagnostic technique for use with electric machine insulation. *IEEE Transactions on Dielectrics and Electrical Insulation*, 9, 1186–1193.

Márquez, F. P. G. Principal Component Analysis Applied to Filtered Signals for Maintenance Management. Quality and Reliability Engineering International. John Wiley & Sons, Ltd. Published Online: 29 Sep 2009.

Marquez, F. P. G., Tobias, A. M., Papaelias, M., Pérez, J. M. P. (2012). Condition Monitoring of Wind Turbines: *Techniques and Methods, Renewable Energy*. Vol. 46, 169–178.

Matić, D., Kulić, F., Bugarski, V. (2010). Survey of Methods for Online Broken Bar Induction Motor Fault Detection. *Journal on Processing and Energy in Agriculture*, vol. 14, No 2, 90-92.

Samanta, B. (2004). Gear fault detection using artificial neural networks and support vector machines with genetic algorithms. *Mechanical Systems and Signal Processing*, 18(3), 625–644.

Stepanić, P., Latinović, I., Đurović, Ž. (2009). A new approach to detection of defects in rolling element bearings based on statistical pattern recognition, *The International Journal of Advanced Manufacturing Technology*, 45, 91-100.

Stone, G. C., Sedding, H. G., Costello, M. J. (1996). Application of partial discharge testing to motor and generator stator winding maintenance. *IEEE Transactions on Industry Application,* 32, 459–464.

Tandon, N., Choudhury, A. (1999). A review of vibration and acoustic measurement methods for the detection of defects in rolling element bearings. *Tribology International*, 32, 469-480.

Thollon, F., Jammal, A., Grellet, G. (1993). Asynchronous motor cage fault detection through electromagnetic torque measurement. *European Transactions on Electrical Power*, 3, 375–378.

Thomson, W. T., Fenger, M. (2001). Current signature analysis to detect induction motor faults. *IEEE Transactions on Industry Applications*, 7, 26–34.

Yang, B. S., Hwang, W. W., Kim, D. J., Tan, A. C. (2005). Condition classification of small reciprocating compressor for refrigerators using artificial neural networks and support vector machines. *Mechanical System and Signal Processing*, 19, 371–390.

Yang, B. S., Hwang, W. W., Ko, M. H., Lee, S. J. (2005). Cavitation detection of butterfly valve using support vector machines. *Journal of Sound Vibration*, 287(1–2), 25–43.

Yang, H., Mathew, J. Ma, L. (2003). Vibration feature extraction techniques for fault diagnosis of rotating machinery: a literature survey. *Asia-Pacific Vibration Conference*, Gold Coast, Australia.

Vapnik, V. N. (1995). The nature of statistical learning theory. Berlin: Springer.

In: Fault Detection
Editors: F. P. G. Márquez and M. Papaelias
ISBN: 978-1-62808-999-8

Chapter 5

FAILURE DETECTION AND ACCOMMODATION APPROACHES FOR THE AIRSPEED SENSOR ON A SMALL UAV

Srikanth Gururajan[1], Matthew Rhudy[1], Mario L. Fravolini[2], Haiyang Chao[1] and Marcello R. Napolitano[1]
[1]West Virginia University, Morgantown, West Virginia , US
[2]University of Perugia, Italy

ABSTRACT

Over the last two decades, significant research efforts have been directed towards increasing operational safety of aircraft through fault tolerant flight control schemes, especially on malfunctions of the actuators or the propulsion system. However, issues related to sensor failures have been considered to be of lower priority. Nevertheless, recent events have highlighted the criticality of "common mode" failures on the Air Data System (ADS) and that physical redundancy is not a foolproof option since all the redundant sensors are subjected to the same environmental conditions in flight.

Analytical Redundancy (AR) can be considered to be a computational approach to estimating the behavior of a parameter of a dynamic system, based on the relationship between its measurement and other parameters of the dynamic system. In this research, multiple analytical redundancy-based approaches to the SFA task for the ADS are evaluated - model free and model based.

The model free approach is based on an Unscented Kalman Filter (UKF) and two different neural networks – a Multilayer Feed forward (MLP) and the Extended Minimal Resource Allocating Network (EMRAN).

In the model based approach, assuming the availability of the complete state of the UAV, the angle of attack equation is rearranged as a quadratic equation in V which is solved to provide estimates of the airspeed. The parameters governing the airspeed equation are identified via Least Squares (LS) optimization from a record of flight data. The performance of both approaches is evaluated in terms of the statistics of the tracking error in the estimation of the airspeed, under nominal conditions. For the SFDI task, two approaches are evaluated based on the based on the combined use of a voting scheme between 3 pitot tubes with either a Cumulative Sum (CUSUM) filter or a Floating

Limiter that monitors a residual signal and declares a failure when it crosses a pre-defined threshold. Two distinct failure types – a Step Bias and a Ramp Bias (fast or slow) in the measured airspeed have been considered.

Additionally, the performance of a sequential Generalized Likelihood Ratio Test (GLRT) was also evaluated for detecting failures on the ADS. To apply sequential GLRT, a residual whitening filter was designed to enable rigorous statistical tests to be performed on the Gaussian whitened signal. The experimental results confirm the capabilities of the proposed method to cope with the important problem of the airspeed sensor fault detection.

The study has been performed using a comprehensive set of flight data from a jet powered WVU research platform was designed, manufactured, instrumented, and flown at the WVU Flight Control Systems Laboratory (FCSL). Results from preliminary experiments indicate that the MLP produced the best estimation results, with the lowest standard deviation of the estimation error and the UKF exhibiting the worst performance. Additionally, based on an ADS without redundancy, the failure detection performance of the Floating Limiter was better than that of the CUSUM for Step Bias failures, while the CUSUM filter performed better for the Slow Ramp Bias failures.

1. INTRODUCTION

Over the last two decades, significant research efforts have been directed towards increasing operational safety of aircraft through numerous approaches to fault tolerant flight control schemes. Extensive research programs led mainly by NASA have focused on designing and validating specific approaches for the design of a number of fault tolerant flight control schemes [1, 2], primarily focused on malfunctions of the actuators of the control surfaces or failures of the propulsion system. However, issues related to sensor failures have been considered to be of lower priority, mainly due to the fact that triple or quadruple physical redundancy of the sensors, along with the implementation of Built-In Testing (BIT) and/or voting schemes, have historically shown to be reliable approaches to cope with failures for the majority of the sensors of the flight control system [3, 4]. Thus, physical redundancy in the sensors to date has been the only accepted approach for manned aircraft.

Nevertheless, recent events have shown that physical redundancy is not a foolproof option for failures on Air Data Systems (ADS) providing airspeed measurements since all the redundant sensors are subjected to the same environmental conditions in flight. Therefore, any failure on one of the sensors is likely to occur on the other sensors under certain weather conditions; this class of failure is known in the literature as "common mode" failure [3-6]. Since dynamic pressure measurements are used by the flight computer for the calculation of the airspeed and since the gains of the control laws are typically scheduled as a function of the airspeed, it is clear that incorrect airspeed measurements can potentially lead to catastrophic conditions. The recent crash of the Air France Flight 447 [5] has increased the awareness of the criticality of ADS failures, which has also been cited as the cause behind the crashes of X-31 research aircraft [6] and the Aero Peru 757 [7]. In summary, the aviation safety database has technical evidence of eleven failures for all Pitot tubes in the ADS over the past 30 years leading to significant damages and/or fatalities [38].

Thus, even in the case of physical redundancy, if one Pitot tube fails due to environmental factors, it is statistically likely that the other Pitot tubes will experience the same failure. Under such circumstances, the use of Analytical Redundancy (AR) approaches to the estimation of airspeed could potentially alleviate the problems associated with the "common mode" failures. In addition, in cases where physical redundancy is not available, analytical redundancy methods can be used within the framework of a failure detection and identification approach to monitor the status of the airspeed sensor. Conceptually, this Sensor Failure Detection Identification and Accommodation (SFDIA) approach can be divided into two distinct and relatively independent tasks - a Sensor Failure Detection and Identification (SFDI) task and a Sensor Failure Accommodation (SFA) task. The SFDI task is designed to monitor the sensor systems without physical redundancy to detect signs of incipient failures followed by the identification and isolation of such failures; following a positive SFDI, the SFA task is responsible for providing an alternative estimate of the flight parameter, in lieu of the measurements from the failed sensor.

The majority of AR-based SFDI techniques rely on the availability of an accurate system modeling, typically featuring some versions of statistical filters [38-40] and nonlinear observers [41, 42]. McLaren [14] developed an airspeed estimation algorithm using Kalman filtering techniques with geometrical concepts based on the velocity triangle which defines the relationship between airspeed, wind speed, and groundspeed. With these implementations, there are important issues related to robustness to non-linearity, low signal-to-noise ratios (S/N), and modeling discrepancies between the actual system and the filter model. An alternative approach to classical analytical redundancy-based SFDI has been offered by the use of on-line learning estimators designed using Neural Network estimators [43, 44]. The main advantage of these approaches is the capacity of learning specific tasks exploiting the experience derived by the continuous interaction with the system under control. Most of the approaches documented in recent research literature implement artificial neural networks (ANN) based analytical redundancy approaches for the detection, identification and accommodation of failures on the rate gyroscopes [8-13].

The application of ANN-based estimation for ADS failures is very limited in the technical literature. Samy [15] studied the use of NNs for modeling the non-linear relationships between air data and pressure data and its use within a comparative study of physical and analytical failure accommodation approaches on a Flush Air Data System (FADS) on a small UAV. Fravolini et al. [16] compared two model based approaches to infer current values of airspeed. The first approach assumed the availability of a perfect aircraft model; within the second approach, the parameters governing the equation of the aircraft dynamics were instead identified via a least squares optimization from experimental flight data and using the measured values of the entire state of the aircraft, except the airspeed; the resulting quadratic equation in the airspeed, derived from the angle of attack relationship was solved to provide real-time estimates of airspeed. Hansen et. al. [52] generated residuals between the Pitot tube measured airspeed and the velocity measured by an onboard GPS receiver as well as the airspeed estimated from the throttle setting in conjunction with a Generalized Likelihood Ratio Test (GLRT) to detect failures on the Pitot tube onboard a small UAV (Banshee) and was able to detect the faults within an acceptable time window.

This chapter presents the results of research efforts investigating the use of analytical redundancy approaches to the problem of airspeed estimation following simulated failures on the Pitot tube. The experiments are based on real flight data collected from the semi-scale WVU-YF22 research platform. This chapter covers two distinct objectives. The first objective is the performance evaluation of various approaches to airspeed estimation for use within a Sensor Failure Accommodation (SFA) task as "virtual sensors" in place of a failed airspeed sensor, including a nonlinear Kalman filtering approach, two ANN-based analytical redundancy approaches and a model-based approach using the nonlinear equations governing the dynamics of the angle of attack of the aircraft, using real data from the flights of a small UAV. The second objective is to validate the performance of three distinct approaches to the task of Sensor Failure Detection and Identification (SFDI) namely (i) fixed failure detection thresholds within a *CUSUM* filter, (ii) an adaptive threshold approach based on the concept of a Floating Limiter, and (iii) Sequential Generalized Likelihood Ratio Test.

The rest of the chapter is organized as follows. Section II describes the WVU-YF22 research platform and its instrumentation followed by a brief description of the flight data used for this effort. The general Sensor Failure Detection Identification and Accommodation problem is discussed in Section III. Section IV outlines the analytical redundancy approaches evaluated, namely the Kalman filtering-based approach, the two different Artificial Neural Networks (ANNs) – Multi-Layer Perceptron (MLP), the Extended Minimal Resource Allocating Networks (EMRAN) for the estimating the airspeed and the model based approach. Section V describes the classes of the simulated failures on the airspeed sensor and the failure detection methodologies, based on the CUSUM filter with fixed thresholds and a Floating Limiter approach with adaptive thresholds. The results of the evaluation are summarized in Section VI.

2. WVU-YF22 Small UAV Test-Bed and Flight Data

The experiments and the results described in this chapter are based on comprehensive data from flight tests of a fleet of 3 jet powered small research UAVs, the WVU YF-22, shown in Figure 1. These research platforms were designed, manufactured, instrumented, and flown at the WVU Flight Control Systems Laboratory (FCSL) for experimental demonstration of GPS based 3-aircraft formation flight [32].

Table 1. Specifications of the WVU-YF 22 Unmanned Research Aircraft

Physical Specifications	
Wingspan	78”
Length	120” (with air data probe)
Height	24” (ground up to top of vertical tail)
Wing Area	14.7 ft^2
Fuel capacity	7 lbs.
Flight Duration	9 min (typical); 12 minutes (max)
Airspeed	Approx. 90 mph (cruise); 130 mph (max)
Take off Speed	Approx. 60 mph.
R/C Radio	JR 10X, 10 channel SPCM
Engine/Thrust	RAM 1000 (28 lbs.)

This unmanned research aircraft has a take-off weight of approximately 23 Kg, including an approximately 5 Kg electronic payload based on a PC-104 format flight computer, a complete suite of sensors (Inertial Measurement Unit (IMU), Vertical Gyroscope and ADS), a

GPS receiver, and a Radio Frequency (RF) modem for data communication. Each aircraft features a set of conventional control surfaces – ailerons, stabilators, flaps, and rudders – that are actuated by digital servos. A subset of the physical specifications of the WVU-YF22 is given below in Table 1. Interested readers are referred to Napolitano [17] and Campa [18] for a comprehensive description of the characteristics of the WVU-YF22 research aircraft.

Figure 1. WVU–YF22 Unmanned Research Aircraft Fleet.

Onboard Computer and Data Acquisition Software

The WVU YF-22 onboard avionics is developed around a PC-104 format computer and consists of a low-power computer (MSI-CM588) with a 6x86 300 MHz processor, a Diamond-MM-32-AT Data Acquisition (DAQ) module, with 32 analog input channels with 16-bit resolution and 24 digital I/O channels and a power supply module interfaced with two customized circuit boards – the controller board and the interface board and are shown in Figures 2 and 3. A 64 MB compact flash card, interfaced with the computer through an IDE adapter contains the real time enabled Linux operating system and flight control laws.

Figure 2. On-Board Avionics – Installed.

Figure 3. On-Board Instrumentation Package.

The customized controller board functionality includes the following:

- Receiving control signals from the onboard computer (OBC), and translating them into Pulse Width Modulation (PWM) signals
- Receiving PWM control signals from the radio receiver;
- Dispatching the control signals from the OBC or the radio receiver to the individual servos (according to the current operation mode of the aircraft).

Software executing the designed flight control schemes (FCS) was implemented on the onboard computer.

The operating system was based on a Linux kernel (Version 2.6.9), patched with the Real Time Application Interface (RTAI, Version 3.2), allowing for the execution of the flight control software with strict timing constraints. Due to the constraints of the on-board memory, the RTAI patched kernel was compiled with a minimum amount of features and Busybox® software, which provides the required Linux utilities.

The FCS was designed and implemented using the Matlab/Simulink® environment to perform data acquisition, communication, execution of control laws, and implementation of the on-board computer-generated control commands. This enabled the OBC to collect and store information from the aircraft sensors during the flight test, respond to pilot commands, and utilize autonomous capabilities, all of which were integral components for performing this parameter identification study.

The data used for the training and validation of the three proposed analytical approaches were collected from flight testing experiments conducted at the WVU Jackson's Mill flight testing facility.

In each flight, the aircraft follows a NASCAR-type loop flight path consisting of two straight sections with two left hand turns; each flight lasted approximately 9 minutes, including the take-off and landing segments.

The WVU-YF22 in flight and a bird's eye view of a typical flight path pattern, overlaid on a Google® Map of the WVU flight testing facility are shown below in Figures 4 and 5. Additional details on the WVU-YF22 flight testing experiments can be found in Gururajan et. al. [19]

Figure 4. WVU-YF 22 in flight.

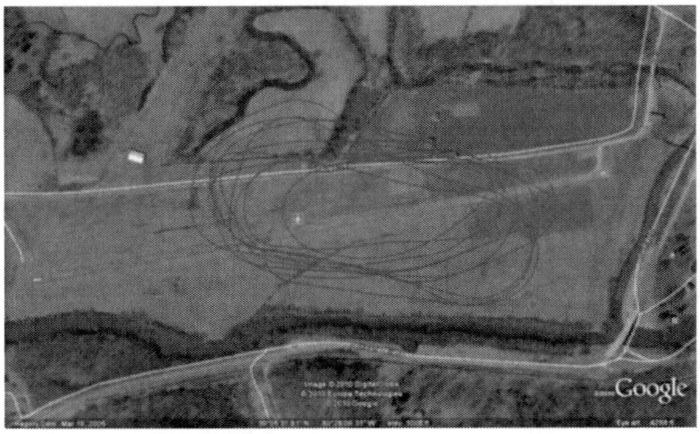

Figure 5. A typical flight path of WVU-YF22, overlaid on Google Earth image.

Sensors and Communication Hardware

The WVU YF-22 aircraft is instrumented with a complete suite of sensors for measuring a variety of flight data parameters. An IMU provides measurements of the linear accelerations and angular rates, and a mechanical gyroscope provides measurements for the pitch and bank angles. A GPS receiver provides position and velocity information in three dimensions with respect to an Earth-fixed reference frame, and the air data probe provides measurements of the angle of attack and sideslip angle. It was experimentally evaluated that the noise for all of the sensors could be approximated to follow a Gaussian probability density function.

- Inertial Measurement Unit (Crossbow IMU400), providing 12-bit measurements for the accelerations a_x, a_y, a_z (range ±4 g), and the angular rates p, q, and r (range ±90°/sec).
- Vertical gyro (Goodrich-VG34), providing measurements for the pitch and roll Euler's angles (θ and φ) with ranges of ±60° and ±90° respectively.
- GPS receiver (Novatel-OEM4), providing measurements for x, y, z, V_x, V_y, V_z with respect to an earth reference frame.

- Potentiometers for the primary control surfaces (10 kΩ each, under a 12 V supply), providing measurements for i_H, δ_A, δ_R, with ranges of [-1, 8]°, [-10, 10]°, and [-7, 7]° respectively.
- Air Data Probe, (SpaceAge© Inc. Mini Air Data Boom), providing measurements of flow angles α and β, with ranges of ±30°.
- Absolute and Differential pressure sensors (SenSym ASCX15AN and ASCX01DN), with ranges of [0-15] and [0-1] psi respectively. Both sensors were connected to the nose probe providing measurements for altitude and airspeed.
- Temperature sensor (Thermistor under a 5V supply).

3. Architecture of Failure Detection, Identification, and Accommodation (SFDIA) for Airspeed Sensor

A general block diagram of the proposed airspeed SFDIA logic is shown in Figure 6, where V_T is the true airspeed as measured by the Pitot tubes.

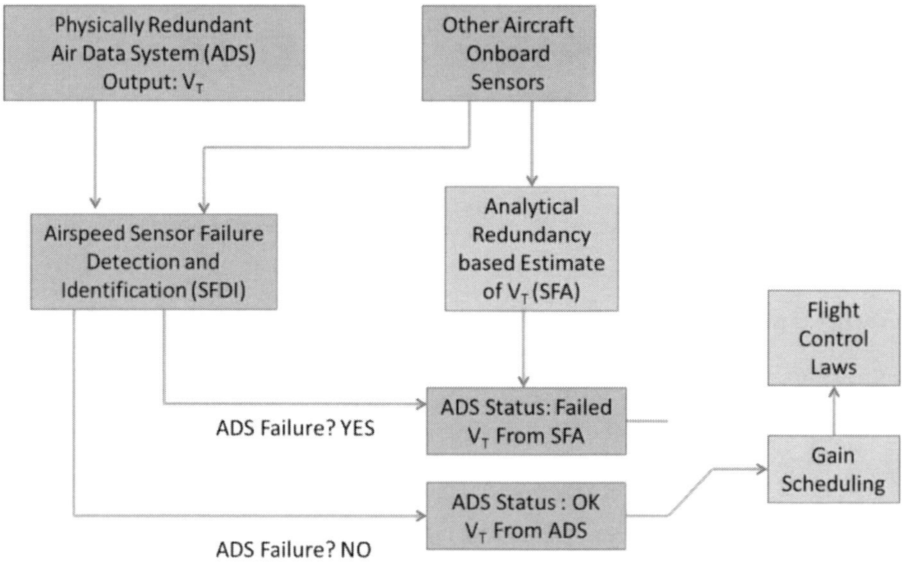

Figure 6. General Framework for Airspeed SFDIA.

The approach is designed in such a way that it can be conceptually separated into the "Sensor Failure Detection and Identification" (SFDI) task and the "Sensor Failure Accommodation" (SFA) task. This design allows for the evaluation of various approaches to the two tasks independently. In this chapter, we discuss and evaluate the performance of three distinct approaches to the task of SFDI. The first approach is based on a fixed failure detection threshold using a *CUSUM* filter; the second is based on an adaptive failure detection threshold using a Floating Limiter; finally, the third approach is based on the Generalized Likelihood Ratio Test (GLRT). In all cases, the residuals, which are defined as the difference between the airspeed measured by the Pitot tube and the output of the analytical method, are monitored. If a fault occurs on the Pitot tube, then the magnitude of this residual exceeds a threshold (either fixed or adaptive), thus triggering failure detection. This approach entails the development of the analytical redundancy solutions as a precursor to addressing the issue of fault detection. It should be emphasized that while within the scope of this discussion, the identification task is trivial since only the airspeed sensor is here assumed to be subjected to failures and the rest of the sensors onboard the aircraft are assumed to be fully functional. Without any loss of generality the overall architecture of the SFDIA can be extended to as many sensors as needed, in conjunction with a bank of residual monitors to isolate the faulty sensor.

The SFDI task is based on monitoring the characteristics of a residual signal. Towards that end, three different analytically redundancy-based approaches to the problem of Sensor Failure Accommodation (SFA) are also evaluated in terms of their performance (mean and standard deviation) in estimating the airspeed, using real flight data. The first approach is based on the implementation of an Unscented Kalman Filter (UKF); the remaining two approaches are based on the implementation of Artificial Neural Networks (ANNs), specifically a Multi-Layer Perceptron (MLP) and an Extended Minimal Resource Allocating Network (EMRAN). These above approaches, although different, share the common advantage of not requiring any dynamic model of the aircraft; this issue is particularly critical considering that the large flight envelope of commercial aircraft will typically require the implementation of a data depository with a large number of nonlinear or linearized mathematical models. This chapter also discusses the results of a separate, but related study featuring a model-based approach to the determination of the airspeed computed as function of the dynamic pressure.

4. Analytical Redundancy Methods for Airspeed Estimation

In order to detect, identify, and accommodate for airspeed sensor failures, four analytical redundancy approaches were investigated. Of the four methods, three can be classified as "Model Free" and the fourth as a "Model Based" method. All these methods take information from other sensors onboard the aircraft to estimate the airspeed, without using the measurements from the airspeed sensor. In particular, three-axis acceleration and angular rate measurements from the IMU, roll and pitch angle measurements from the vertical gyroscope, and angle of attack and sideslip angle measurements from the ADS were used as inputs, as outlined in Figure 7. In the case of the "Model Free" estimators, the inputs are independent from the aircraft dynamics, and therefore do not require a mathematical model of the dynamic system. This is a very desirable feature for an airspeed estimator since it can be applied to any aircraft under varying flight conditions without having to modify the parameters of the estimator.

The following paragraphs describe the AR approaches. First, a nonlinear Kalman filtering approach is presented which exploits the known kinematic relationships between some key aircraft measurements; next, two different ANN approaches are presented which use the same inputs as the nonlinear Kalman filtering approach, but do not use any explicit functional relationship from aircraft dynamics. Each of these three approaches can be executed in real-time, thus allowing for both airspeed sensor failure detection (the estimation is used for the generation of the residual) and accommodation (the estimation is used as virtual sensor in lieu of the failed one).

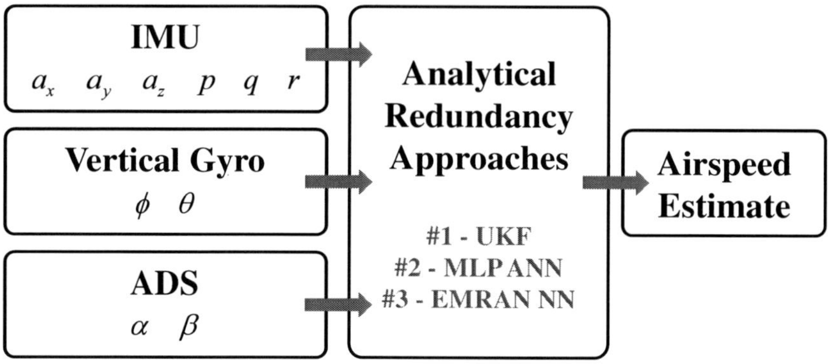

Figure 7. Analytical Redundancy Estimation Diagram.

5. Nonlinear Kalman Filtering Method for Airspeed Estimation

To properly exploit the kinematic relationships between aircraft dynamic variables, a nonlinear Kalman filtering approach was developed. Although the total airspeed is the desired output for the estimator, this method considers the estimation of the aircraft body components of airspeed velocity, that is u,v,w. The components of the velocity vector are considered in order to properly relate the three-axis accelerations with the dynamics of the airspeed. Additionally, the roll angle, ϕ, and pitch angle, θ, are considered as states to capture the rotational kinematics, thus giving the state vector:

$$\mathbf{x} = \begin{bmatrix} u & v & w & \phi & \theta \end{bmatrix}^T \tag{1}$$

Using these states, the airspeed can be calculated using:

$$V_T = \sqrt{u^2 + v^2 + w^2} \tag{2}$$

To aid in the prediction of these states, an input vector containing aircraft body acceleration, a_x, a_y, a_z, and angular rates, p, q, r, is used:

$$\mathbf{u} = \begin{bmatrix} a_x & a_y & a_z & p & q & r \end{bmatrix}^T \tag{3}$$

Using the above definitions, the state dynamics can be defined using the three so-called "Conservation of Linear Momentum Equations (CLMEs)" and two kinematic equations [26]

$$\begin{aligned}
\dot{u} &= rv - qw - g\sin\theta + a_x \\
\dot{v} &= pw - ru + g\cos\theta\sin\phi + a_y \\
\dot{w} &= qu - pv + g\cos\theta\cos\phi + a_z \\
\dot{\phi} &= p + q\sin\phi\tan\theta + r\cos\phi\tan\theta \\
\dot{\theta} &= q\cos\phi - r\sin\phi
\end{aligned} \tag{4}$$

Additionally, a lever arm position correction [26] was applied to adjust the accelerometer measurements:

$$\begin{bmatrix} a_x \\ a_y \\ a_z \end{bmatrix} = \begin{bmatrix} a_x \\ a_y \\ a_z \end{bmatrix}_{IMU} + \begin{bmatrix} \left(q^2+r^2\right) & -\left(pq-\dot{r}\right) & -\left(pr-\dot{q}\right) \\ -\left(pq+\dot{r}\right) & \left(p^2+r^2\right) & -\left(qr-\dot{p}\right) \\ -\left(pr-\dot{q}\right) & -\left(qr+\dot{p}\right) & \left(p^2+q^2\right) \end{bmatrix} \begin{bmatrix} x_a \\ y_a \\ z_a \end{bmatrix} \tag{5}$$

where (x_a, y_a, z_a) is the *a priori* known location of the accelerometers relative to the Center of Gravity (CG) of the aircraft.

The outputs (available measurements) for this formulation are given by the angle of attack, α, sideslip angle, β, roll angle, ϕ, and pitch angle, θ, thus defining the output vector:

$$y = \begin{bmatrix} \alpha & \beta & \phi & \theta \end{bmatrix}^T \tag{6}$$

The measurement equations are then defined by the following equations for relative wind angles (also with lever arm position correction) [26] and direct measurement for attitude states from the vertical gyroscope:

$$\begin{aligned}
\alpha &= \tan^{-1}\left(\frac{w - qx_\alpha + py_\alpha}{u}\right) \\
\beta &= \sin^{-1}\left(\frac{v + rx_\beta - pz_\beta}{\sqrt{u^2+v^2+w^2}}\right) \\
\phi &= \phi_{VG} \\
\theta &= \theta_{VG}
\end{aligned} \tag{7}$$

where (x_α, y_α) and (x_β, z_β) are the ‘a priori’ known locations of the angle of attack and sideslip angle vanes with respect to the CG of the aircraft. Using the above relationships, the state space formulation can be written as:

$$\dot{\mathbf{x}} = \mathbf{f}_c(\mathbf{x}, \mathbf{u})$$
$$\mathbf{y} = \mathbf{h}(\mathbf{x}, \mathbf{u}) \quad (8)$$

Note that the state dynamics are nonlinear in nature; therefore, the use of a nonlinear Kalman filter such as the Extended Kalman Filter (EKF) [27] or Unscented Kalman Filter (UKF) [28] is required.

The UKF was selected for this analysis due to ease of implementation [29]. The state dynamics are discretized using a first order discretization [30], i.e.:

$$\mathbf{x}_k = \mathbf{x}_{k-1} + T_s \mathbf{f}_c(\mathbf{x}_{k-1}, \mathbf{u}_k) = \mathbf{f}(\mathbf{x}_{k-1}, \mathbf{u}_k) \quad (9)$$

where T_s is the sampling time of the measurements selected to be 0.02 s. This sampling time is sufficiently small for accurate sampling based on the knowledge of the highest rigid body natural frequency of the aircraft.

It is important to note that this state model describes the kinematics of any rigid body, and is independent of the specific aircraft model. Additionally, since this is a Kalman filtering-based approach, the uncertainty in the equations can also be considered. Taking advantage of the fact that the state dynamic equations are derived from kinematics, no modeling errors are expected for this case.

However, a certain level of uncertainty is expected in the input vector from the IMU; therefore the process noise is considered as additive to the inputs, i.e.:

$$\mathbf{x}_k = \mathbf{f}(\mathbf{x}_{k-1}, \mathbf{u}_k + \mathbf{w}_k) \quad (10)$$

where w is the zero-mean Gaussian process noise vector with covariance matrix, Q, determined from static measurements of the IMU, as in:

$$\mathbf{Q} = diag\left[\sigma^2(a_x) \quad \sigma^2(a_y) \quad \sigma^2(a_z) \quad \sigma^2(p) \quad \sigma^2(q) \quad \sigma^2(r)\right] \quad (11)$$

For these noise assumptions, a non-additive noise model is used for the UKF, which requires the augmentation of the state vector with process noise components, e.g. see Rhudy et. al.[29] or van Der Merwe et. al.[31]. The measurement noise model is considered as additive noise, i.e.:

$$\mathbf{y}_k = \mathbf{h}(\mathbf{x}_k, \mathbf{u}_k) + \mathbf{v}_k \quad (12)$$

where v is the zero-mean Gaussian measurement noise vector with covariance matrix, R, determined from static measurements of the vanes and vertical gyroscope, as in:

$$\mathbf{R} = diag\left[\sigma^2(\alpha) \quad \sigma^2(\beta) \quad \sigma^2(\phi) \quad \sigma^2(\theta)\right] \tag{13}$$

Using these definitions, the UKF equations can be implemented, and are summarized as follows [28].

For this application of the UKF, due to the non-additive noise assumption in the state dynamics, the state vector is augmented to include navigation states, x, and process noise states, w, leading to the augmented state vector, $\mathbf{x}^a = \left[\mathbf{x}^T \quad \mathbf{w}^T\right]^T$ where the superscript a denotes augmentation, and the process noise states are assumed to be zero mean Gaussian. The total dimension of the augmented state vector is denoted by l. The process noise states are initialized to zero and remain constant with time. The effect of the process noise, however, is implemented implicitly by the use of the unscented transformation [28]. Due to the state augmentation, the state covariance matrix is augmented accordingly to include the noise characteristics of the process noise states, as in

$$\mathbf{P}_k^a = \begin{bmatrix} \mathbf{P}_k & \mathbf{0} \\ \mathbf{0} & \mathbf{Q}_k \end{bmatrix} \tag{14}$$

The UKF uses the unscented transformation to obtain the estimates of mean and covariance. The first step of the UKF implementation is to calculate $2l+1$ sigma points based on the square-root of the augmented state covariance matrix:

$$\boldsymbol{\chi}_{k-1}^a = \begin{bmatrix} \boldsymbol{\chi}_{k-1}^x \\ \boldsymbol{\chi}_{k-1}^w \end{bmatrix} = \left[\hat{\mathbf{x}}_{k-1|k-1}^a \quad \hat{\mathbf{x}}_{k-1|k-1}^a + \eta\sqrt{\mathbf{P}_{k-1|k-1}^a} \quad \hat{\mathbf{x}}_{k-1|k-1}^a - \eta\sqrt{\mathbf{P}_{k-1|k-1}^a}\right] \tag{15}$$

where χ is a matrix of sigma points, and η is the sigma point spread parameter, given by [28]:

$$\eta = \sqrt{l + \lambda} \tag{16}$$

$$\lambda = l\left(\alpha^2 - 1\right) \tag{17}$$

where λ is the compound sigma point parameter, and α is the primary sigma point scaling parameter, which is suggested to vary between 0.001 and 1.

The prediction step for the UKF consists of passing the sigma points associated with each state through the state prediction equations:

$${}^i\boldsymbol{\chi}_{k|k-1}^x = \mathbf{f}({}^i\boldsymbol{\chi}_{k-1}^x, \mathbf{u}_k + {}^i\boldsymbol{\chi}_{k-1}^w) \qquad i = 0,1,\ldots,2l \tag{18}$$

where the superscript i denotes the $(i+1)^{\text{th}}$ column of the matrix. Next, the state mean and covariance are predicted using a weighted average of the transformed sigma points using:

$$\hat{\mathbf{x}}_{k|k-1} = \sum_{i=0}^{2l} \mathbf{w}_i^m \, {}^i\boldsymbol{\chi}_{k|k-1}^x \tag{19}$$

$$\mathbf{P}_{k|k-1} = \sum_{i=0}^{2l} \mathbf{w}_i^c \left({}^i\boldsymbol{\chi}_{k|k-1}^x - \hat{\mathbf{x}}_{k|k-1} \right)\left({}^i\boldsymbol{\chi}_{k|k-1}^x - \hat{\mathbf{x}}_{k|k-1} \right)^T \tag{20}$$

where w_m and w_c are weight vectors [28]. These are the predicted estimates of the mean and covariance for the time-step k. Next, the transformed sigma-points are sent through the observation function. Since the measurement noise is additive and zero-mean, the measurement noise is omitted from the observation function, as in

$${}^i\boldsymbol{\psi}_{k|k-1} = \mathbf{h}\left({}^i\boldsymbol{\chi}_{k|k-1}^x, \mathbf{u}_k \right), \quad i = 0,1,\ldots,2L \tag{21}$$

where ψ is a matrix of output sigma-points. These output sigma-points are then used to calculate the predicted output, output covariance matrix, and cross-covariance between the state and output, using

$$\begin{aligned}
\hat{\mathbf{y}}_{k|k-1} &= \sum_{i=0}^{2L} \mathbf{w}_i^m \, {}^i\boldsymbol{\psi}_{k|k-1} \\
\mathbf{P}_k^{yy} &= \mathbf{R}_k + \sum_{i=0}^{2L} \mathbf{w}_i^c \left({}^i\boldsymbol{\psi}_{k|k-1} - \hat{\mathbf{y}}_{k|k-1} \right)\left({}^i\boldsymbol{\psi}_{k|k-1} - \hat{\mathbf{y}}_{k|k-1} \right)^T \\
\mathbf{P}_k^{xy} &= \sum_{i=0}^{2L} \mathbf{w}_i^c \left({}^i\boldsymbol{\chi}_{k|k-1}^x - \hat{\mathbf{x}}_{k|k-1} \right)\left({}^i\boldsymbol{\psi}_{k|k-1} - \hat{\mathbf{y}}_{k|k-1} \right)^T
\end{aligned} \tag{22}$$

Note that the assumed measurement noise covariance matrix, R, is added to the output covariance matrix due to the additive noise assumption. These covariance matrices are used to calculate the Kalman gain matrix, K, using

$$\mathbf{K}_k = \mathbf{P}_k^{xy} \left(\mathbf{P}_k^{yy} \right)^{-1}$$

This Kalman gain is then used to update both the state and covariance estimates, as in

$$\begin{aligned}
\hat{\mathbf{x}}_k &= \hat{\mathbf{x}}_{k|k-1} + \mathbf{K}_k \left(\mathbf{y}_k - \hat{\mathbf{y}}_{k|k-1} \right) \\
\mathbf{P}_k &= \mathbf{P}_{k|k-1} - \mathbf{K}_k \mathbf{P}_k^{yy} \mathbf{K}_k^T
\end{aligned} \tag{24}$$

where y_k is the measurement vector provided from an independent measurement, and $\hat{\mathbf{x}}_k$ and P_k are the updated state and covariance estimates.

6. Multi-Layer Perceptron (MLP) Artificial Neural Network (ANN)

As an alternative approach, Artificial Neural Networks (ANNs) were considered. From classic neural network literature, ANNs can be defined as "*A massively parallel distributed processor that has a natural propensity for storing experimental knowledge and making it available for use*" [20]. Relying on the property of being universal function approximators, ANNs can be used to estimate and compensate for completely unknown functions within uncertain dynamic systems. Also, neural networks are by definition multiple input multiple output systems, and as a result are ideally suited to be applied to multivariable systems. Hornik [37] showed that a standard feedforward network with at least one hidden layer is capable of approximating any function to any desired degree of accuracy, provided enough hidden layer neurons are available. As a consequence, the ANNs do not require exact mathematical models to learn/estimate the dynamics of a system. Also, these methods are not correlated in time, i.e. the estimate at the current time step does not depend on the previous estimate, which prevents the possibility for error accumulation that occurs with recursive methods such as the UKF. In the following sections, two different ANN approaches are presented as alternatives to the UKF approach.

The first ANN is a *Multilayer Feed-forward* network with supervised learning, also known as Multi-Layer Perceptron (MLP). This ANN architecture, provided by the Matlab Neural Network toolbox (*newff*), was selected due to is flexibility for a number of applications, including function fitting and pattern recognition. The architecture of the MLP is shown below in Figure 8. Prior to being used within the SFA task for the purpose of providing an on-line estimate of the airspeed, the MLP was trained to learn the functional relationship between the airspeed and a set of correlated measurements provided by the other aircraft sensors.

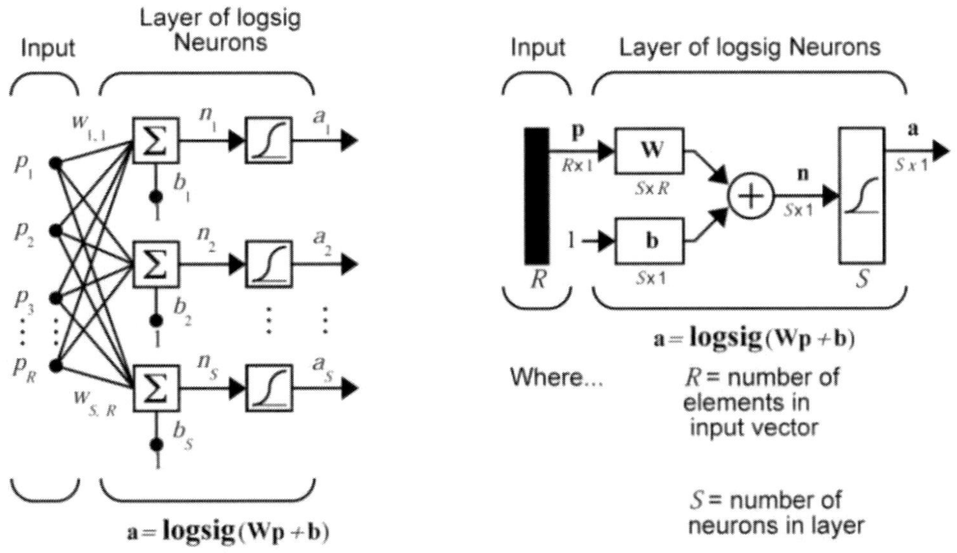

Figure 8. Structure of Matlab Feed-forward Neural Network [21].

A MLP ANN with one hidden layer with three neurons and a sigmoid activation function was employed to estimate the function $V_{nn}(t) = f_{nn}(X(t), u(t))$ where V_{nn} is the NN estimation of the airspeed provided by the ADS.

7. Extended Minimal Resource Allocating Network (EMRAN)

The second ANN used is the *Extended Minimal Resource Allocating Networks* (EMRAN) [22]. The EMRAN is a more powerful version of the original MRAN architecture. The network architecture, illustrated below in Figure 9, features a growing and pruning scheme; additionally, the weights of the connections are updated following a "winner takes all" strategy [23].

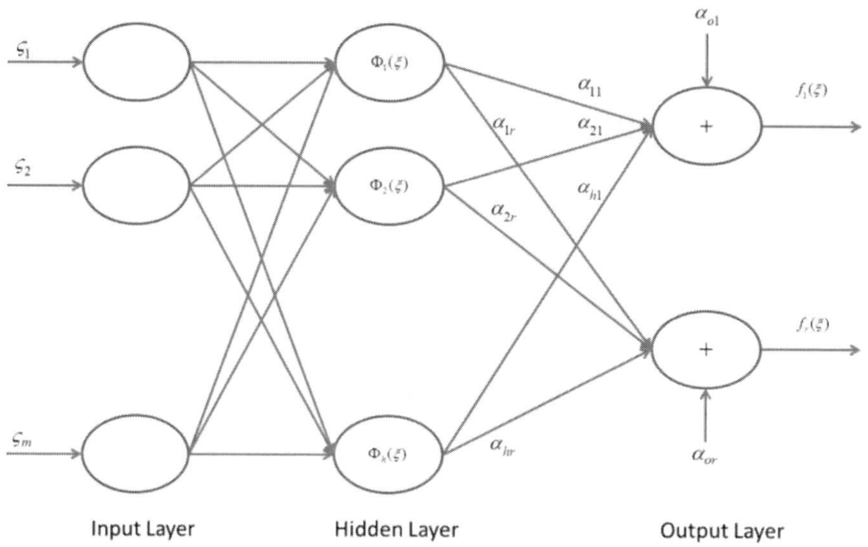

Figure 9. Structure of RBF.

The EMRAN architecture allows only the parameters of the most active neurons to be updated, while all the others are left unchanged. Essentially, the EMRAN algorithm allocates neurons in order to decrease the estimation error in regions of the state space where the mapping accuracy is poor. This strategy results in a significant reduction of the number of parameters to be updated on-line, thus reducing the computational burden, and therefore making this architecture particularly suitable for on-line applications [24].

For Gaussian basis functions, the NN output is computed with the expression:

$$\hat{y}(x,\xi) = \sum_{i=1}^{M} w_i e^{\left(\frac{|x-\mu_i|^2}{2\sigma_i^2}\right)} \tag{25}$$

where x is the input vector, ξ is the set of parameters to be tuned by the learning algorithm, including the weight w, the Gaussian center positions μ, and the variance σ. In the EMRAN approach, a new neuron is introduced only if 3 distinct criteria are satisfied, that is if the estimation error exceeds a threshold E_1 as in (26), if the windowed estimation error exceeds a threshold E_2 as in (27), and if the distance from the input to the nearest center is larger than a threshold E_3 as in (28)

$$e_1(k) = y(k) - \hat{y}(k) > E_1 \tag{26}$$

$$\frac{1}{N}\sqrt{\sum_{i=1}^{M} e(k-i)} > E_2 \tag{27}$$

$$d_{BMU} = \min_{j=1,\ldots,neu} \left(\left\| c(k) - x(k) \right\|\right) > E_3 \tag{28}$$

If all the above criteria are satisfied, a new neuron will be created at time instant k with center, variance, and weight given by:

$$\mu_{M+1}(k) = x(k) \tag{29}$$

$$\sigma_{M+1} = \lambda \inf_{j=1}^{M} \left\| x(k) - \mu_j(k) \right\| \tag{30}$$

$$w_{M+1}(k) = e(k) = y(k) - \hat{y}(k) \tag{31}$$

If all or any of the above criteria are not satisfied, no new neurons are created. The parameters of the current neurons are updated using a simple delta rule using the relationship:

$$\theta(k+1) = \theta(k) - \eta \left.\frac{\partial \hat{y}(k)}{\partial \theta(k)}\right|_{(k)} \cdot e(k) \tag{32}$$

where, $e(k)$ is the estimation error and η is the user-defined learning rate.

7. Model-Based Approach

The airspeed $V_T(t)$ of the UAV is typically computed as function of the aerodynamic dynamic pressure $q_{dyn}(t)$ exploiting the Bernoulli's equation:

$$q_{dyn}(t) = \frac{1}{2}\rho(t)V_T^{\,2}(t) \tag{33}$$

where $\rho(t)$ is the air density that was here estimated from the static pressure signal using the COESA Atmosphere Model. From (33) it is evident that a fault on the airspeed velocity sensor can be traced essentially to a malfunction of the Pitot tube that provides the direct measurement of the dynamic pressure $q_{dyn}(t)$. The approach proposed here for the estimation of the airspeed V_T relies on the redundant measurements from other sensors onboard the aircraft and on the modeling information of the aircraft systems. In particular, based on the well-known nonlinear model of an aircraft system[45] characterized by 12 nonlinear equations depending on the state variables $X = [V_T \quad \alpha \quad \beta \quad p \quad q \quad r \quad \theta \quad \varphi \quad \phi \quad h \quad x \quad y]$ and on the 4 control commands $U = [X_T \quad \delta_e \quad \delta_a \quad \delta_r]$. In the specific case of the WVU-YF22 UAV, the X and U signals are all measured and available for SFDI and SFA purposes [46].

Although the airspeed V_T enters implicitly in all the 12 nonlinear dynamics equations, this parameter is strongly correlated with the equation of the angle of attack α. Therefore, this equation is particularly suitable for extracting information about the airspeed itself. From Stevens [45], the nonlinear equation of the angle of attack can be expressed as follows:

$$\dot{\alpha} = \frac{-F_X \sin(\alpha) + F_Z \cos(\alpha)}{mV\cos(\beta)} - q + (p\cos(\alpha) + r\sin(\alpha))\tan(\beta) \tag{34}$$

where m is the aircraft mass. In (23), F_x and F_z represent the total forces acting along the x and z axis of the aircraft, expressed in the body-axes reference frame. These forces in (23) are expressed as:

$$\begin{aligned} F_X &= X_a + X_T + X_{gr} \\ F_Z &= Z_a + Z_T + Z_{gr} \end{aligned} \tag{35}$$

where the T subscript indicates the contribution from the engine thrust force while the aerodynamic forces generated by the control surface deflections are defined as follows

$$\begin{aligned} X_a &= C_{Xa} q_{dyn} S \\ Z_a &= C_{Za} q_{dyn} S \end{aligned} \tag{36}$$

where S is the aircraft wing surface area. In (25) the so called aerodynamic coefficients, expressed by affine functions of the aircraft states and inputs are used to model the aerodynamic forces and moments acting on the aircraft as:

$$C_{Xa} = C_{X0} + C_{X\alpha}\alpha + C_{Xde}\delta_e$$
$$C_{Za} = C_{Z0} + C_{Z\alpha}\alpha + C_{Zq}\frac{q\bar{c}}{2V} + C_{Zde}\delta_E \tag{37}$$

Finally, the body axes components of the gravity force are:

$$X_{gr} = -mg_0 \sin(\theta)$$
$$Z_{gr} = mg_0 \cos(\theta)\cos(\varphi) \tag{38}$$

where m is the aircraft mass and g_0 is the gravitational acceleration at sea level. Substituting relations (35-38) into (34), we get the following quadratic expression in the airspeed $V_T(t)$:

$$\begin{aligned} &0.5\rho S * \left(-C_{zde}\cdot\delta_e - C_{z0} - C_{za}\cdot\alpha\right)V_T^{\,2} \\ &+\begin{pmatrix} q\cdot m\cdot cos\beta - sin\beta\cdot m\cdot p\cdot cos\alpha - \\ sin\beta\cdot m\cdot r\cdot sin\alpha - 0.25\rho S\cdot C_{zq}q\cdot\bar{c} - \dot{\alpha}\cdot m\cdot cos\beta \end{pmatrix}V_T \\ &-X_t\cdot sin\alpha + m\cdot g\cdot sin\alpha\cdot sin\theta + m\cdot g\cdot cos\alpha\cdot cos\theta\cdot cos\varphi = 0 \end{aligned} \tag{39}$$

For the sake of brevity, the time dependency of the state variables is not shown. This quadratic expression in $V_T(t)$ serves as the basis of the proposed airspeed estimation algorithm that will be introduced in Section V.

8. Parameter Estimation

In this study, it was assumed that the geometric, inertial and aerodynamic parameters that appear in (39) are unknown and need to be identified. Within the scope of this research, this identification was performed offline using recorded flight data from the WVU YF-22. For estimation purposes, it is necessary to recast equation (39) in a form that is suitable for parameters identification and defining the unknown coefficients as W_i. Equation (28) can be rewritten as follows:

$$\begin{aligned} &W_1\cdot V_T^{\,2} + W_2\cdot\alpha\cdot V_T^{\,2} + W_3\cdot V_T\cdot q + W_4\cdot\delta_e\cdot V_T^{\,2} + W_5\cdot sin\alpha\cdot X_t + \\ &\qquad W_6\cdot q\cdot sin\alpha + W_7\cdot r\cdot sin\alpha + W_8\cdot p\cdot cos\alpha = \\ &-[9.806\cdot sin\alpha\cdot sin\theta + 9.806\cdot cos\alpha\cdot cos\theta\cdot cos\phi + cos\alpha\cdot q \\ &\qquad + q\cdot V_T\cdot cos\beta - sin\beta\cdot V_T\cdot p\cdot cos\alpha \\ &\qquad - sin\beta\cdot V_T\cdot r\cdot sin\alpha - \dot{\alpha}\cdot V_T\cdot cos\beta] \end{aligned} \tag{40}$$

It can be observed that the left hand side of (40) is a linear combination of unknown coefficients W_i and known nonlinear functions of the measured signals, while the right hand side is a linear combination of known nonlinear functions. At any given time instant *k,* (40) can be arranged in a compact vector form as follows:

$$x(k) \cdot W^T = y(k) \tag{41}$$

where $x(k) \in \mathbb{R}^{8\times1}$ is a row vector, $y(k)$ is a scalar, and $W \in \mathbb{R}^{8\times1}$ is the unknown vector. Now, since the above relation holds $\forall k$ a collection of N samples of (41), considered over a pre-defined time window, is grouped together to constitute a redundant system of equations:

$$\begin{bmatrix} x(k_1) \\ x(k_2) \\ \vdots \\ x(k_N) \end{bmatrix} \begin{bmatrix} W_1 \\ W_2 \\ \vdots \\ W_8 \end{bmatrix} = \begin{bmatrix} y(k_1) \\ y(k_2) \\ \vdots \\ y(k_N) \end{bmatrix} \tag{42}$$

The above relationship can be expressed in matrix form as $X \cdot W^T = Y$ where $X \in \mathbb{R}^{N\times8}$ and $Y \in \mathbb{R}^{N\times1}$. Since the equation (42) is linear in W, the well-known Least Squares (LS) method was applied to estimate W, resulting in:

$$W^T = \left(X^T X\right)^{-1} X^T Y \tag{43}$$

In the current effort, the coefficients W were estimated offline using a set of flight data at fault free conditions.

Online Estimation of Airspeed

The in-flight estimates of the velocity $V_T(k)$ at time instant k is based on the online solution of the quadratic equation (39) where the unknown coefficients W_i are substituted by the LS estimates obtained from (43). In particular, at a given time instant *k,* the measurements of the variables $\delta_e(k), \alpha(k), \dot{\alpha}(k), \beta(k), \theta(k), \varphi(k), r(k), p(k), q(k), X_t(k)$ and the coefficients W_i i=1…8 are substituted in (39) so that it results in a quadratic equation in the only unknown $V_T(t)$ as:

$$eq_\alpha(k) = a_1(k)V_T^2(k) + a_2(k)V_T(k) + a_3(k) = 0 \tag{44}$$

where the coefficients $a_1(k), a_2(k), a_3(k)$ are known.

The solution of the quadratic equation $eq_{\alpha}(k)=0$ provides the estimates of $V_T(k)$ .In the course of our experiments, we verified that the $eq_{\alpha}(k)=0$ always provides both a negative and positive solution.

As this study is directed towards the application of AR methods to a fixed wing aircraft in forward motion, it is trivial to discard the negative solution and to associate the positive solution to the estimate of $V_T(k)$, that is:

$$\hat{V}_T(t)=V_{T+}(t) \tag{45}$$

9. ANN Training and Validation Procedure

The inputs for both MLP and EMRAN architectures for estimating the airspeed were set to be $X=[p,q,r,a_x,a_y,a_z,\alpha,\beta,\phi,\theta]$. Once this set of inputs was identified, the general procedure that was followed in training and validating the performance of the MLP ANN and EMRAN NN was:

Step 1: Create and configure the network
Step 2: Initialize the interconnection weights and biases

For the MLP ANN the weights were initialized using the Widrow-Nyugen initialization procedure [25].

This algorithm initializes the weights in such a way that it distributes the active region of each neuron in the hidden layer approximately evenly across the layer's input space and helps in decreasing the training time. It should be emphasized that these weights are different with each initialization of the network.

For the EMRAN ANN, the network starts without hidden layer neurons; new neurons are introduced based on the three error criteria described above.

Step 3: Process the input data and train the network for a user-defined number of instants.
Step 4: Validate the performance of the network in terms of a Tracking Error defined as

$$TE=V_{Pitot}-V_{NN\_Est}$$

10. Modeling of Airspeed Sensor Faults and Failure Detection Methods

Following the description of different approaches to estimate the airspeed in the previous sections, the different failure types that were artificially injected on the airspeed sensor and fault detection methods that were considered in this chapter are discussed in detail in the subsequent sections.

Air Data System Faults and Residual Signals

The fault on the airspeed sensor was modeled as an additive bias of amplitude ΔV occurring at time t_f and the resulting faulty signal is:

$$V_{faulty}(t) = V(t) + \Delta V(t - t_f) \tag{46}$$

Two distinct fault scenarios that are commonly used in Fault Detection and Identification literature [9] were tested: a Sudden Bias (SB) failure, here referred to as Type 1, and a Slow Ramp Bias (SRB) failure, here referred to as Type 2 failure, as illustrated in Figures 10 and 11. The SB fault is essentially a sudden step bias on the airspeed measurement at the instant of failure. In the SRB scenario the fault was designed to reach its final amplitude after certain $\Delta T_{Failure}$ with the failure amplitude increasing linearly to its final value. The value of $\Delta T_{Failure} = 100$ sec was selected for this part of the study.

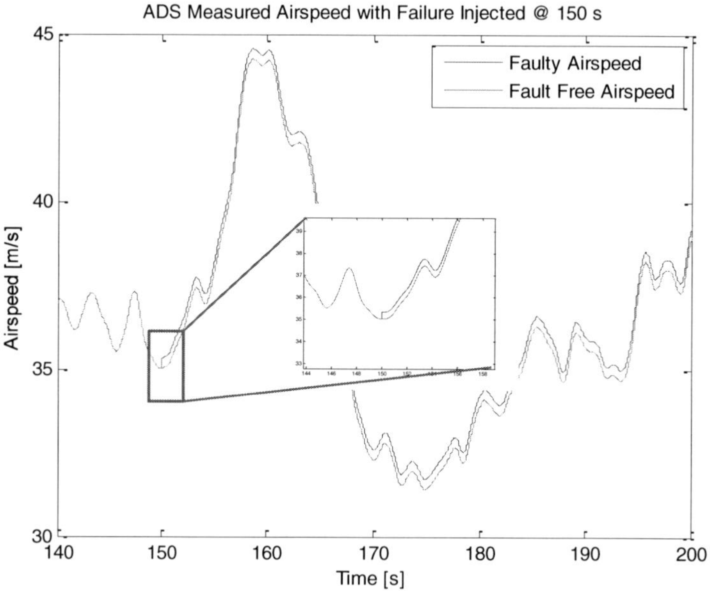

Figure 10. ADS Measured Airspeed with SB Failure, Injected at 150s.

The residual signal used for the purpose of SFD is here defined in (47) as the difference between the measured airspeed from the ADS and the estimated airspeed.

$$r(t) = V_{ADS}(t) - \hat{V}_{AR}(t) \tag{47}$$

A typical residual signal for a SRB failure is shown in Figure 12 for a simulated failure of magnitude 2.5 m/s injected into the flight data.

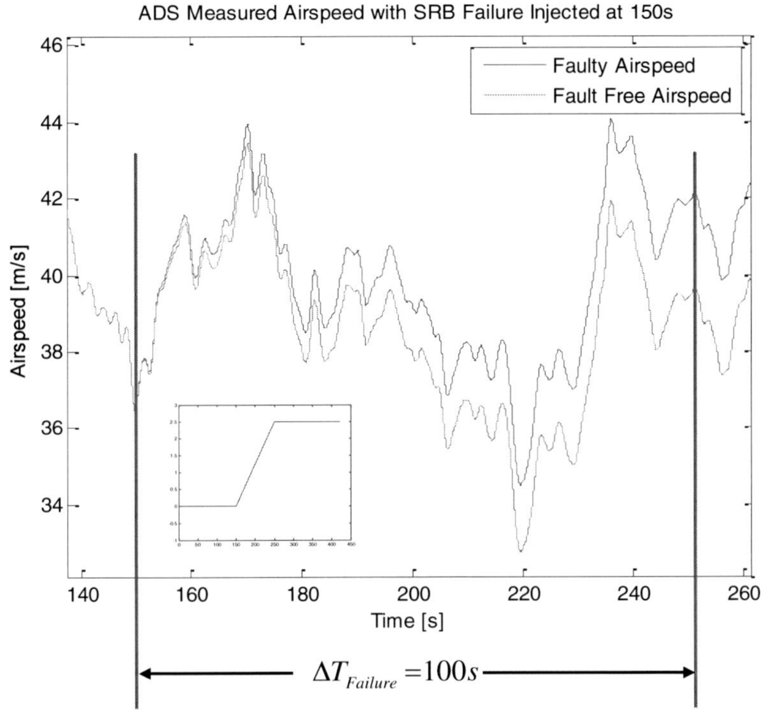

Figure 11. ADS Measured Airspeed with SRB Failure, Injected at 150s (SRB failure shown in inset).

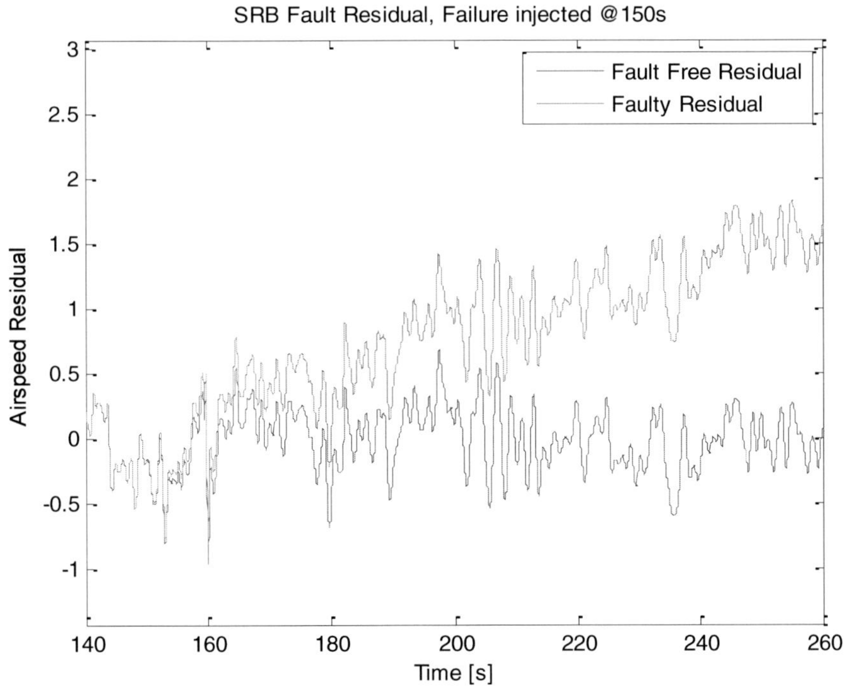

Figure 12. SRB Fault Residual, Fault Injected @ 150s.

Fault Modeling and Residual Signal – Model Based Approach

In the context of the concurrent model-based approach, the failure on the airspeed sensor was modeled similarly as an additive fault ΔV that occurs at time instant k_f; the resulting faulty signal is:

$$V_{faulty}(k) = V(k) + \Delta V(k - k_f) \tag{48}$$

To allow the study of both hard and soft sensor failures the ΔV fault was derived as the output of unit gain low pass filter with time constant τ that is:

$$\Delta V(k) = \frac{1}{1 + s\tau} \overline{\Delta V} \delta_{-1}(k) \tag{49}$$

where $\delta_{-1}(k)$ is a (unitary) step fault function and $\overline{\Delta V}$ is the actual steady state fault amplitude. By varying τ, it is possible to simulate fast ("hard") and slowly rising ("soft") failures.

The residual signal that is used for SFD purpose is defined as the difference between the measured airspeed coming from the sensor and the estimation that is:

$$r(k) = \left(V(k) + \Delta V(k - k_f)\right) - \hat{V}(k) \tag{50}$$

As with the other approaches, in fault free conditions, the residual $r(k)$ should statistically approach zero, while in real world applications, due to the presence of modeling uncertainties and noise $r(k) \neq 0$ even without failures. Therefore, the development of a robust decision algorithm able to detect the occurrence of a fault while minimizing the rate of false alarms is of great interest. In the experiments we have found that the raw residual signal has a significant autocorrelation length that is introduced, mainly, by modeling uncertainties in the low frequency range. On the other hand, since white noise (completely uncorrelated residual signal) is one of the requirements for most statistical change detectors to perform optimally, a whitening filter was designed to remove correlation in the residual signal.

The Cumulative Sum (*CUSUM*) Filter with a Static Threshold

A filter extensively used for the detection of moderate persistent shifts in the mean value of residuals is the two-sided *CUSUM* filter [35]. This filter is typically used to detect both positive and negative changes of amplitude ν in the mean value μ_0 of the residual signal $r(k)$ caused by the occurrence of a fault. In this effort we limited the analysis to faults with positive amplitude that is, following the fault, the mean value of the residual is expected to be: $\mu_f = \mu_0 + \nu$. The *CUSUM* filter is defined by the following equation:

$$S^{+}(k)=\sup\left(0,S^{+}(k-1)-\frac{\nu(\nu-2r(k))}{2\sigma_{0}^{2}}\right) \tag{51}$$

where $S^{+}(k)$ is the *CUSUM* state variable and $\sigma_0^{\ 2}$ is the variance of the airspeed residual $r(k)$ in fault free condition. The detection of a fault occurs when $S^{+}(k)$ exceeds a predefined detection threshold. The parameters of the *CUSUM* filters were determined from the statistics of the residual signal in fault free condition, specifically, the standard deviation σ_0 and the mean μ_0. The values of the parameter ν (that is the value of the expected fault amplitude $\Delta V = \nu$) strongly affects the sensitivity of the *CUSUM* signals to faults having generic amplitudes. In this study, this parameter was fixed to a baseline value equal to the variance of the estimation error for the airspeed channel. The performance of the *CUSUM* filter in detecting failures on the airspeed sensor is dependent on the value of the threshold. The process used to identify this threshold used in the work is described in Section VI. E.

Floating Limiter (FL) with Adaptive Thresholds

The Floating Limiter architecture was originally designed to provide protection against undetected "hard overs" being sent through a single string computer to a control surface. This concept was first implemented on an F-8 digital fly-by-wire aircraft, under a joint NASA/RAE program known as the Cooperative Advanced Digital Research Experiment (CADRE). Conceptually, the "Floating Limiter" is designed to feature two boundaries (upper and lower) that fluctuate around a given input signal and drift with it at a rate equal to the rate of the signal itself, but under a pre-defined limit. The signal is free to vary at any rate within these upper and lower boundaries, but if it breaches either boundary, it triggers specific logic schemes. The Floating Limiter concept has been successfully implemented within fault tolerant flight control schemes for safe transitions between conventional control laws to control laws under failure conditions [33]. Additionally, the concept has also been adapted to serve as a better alternative to the use of fixed thresholds in the detection of failures onboard aircraft [34]. A conceptual block diagram of the floating limiter illustrating the logic behind the computation of the fault detection thresholds is shown in Figure 13.

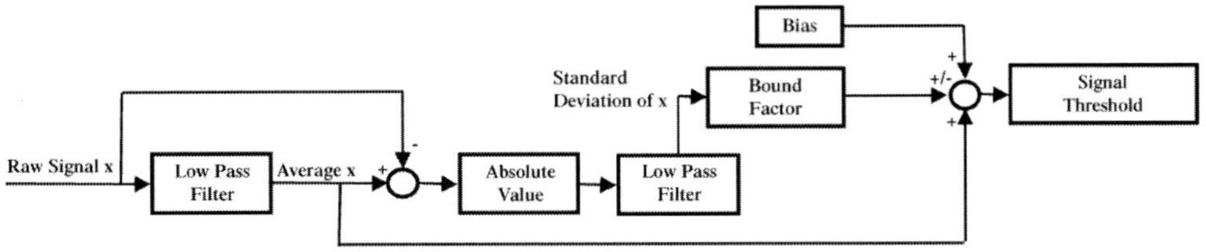

Figure 13. Conceptual representation of Floating Limiter.

The upper and lower bounds of the FL are computed based on the statistics (mean and standard deviation) of the signal over a moving time window (defined to be 1s in this study) and is given by:

$$X_{Thr} = \overline{X} \pm \beta \cdot \sigma(X) \pm b \tag{52}$$

where $\overline{X}$ is the average of the signal, $\sigma(X)$ is its standard deviation, β is a bound factor and b is a bias and were determined based on a parametric study. The details of the study are discussed in Section VI.D. Within the context of this effort, the signal *X(t)* that is monitored is the residual $r(t)$, defined as the difference between the actual airspeed computed from Pitot tube measurements and its corresponding estimate from any of the three analytical redundancy methods. Under nominal, fault free conditions, the residual stays within the boundaries defined by the upper and lower bounds of the floating limiter; on the other hand, following the occurrence of a failure, the residual breaches either of the bounds, which triggers the failure detection. The design parameters in (52) are positive constants that depend on the nature of the signal itself and are calculated over a time window δ and can be tuned to achieve desirable failure detection performance. The tuning process adopted in the work presented in this chapter is described in Section IX. The approximations of the average and standard deviations of the residual signal is calculated over a period of 1s (50 samples) using an autoregressive moving average filter given by

$$D(Z) = \frac{z^3 + z^2 + z + 1}{50(z^3 - 0.3z^2 + 0.3z - 0.3)} \tag{53}$$

Model Based Approach – Residual Whitening and Fault Detection Methodology

Residual Whitening

Linear AR modeling provides an easy approach to remove correlation from the residual [47, 48,49]. A sequence *r*(*k*) is an Auto Regressive (AR) process of order *N* if it is generated by the model:

$$r(k) = \sum_{i=1}^{M} d_i \cdot r(k-i) + \varepsilon(k) \tag{54}$$

with k designating the sampling time, d_i the AR modeling parameters, M the model order, $\varepsilon(k)$ a white noise with zero mean and variance σ^2 and the sum $r_{FIR}(k) = \sum_{i=1}^{M} d_i \cdot r(k-i)$ represents the LS one step ahead prediction of *r*(*k*) based on the previous M samples. In practice, given a random sequence *r*(*k*) (with possible autocorrelation), an AR model can be fitted to estimate the overdetermined parameters d_i which minimize the empirical LS prediction error $\sum_L \varepsilon(k)^2 = \sum_L (r(k) - r_{FIR}(k))^2$.

For Gaussian sequences, the resulting prediction error sequence: $\varepsilon(k) = r(k) - r_{FIR}(k)$ is white, implying that the filter $W_{FIR}(z) = 1 - d_1 z^{-1} - d_2 z^{-2} - ... - d_N z^{-M}$ is a whitening filter for $r(k)$ (where z^{-1} is discrete time one step delay operator in the Z domain). This implies that the whitened residual $\varepsilon(k)$ is simply the output of the $W_{FIR}(z)$ filter and can be computed as follows:

$$\varepsilon(z) = W_{FIR}(z) r(z) \tag{55}$$

The whitened residual $\varepsilon(k)$ is the input signal of the GLRT detector that is introduced in the next section.

The Neyman-Pearson Approach to Fault Detection

The Neyman-Pearson approach [50] is a probabilistic method used to discriminate between two possible hypotheses. In the context of SFDI, H_0 is the hypothesis that the fault is not present and H_1 is the hypothesis that the fault is present. In the present study the purpose of the detector is thus to decide between either H_0 or H_1, based on the observation of the characteristics of the (whitened) residual ε(k). Let $p(\varepsilon | H_1)$ be the probability of a faulty residual under hypothesis H_1 and $p(\varepsilon | H_0)$ the probability of a no fault residual under hypothesis H_0, then the probability likelihood ratio is defined as: $L = p(\varepsilon | H_1) / p(\varepsilon | H_0)$. The higher L is, the higher is the possibility that hypothesis H_1 is true compared to H_0. Therefore, if the likelihood ratio is higher than a specific threshold value γ, then the hypothesis H_1 is accepted to be true and a fault is declared, otherwise the hypothesis H_0 is accepted to be true.

Formally, a fault is declared if

$$L = \frac{p(\varepsilon | H_1)}{p(\varepsilon | H_0)} \geq \gamma \tag{56}$$

The threshold γ is selected as a compromise to provide a desired probability of false alarm and fault detection where the probability of a false P_{FA} is the probability of deciding H_1 when H_0 is true while the probability of detection P_D is the probability of deciding H_1 when H_1 is true.

Generalized Likelihood Ratio Test (GLRT) – Change in Variance

In the present effort, fault detection is performed on-line based upon a statistical decision making procedure that utilizes the whitened residual signal $\varepsilon(k)$. At nominal conditions this signal should be precisely uncorrelated (white) and follow a normal distribution. The occurrence of a fault leads to changes in the residual dynamics, which results in increased residual variance [47]. Therefore, a formal statistical hypothesis testing problem may be formulated in order to detect faults through the detection of increased variance.

To cast this problem as a probabilistic real-time hypothesis testing applied to the stream of samples of the whitened residual let us consider a fixed window (W_0) of n successive samples of the whitened residual signal $\varepsilon(k_0), \varepsilon(k_0+1), ..., \varepsilon(k_0+n-1)$ beginning at sample time k_0; the m samples in W_0 are assumed to be Independent, Identically Distributed $IID(\sigma_0, \mu_0)$. Consider also another moving window (W_1) of m samples $\varepsilon(k), \varepsilon(k-1), ..., \varepsilon(k-m+1)$ beginning at a generic time instant k; assume that also in W_1 the m samples are $IID(\sigma_1, \mu_1)$. The finite length sequences of samples of the $\varepsilon(k)$ signal in the windows W_0 and W_1 are used to test the hypothesis:

$$H_0: \quad \sigma_0 = \sigma_1 \tag{57}$$

against

$$H_1: \quad \sigma_1 > \sigma_0 \tag{58}$$

under the assumption that σ_1, σ_2 and the means μ_1, μ_2 are unknown. As the theoretical variances σ_1^2 and σ_2^2 as well as the means μ_1, μ_2 are assumed unavailable, appropriate estimates are used need to be used given by the sample variances and means respectively, that is:

$$\hat{\sigma}_0^2 = \frac{1}{n-1}\sum_{i=k_0}^{k_0+n-1} (\varepsilon(i) - \hat{\mu}_0)^2 \; ; \; \hat{\mu}_0 = \frac{1}{n}\sum_{i=k_0}^{k_0+n-1} \varepsilon(i) \tag{59}$$

$$\hat{\sigma}_1^2 = \frac{1}{m-1}\sum_{i=k-m+1}^{k} (\varepsilon(i) - \hat{\mu}_1)^2 \; ; \; \hat{\mu}_1 = \frac{1}{m}\sum_{i=k-m+1}^{k} \varepsilon(i) \tag{60}$$

Under the above assumptions it can be shown that the Generalized Likelihood Ratio Test (GLRT) may be expressed by the following ratio:

$$L = \frac{\hat{\sigma}_1^2}{\hat{\sigma}_0^2} = \frac{\sum_{i=k-m+1}^{k} (\varepsilon(i) - \hat{\mu}_1)^2 \Big/ (m-1)}{\sum_{i=k_0}^{k_0+n-1} (\varepsilon(i) - \hat{\mu}_0)^2 \Big/ (n-1)} = \frac{\sigma_1^2 \chi_{m-1}^2 / (m-1)}{\sigma_0^2 \chi_{n-1}^2 / (n-1)} = \frac{\sigma_1^2}{\sigma_0^2} F_{m-1,n-1} \tag{61}$$

where $F_{m-1,n-1}$ is the so called F-distribution with m-1,n-1 degrees of freedom (that is the ratio of 2 chi-square distributions). Now, under the hypothesis H_0 as $\sigma_0 = \sigma_1$, results in the Likelihood Ratio being exactly $L \sim F_{m-1,n-1}$. On the other hand, if H_1 is true (that is in case $\sigma_1^2 > \sigma_0^2$), the function L will tend to be greater than when H_0 is true, so we reject H_0 if this ratio is large. This leads to the following F-test at the risk level α (i.e. false alarm probability equal to α) that is:

$L < \gamma$ (H_0 accepted, that is the fault is not present)

$L \geq \gamma$ (H_1 accepted, that is the fault is present) (62)

where the threshold γ in (62) is the "critical" value of the $F_{m-1,n-1}$ distribution such that the area of the tail of its probability density function is exactly equal to α, that is:

$$\alpha = \int_{\gamma}^{\infty} PDF\left(F_{m-1,n-1}(x)\right)dx \tag{63}$$

For a given risk level α, the threshold γ can be computed from (63). This γ is used in the experimentation as the threshold for the GLRT detector $\hat{\sigma}_1^2(k)/\hat{\sigma}_0^2(k)$ defined in (61).

11. Results

The WVU YF-22 research aircraft has a limited operational flight envelope as it is flown under the Academy of Model Aeronautics (AMA) rules, which has a restriction of 55 lbs take-off weight and a flight ceiling of 400 ft within visual range. However, even within a limited flight envelope, the aircraft aerodynamics is substantially different at take-off and landing conditions due to the use of flaps.

Table 2. Characteristics of Flight Data Segments

		Inputs [p (rad/s), q (rad/s), r (rad/s), ax (g), ay (g), az (g), α (rad), β (rad),	
FDS	Length		
001	447 s	Min	[-0.9714, -0.2839, -0.3302, -0.1633, -0.1686, -3.0070, -0.0268, -
		Max	[0.6321, 0.6926, 0.2299, 0.2164, -0.0504, 0.3341, 0.1677, 0.0388,
003	368 s	Min	[-0.9327 -0.2044 -0.3063 -0.1260 -0.1469 -2.6588 0.0138 -0.0496
		Max	[1.1161 0.5995 0.4304 0.2284 0.0206 -0.3743 0.1813 0.0947
004	452 s	Min	[-0.6049 -0.1566 -0.3326 -0.1535 -0.1375 -2.2525 0.0027 -0.0521
		Max	[0.8597 0.4770 0.1264 0.2036 -0.0262 -0.3165 0.1652 0.0708
005	422 s	Min	[-1.0121 -0.3046 -0.3543 -0.0915 -0.1550 -3.3062 -0.0155 -0.0832
		Max	[0.9812 0.6890 0.1125 0.1999 0.0447 0.2643 0.1408 0.0859

Therefore, to avoid inaccuracies in airspeed estimation during the take-off and landing phases, the data segment from each flight is restricted to flight altitudes over 100 ft. The duration of the flight segments and the maximum and minimum values of each of the inputs are given below in Table 2.

Airspeed Estimation using UKF

The UKF airspeed estimation algorithm was executed for each of the four considered sets of UAV flight data. For each flight, the same noise assumptions were used based on offline static measurements from the IMU, relative wind vanes, and vertical gyroscope, resulting in the following process and measurement noise matrices:

$$\mathbf{Q} = diag\left(5.73\times10^{-4},\ \ 1.06\times10^{-3},\ \ 7.93\times10^{-4},\ \ 2.04\times10^{-7},\ \ 2.47\times10^{-7},\ \ 4.83\times10^{-7}\right) \tag{64}$$

$$\mathbf{R} = diag\left(1.09\times10^{-5},\ \ 1.10\times10^{-5},\ \ 9.44\times10^{-6},\ \ 3.77\times10^{-6}\right) \tag{65}$$

The units for these matrices are the squares of m/s^2 for accelerations, *rad/s* for angular rates, and *rad* for angles. The initial state of the filter is defined by zeros except for the velocity component u, which is initialized to the current value of the airspeed from the Pitot tube. The initial covariance matrix is defined by 10^{-2} times the identity matrix, which was selected to give a reasonable convergence rate of the estimation.

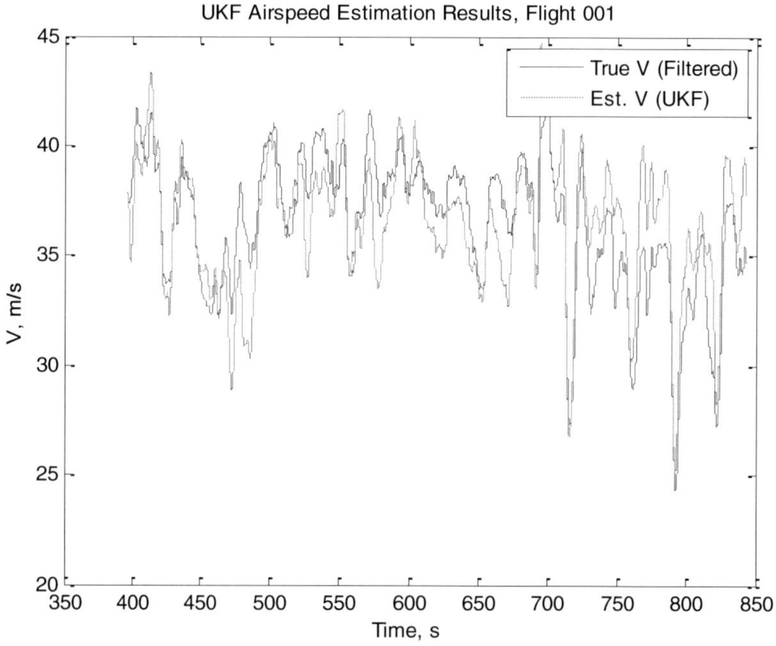

Figure 14. Pitot tube airspeed vs. UKF estimate – Flight Data Set 001.

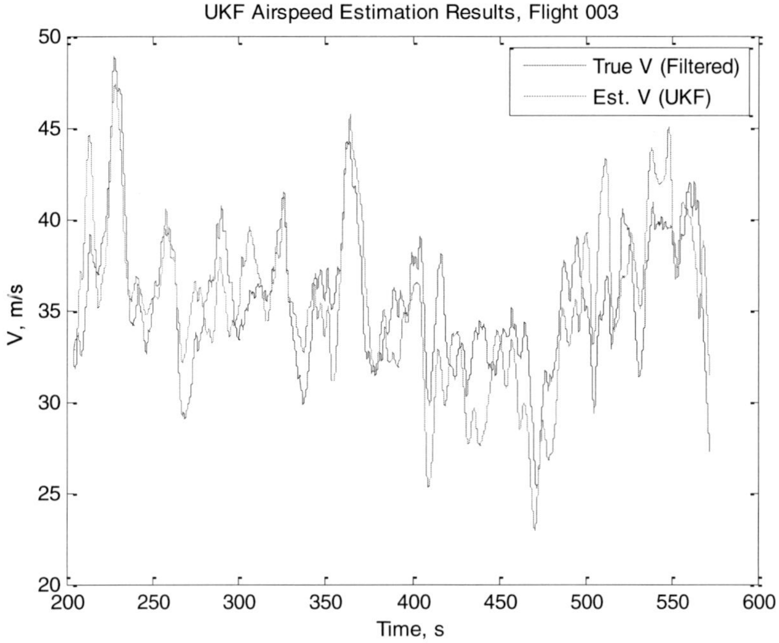

Figure 15. Pitot tube airspeed vs. UKF estimate – Flight Data Set 003.

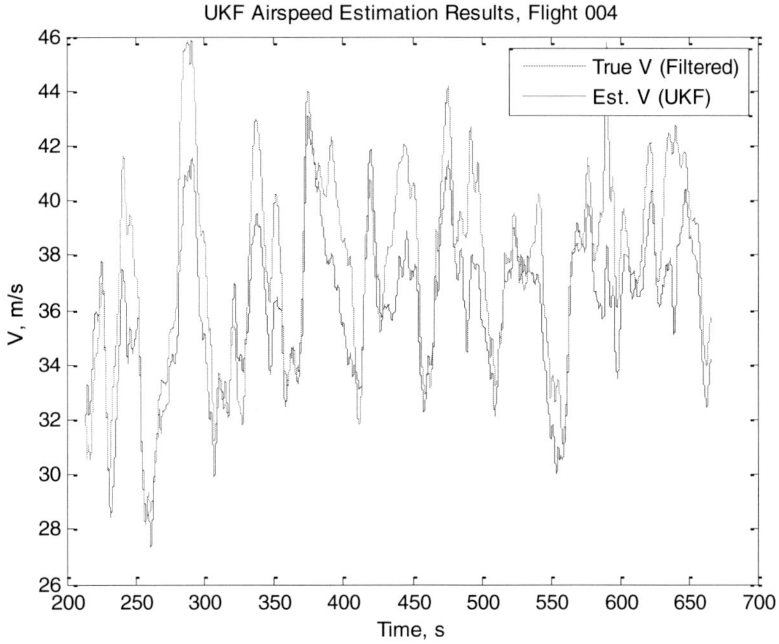

Figure 16. Pitot tube airspeed vs. UKF estimate – Flight Data Set 004.

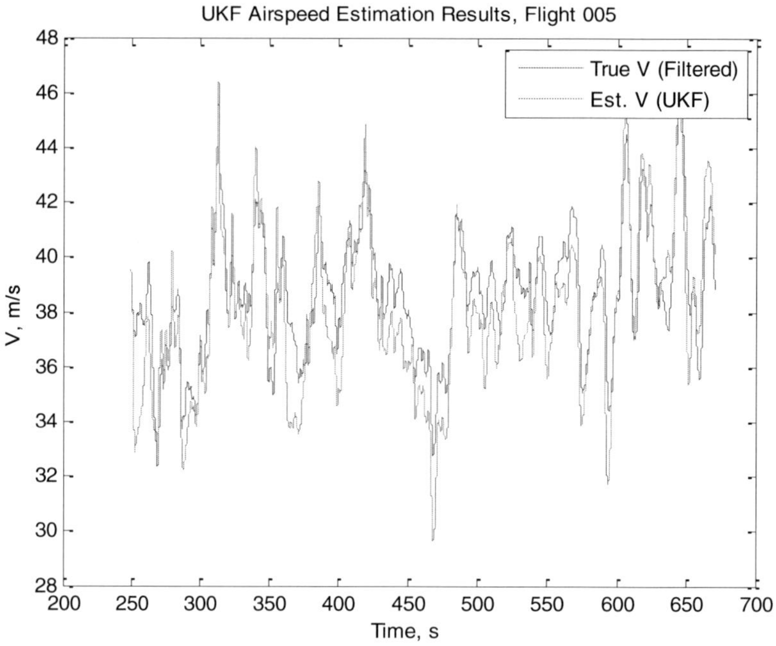

Figure 17. Pitot tube airspeed vs. UKF estimate – Flight Data Set 005.

The UKF primary, secondary, and tertiary scaling parameters were set to 0.5, 2, and 0 respectively, as recommended for Gaussian noise assumptions [28]. Using these settings, the UKF algorithm was executed using the method described in IV. A. for each flight data set. The airspeed estimation results for each of the four data sets are plotted against the "true" airspeed which is the low pass filtered value of the airspeed measured using the Pitot tube, and is shown in Figures 14-17. The statistical results for this method are given in Table 3.

Table 3. Tracking Error Statistics – UKF

	Mean(m/s)	Std (m/s)
FDS 001	0.0641	2.0502
FDS 003	-0.1807	2.8930
FDS 004	1.6308	1.7450
FDS 005	-0.6875	1.3947

Airspeed Estimation using MLP ANN (Training and Validation)

The parameters used to determine the structure of the Matlab NN and used for training with all the four sets of flight data are shown below in Table 4.

To minimize the impact of the differences in the initialization of the interconnection weights between the input layer and hidden layer as well as the hidden layer and output layer on the training and validation performance of the network, for each set of flight data the training procedure was run 100 times, with a new network being initialized at each run. The tracking error statistics (mean (*m/s*) and standard deviation (*m/s*)) from the training phase for each flight data set is shown in Figure 18. The distribution of the standard deviation of the tracking error for the 100 training runs of each data set is shown in Figure 19.The average of the standard deviation from the training phase for each flight data set, tabulated along the diagonal entries in Tables 5 and 5a, is used as the benchmark value for evaluating the validation performance.

Table 4. Parameters of the Multilayer Feed-Forward NN

Parameter	Value
[Ni No]	[10 1]
[Num of Hidden Layer Neurons]	[5]
[Training Goal]	[0.002]
[Max. Number of Epochs]	[500]
[Activation Function]	[log-sigmoidal (hidden layer)] linear(output layer)]
[Interconnection Weights]	Initialized using Widrow-Nyugen Rule

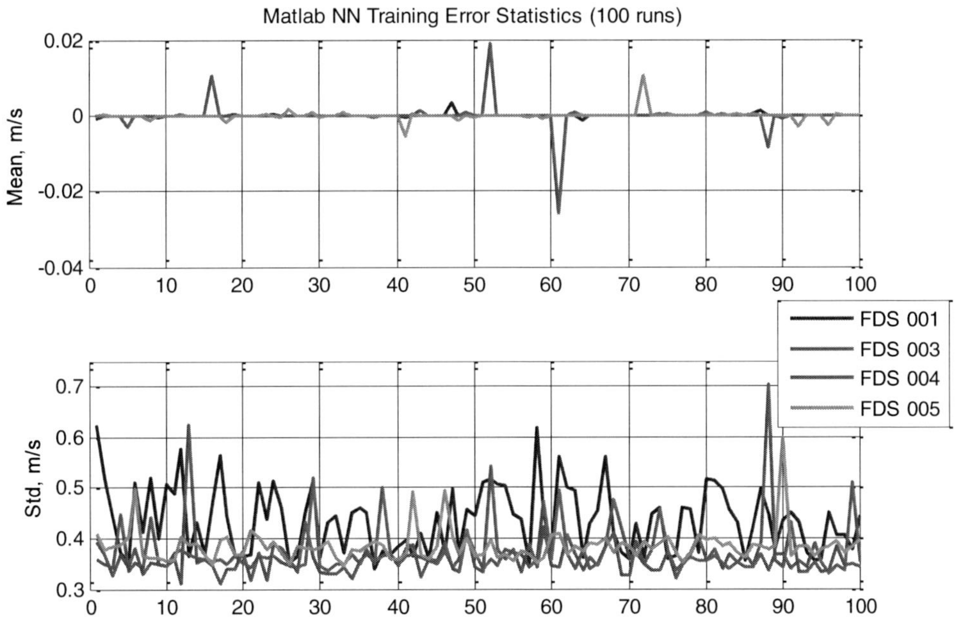

Figure 18. Mean and Std. Dev. from 100 runs of MLP ANN training for each FDS {001, 003, 004, 005}.

Once the MLP ANN was trained with each set of flight data, its training was validated with all the remaining flight data sets. For example, after training a network with Flight Data Set 001, it was validated against FDS 003, FDS 004 and FDS 005 and as with the training, each of the 100 runs of the NN was validated and the average of the prediction error statistics tabulated in Tables 5, with the bolded entries along the diagonal representing the training performance. Considering the fact that the tracking error performance and statistics is typically the best during training phase (and as a consequence in the validation phase, the mean and standard deviations of the tracking error are typically higher), the average performance of the NNs in the validation phase is computed as:

$$perf = -100 * (TE\_Std_{Val} - TE\_Std_{Train}) / TE\_Std_{Train} \tag{66}$$

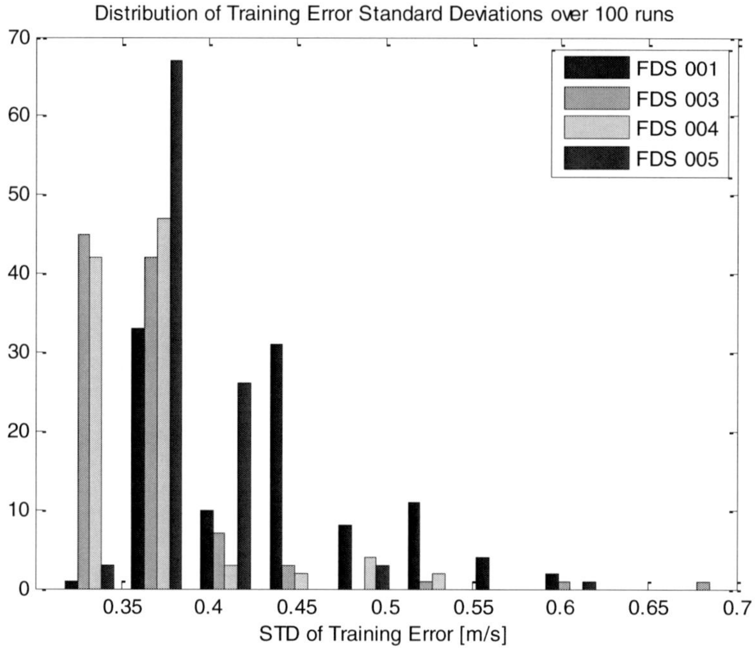

Figure 19. Distribution of Std. Dev. over 100 runs of MLP ANN training for each FDS {001, 003, 004, 005}.

Table 5. Estimation Error Statistics – Training and Validation (MLP ANN)

Data Set	FDS 001 (val)			FDS 003 (val)		
	Mean (m/s)	Std. (m/s)	Perf.(%)	Mean (m/s)	Std. (m/s)	Perf. (%)
FDS 001	1.6948e-005	0.4358	NA	0.6229	0.8351	-91.6
FDS 003	-0.7268	0.7464	-103.7	-6.4628e-	0.3664	NA
FDS 004	1.8179	1.1164	-204.53	2.0883	0.8094	-120.79
FDS 005	0.0875	0.8315	-115.92	1.1513	1.8722	-386.16
	FDS 004 (val)			FDS 005 (val)		
FDS 001	-2.1743	0.9009	-106.7	-0.1303	0.5708	-31
FDS 003	-2.8208	0.9805	-167.6	-0.6518	0.5550	-51.47
FDS 004	1.8118e-005	0.3666	NA	2.1990	0.7802	-112.82
FDS 005	-2.0833	0.8370	-117.35	-4.1971e-	0.3851	NA
FDS 001	-2.1743	0.9009	-106.7	-0.1303	0.5708	-31
FDS 003	-2.8208	0.9805	-167.6	-0.6518	0.5550	-51.47

An analysis of Table 5 reveals that while the training proceeds well with all the flight data sets, the validation performance of the networks trained with data sets 004 and 005 are very poor. The best performing flight data set in the validation phase is FDS 003, with the lowest estimation error standard deviation in training and with an average of approximately 108% higher standard deviation of the estimation error. The MLP ANN estimate of the airspeed during the training and validation phases is shown below in Figures 20 and 21.

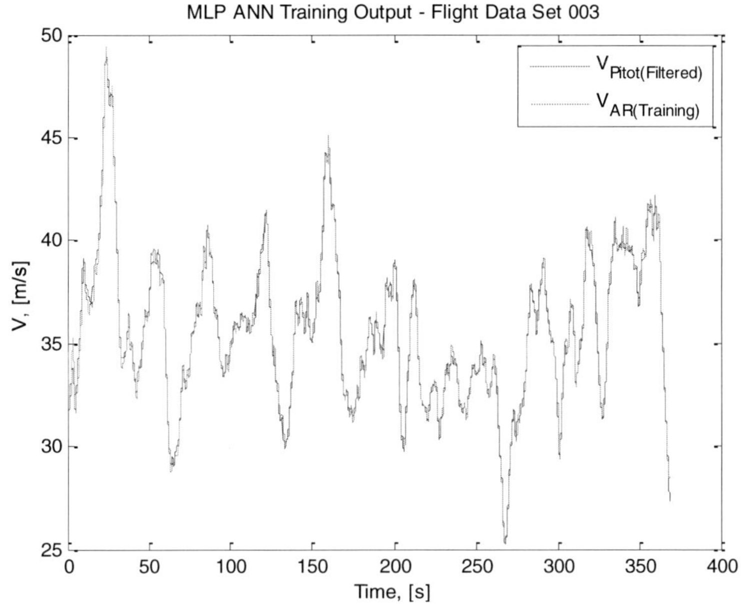

Figure 20. Pitot tube airspeed vs. MLP ANN estimate (trained using FDS 003).

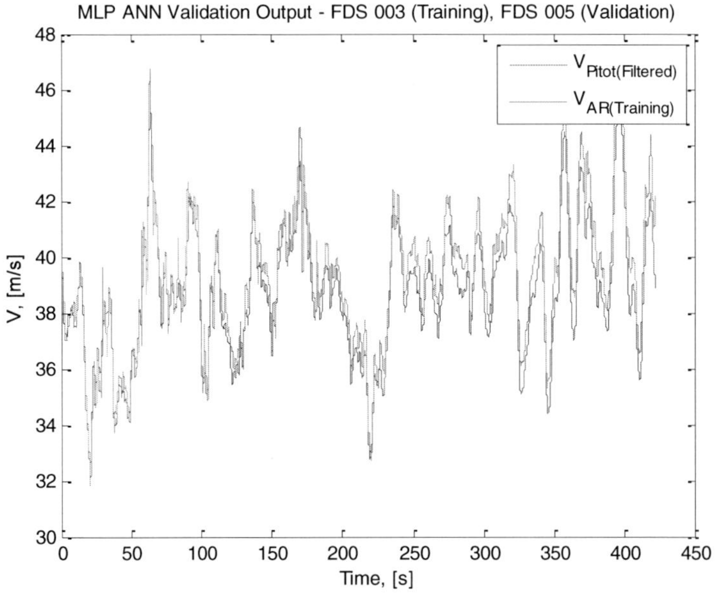

Figure 21. Pitot tube airspeed vs. MLP ANN estimate (trained using FDS 003 and validated using FDS 005).

Airspeed Estimation Using EMRAN ANN (Training and Validation)

The parameters used to determine the structure of the EMRAN NN are given below in Table 6 and used for training with all the four sets of flight data. During the training and validation phases, the inputs to the EMRAN are normalized using the maximum and minimum values of each input from Table 2. It should be emphasized that the learning rate, the error thresholds and the other parameters of the EMRAN network were selected by trial and error to provide a desirable trade-off of performance, network structure (number of hidden layer neurons), and training time.

The EMRAN ANN estimates of the airspeed during the training and validation phases are shown below in Figures 22 and 23. Figure 24 shows the number of neurons during the training phase for each set of flight data.

Table 6. Parameters of the EMRAN NN

Parameter	Value
[Ni, No]	[10 1]
[Nmax, Overlap, Radius, Prune]	[250 0.5 10 1]
[LR_weights, LR_sigma, LR_center]	[0.0001 0.0001 0.0001]
[SF_weights, SF_sigma, SF_center]	[1E-6 1E-6 1E-6]
[Err,_Thr]	[0.02]
[CD_Emax, CD_Emin, CD_Egam]	[0.6 0.6 1]
[FE_Thr, FE_Pole]	[0 0.5]

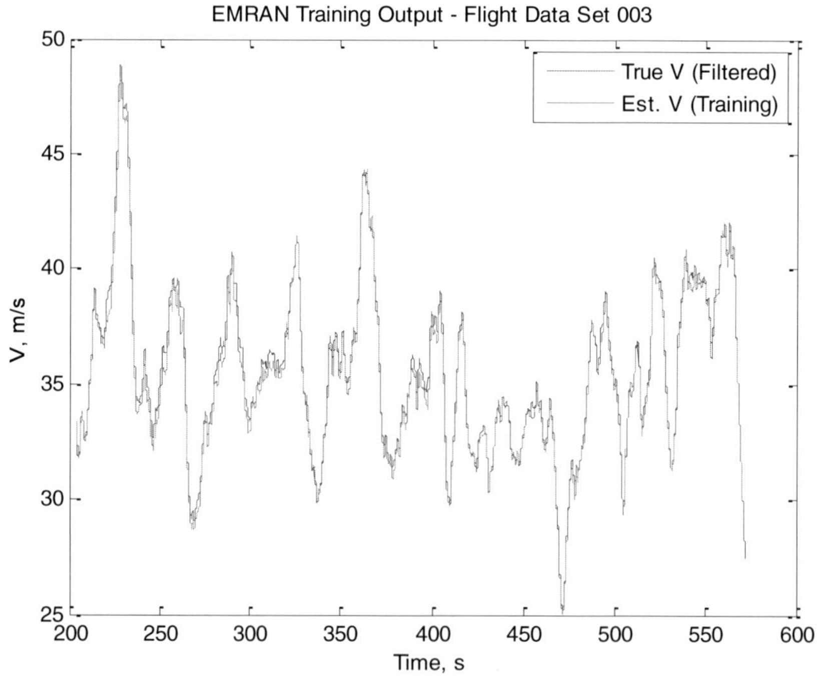

Figure 22. Pitot tube airspeed vs. EMRAN NN estimate (trained using FDS 003).

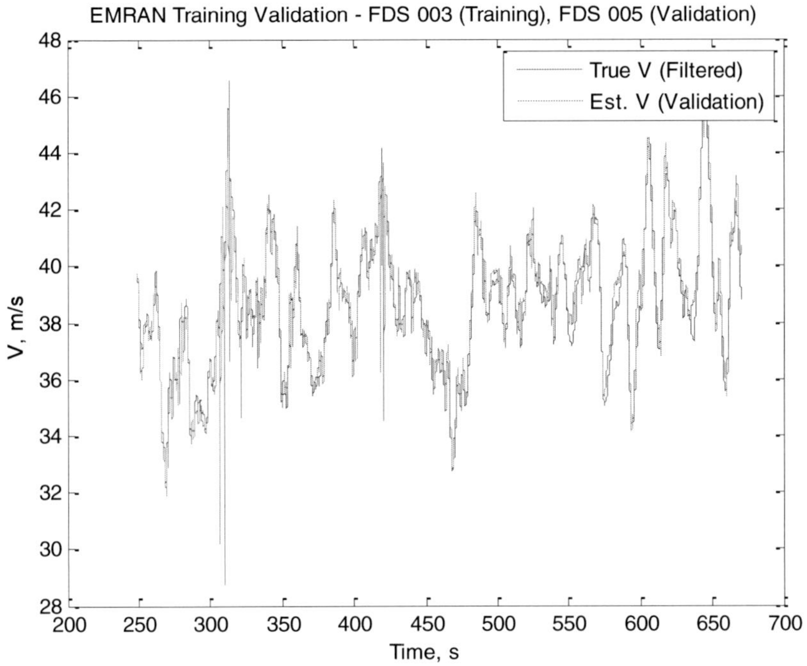

Figure 23. Pitot tube airspeed vs. EMRAN NN estimate (trained using FDS 003 and validated using FDS 005).

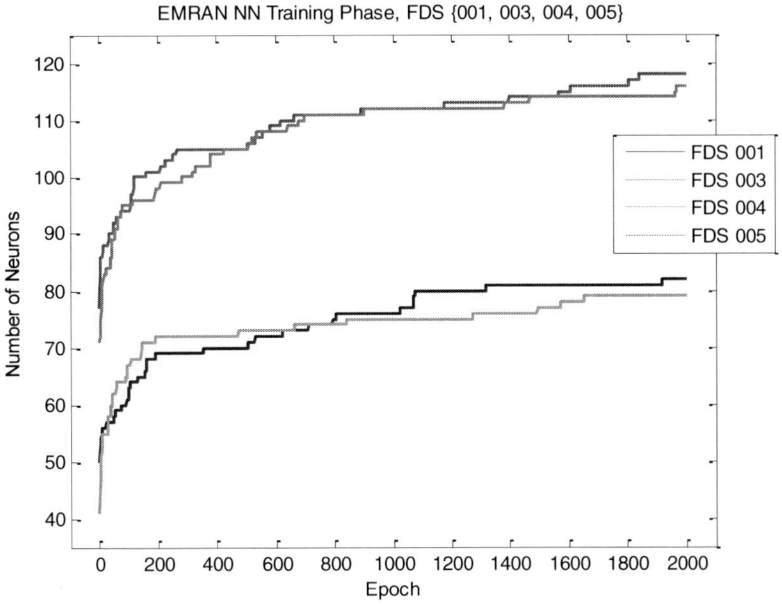

Figure 24. Number of neurons in EMRAN network during training phase for each data set.

The mean and standard deviation of the tracking error from the training phase for each flight data set is tabulated in Table 7 and is used as the benchmark value for evaluating the validation performance.

Table 7. Tracking Error Statistics from Training Phase of EMRAN NN

	FDS 001	FDS 003	FDS 004	FDS 005
No. of active neurons (at end of training)	82	118	79	116
Number of training epochs	2000	2000	2000	2000
Mean of training error (m/s)	-0.0149	0.0136	-0.0777	-0.0482
Standard deviation of training error	0.4845	0.3620	0.2775	0.4664

As with the MLP ANN, once the EMRAN NN was trained with each set of flight data, its training was validated with all the remaining flight data sets. For example, after training the network with Flight Data Set 003, it was validated against FDS 001, FDS 004 and FDS 005. The error statistics and the performance (computed using Eq. 53) from the validation phase are given in Table 8, with the bolded entries along the diagonal representing the training performance.

The analysis of Table 8 reveals that the training proceeds generally well with all the flight data sets; however, the validation performance of the network trained with data set 004 is poor.

The best performing network in the validation phase is the one trained with FDS 003, with an average of approximately 133% higher standard deviation of the estimation error, as compared to the baseline.

Table 8. Estimation Error Statistics from Validation Phase using EMRAN NN

Data Set	FDS 001 (val)			FDS 003 (val)		
	Mean(m/s)	Std(m/s)	Perf.	Mean(m/s)	Std(m/s)	Perf. (%)
FDS 001 (tr)	-0.0149	0.4845	NA	1.9606	1.7830	-268
FDS 003 (tr)	-0.3053	0.9415	-160	0.0136	0.3620	NA
FDS 004 (tr)	2.0772	1.3457	-384.9	3.1129	2.0628	-643.35
FDS 005 (tr)	0.1712	1.1098	-137.95	0.9886	1.5655	-235.65
Data Set	FDS 004 (val)			FDS 005 (val)		
	Mean(m/s)	Std(m/s)	Perf.	Mean(m/s)	Std(m/s)	Perf. (%)
FDS 001 (tr)	-1.1329	0.7190	-48.4	1.3196	1.3744	-183.67
FDS 003 (tr)	-2.2626	0.7056	-94.91	-0.1668	0.9519	-162.96
FDS 004 (tr)	-0.0777	0.2775	NA	2.6669	1.5189	-447.35
FDS 005 (tr)	-1.9170	0.6992	-49.91	-0.0482	0.4664	NA

Comparison of Estimation Performance of UKF, MLP and EMRAN NN

The average of the validation error performance of the three analytical redundancy approaches is given below in Table 9. For each flight data set, the mean and standard deviation of the validation error of the MLP ANN and EMRAN NN is averaged to represent overall performance of the networks. This is representative of a real world scenario wherein a suitable trained network is assumed to be flown onboard the aircraft to estimate the airspeed based on inputs not 'seen' during the training phase. On the other hand, the estimate of airspeed from the UKF implementation is equivalent to the validation outputs of the NNs, since this method does not require a training procedure. An analysis of Table 9 reveals that the MLP ANN is the best performing approach, with the lowest standard deviations of the tracking error, for all cases, except FDS 005, where it is comparable to the EMRAN NN and UKF. Both of the ANN methods tend to have lower standard deviations than the UKF, while the UKF is capable of better estimating the mean. While each of these benefits has some merit, the larger standard deviations of the UKF lead to increased difficulty in failure detection. An advantage of the ANN approaches over the UKF is that the ANNs are capable of modeling the complex nonlinearities and stochastic processes involved in the system, while the UKF is limited to its assumed dynamic structure and noise assumptions. Also, the time correlated nature of the UKF can lead to error accumulation with time, which is not a problem that occurs with the ANN method. An effort was conducted to tune the covariance matrices for the UKF in order to improve the estimation performance. This tuning was done only using FDS 001, and then the same covariance matrix selections were used across all the flight data sets. This represents a comparable case to the ANN methods training on FDS 001 and validating on the others. There was a balance in this tuning procedure of over correcting with the measurement vector which leads to larger mean errors in the UKF estimates, or under correcting which leads to drift in the velocity estimates. The estimates of the airspeed from the three analytical approaches using FDS 003 are shown in Figure 25.

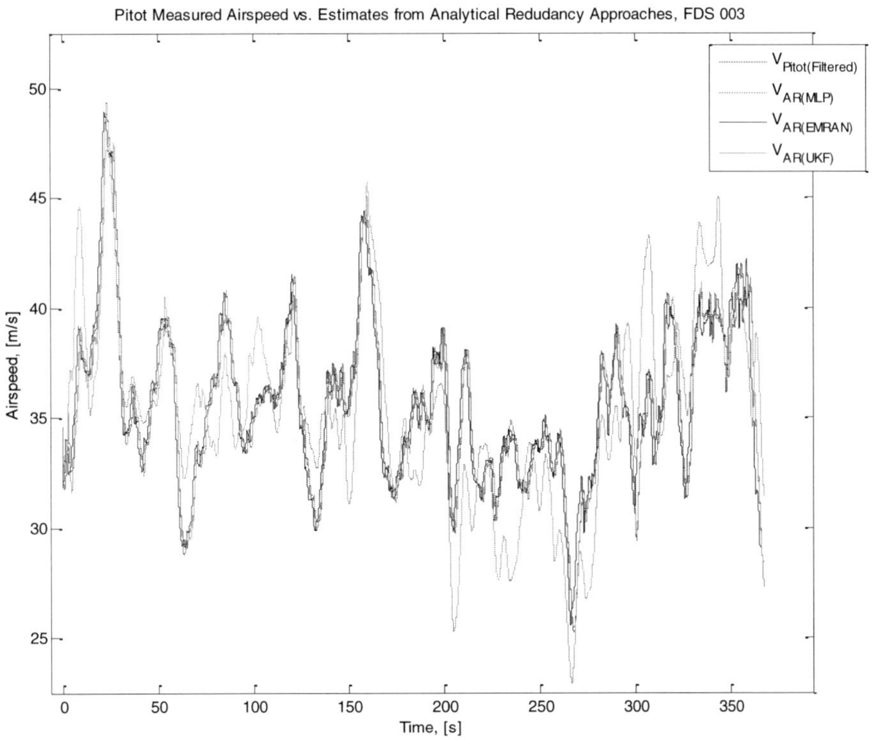

Figure 25. Actual Airspeed vs. Estimates from Analytical Redundancy Approaches.

Table 9. Comparison of validation performance of the three analytical redundancy approaches

Flight Data Set	MLP ANN		EMRAN NN		UKF	
	Mean	Std	Mean	Std	Mean	Std
FDS 001	-0.5599	0.7689	0.7158	1.2921	0.0641	2.0502
FDS 003	-1.3995	0.7618	-0.9116	0.8663	-0.1807	2.8930
FDS 004	2.0355	0.9015	2.6190	1.6425	1.6308	1.7450
FDS 005	-0.2800	1.1787	-0.2524	1.1248	-0.6875	1.3947

Failure Detection Performance of CUSUM Filter and Floating Limiter

Experiments were conducted with flight data from the WVU-YF22 for evaluating the Sensor Failure Detection and Identification performance of the *CUSUM* filter. As with the estimation process, the data was restricted to flight altitudes over 100 ft. Additive failures as described in Section V were introduced on the airspeed sensor at 150 s. Additive failures were introduced as multiple of an entity called the "Failure Basis Unit (FBU)", which was set to 0.5 m/s. As mentioned in Section VI, the values of the parameter ν (that is the value of the expected fault amplitude $\Delta V = \nu$) strongly affects the sensitivity of the *CUSUM* filter to faults having generic amplitudes. In this study this parameter was fixed to a value equal to the standard deviation of the airspeed estimation error using the best performing data set, FDS 003. For the MLP ANN, the network trained with FDS 003 (run 17) was selected as the baseline network and similarly for the EMRAN NN, the network structure trained with FDS 003 was chosen.

To maintain consistency across all three methodologies, the UKF with FDS 003 was selected. The fault detection threshold for the *CUSUM* signal $S^{+}(k)$ was selected as a compromise between the need for fast fault detection and the need of low false alarm rates. An acceptable trade-off was found by setting the threshold equal to 150% of the maximum value of the *CUSUM* signal that is $CUSUM\_th = 1.5 * \max\left(S^{+}(k)\right)$ observed in fault free conditions. The *CUSUM* value for FDS 003 in fault free conditions for the MLP ANN is shown below in Figure 26. Table 10 lists all the parameters used in the *CUSUM* filter. The failure detection performance of the *CUSUM* filter and the FL were evaluated with two data sets FDS 001 and FDS 005.

Table 10. Parameters of the CUSUM Filter

	MLP ANN		EMRAN NN		UKF	
	FDS 001	FDS 005	FDS 001	FDS 005	FDS 001	FDS 005
ν	0.7464	0.5550	0.9415	0.9519	2.0502	1.3947
$CUSUM_{max}$	1172		1071		2228	
$CUSUM_{Thr}$	1758		1606		3342	

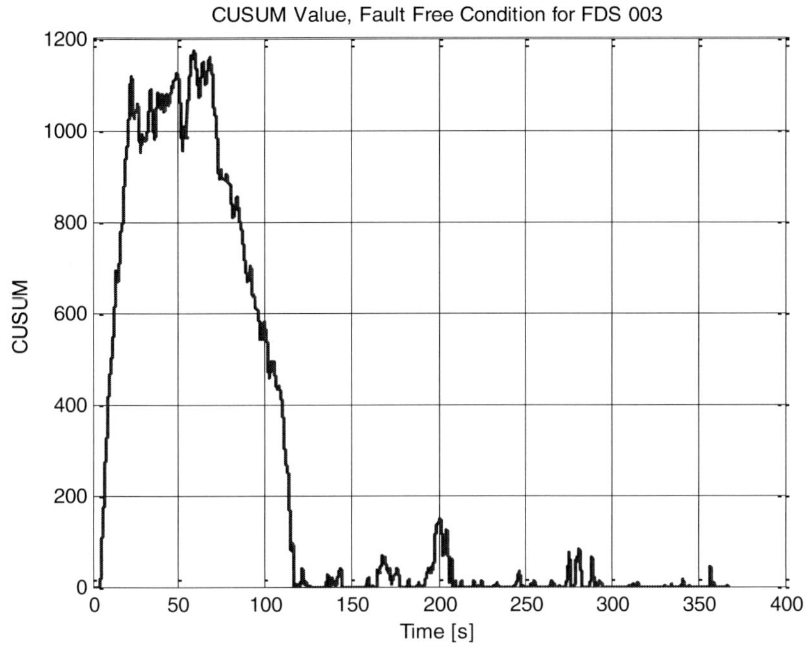

Figure 26. CUSUM Value for a Fault Free condition (MLP ANN estimate of Airspeed using FDS 003).

To assess the performance of the *CUSUM*-based SFD scheme, a parametric study was performed by varying the amplitude of the faults as multiples of the FBU from 1 FBU to 40 FBUs for the different classes of considered faults. Tables 12-15 reports the results for both Type 1 and Type 2 failures by varying the magnitudes from 1 FBU to 40 FBUs, in terms of the fault detection delay.

As with the *CUSUM* filter, in the Floating Limiter approach to fault detection, the networks and the UKF trained with the data from FDS 003 was used as the baseline while data sets FDS 001 and FDS 005 were used in the evaluation process. As described in Section V, there are three parameters of the FL that could be tuned to achieve desirable fault detection performance. In the work presented in this chapter, the parameter 'β' was constant and fixed at 3, the time window 'δ' was set to 1s and the parameter 'b' was tuned. A parametric study was conducted by varying the value of 'b' in increments of 0.001, to determine its optimal value for each of the three analytical redundancy approaches.

The tuning was performed on the "fault free" data from FDS 003, and the parameter 'b' varied until the residual signal stopped exceeding either the upper or lower bounds and no failure detection flags were raised. This resulted in different values of the tuning parameter 'b' for each of the three approaches, which are tabulated in Table 11. The results from this parametric study for various values of 'b' are shown in Figures 27-30. Figure 29 shows the case where the parameter 'b' is such that the fault free residual $r(t)$ does not breach the upper bound for the entire data set.

Figure 30 shows the case where the residual does not breach the lower bound. For each of the three analytical redundancy methods, the higher of the two values of 'b', highlighted in green in Table 11 was selected to be used in the evaluation of FDI for data sets FDS 001 and FDS 005.

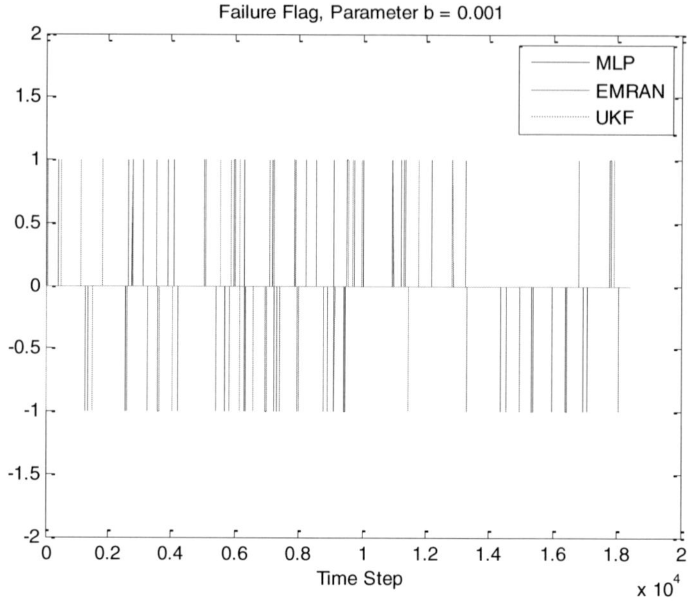

Figure 27. Tuning of floating limiter parameter 'b'; b = 0.001.

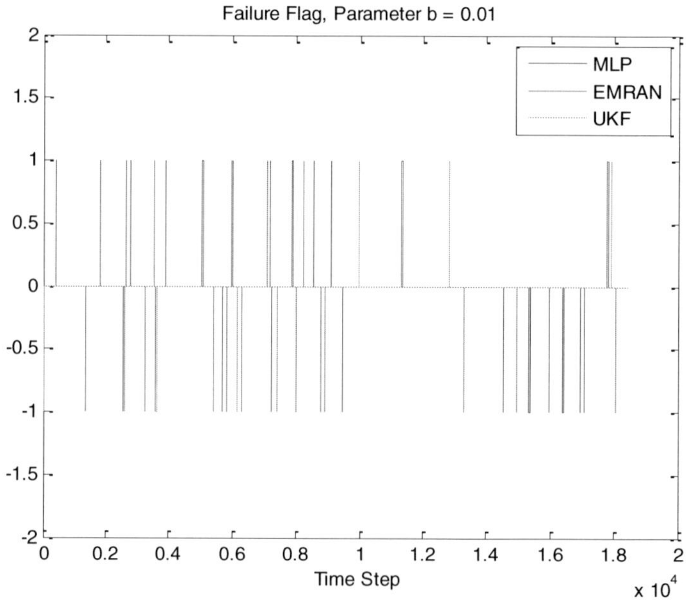

Figure 28. Tuning of floating limiter parameter 'b'; b = 0.01.

Table 11. Parameters of the Floating Limiter (δ = 1s)

	MLP ANN		EMRAN NN		UKF	
	β	b	β	b	β	b
No Upper	3	0.110	3	0.027	3	0.037
No Lower	3	0.293	3	0.058	3	0

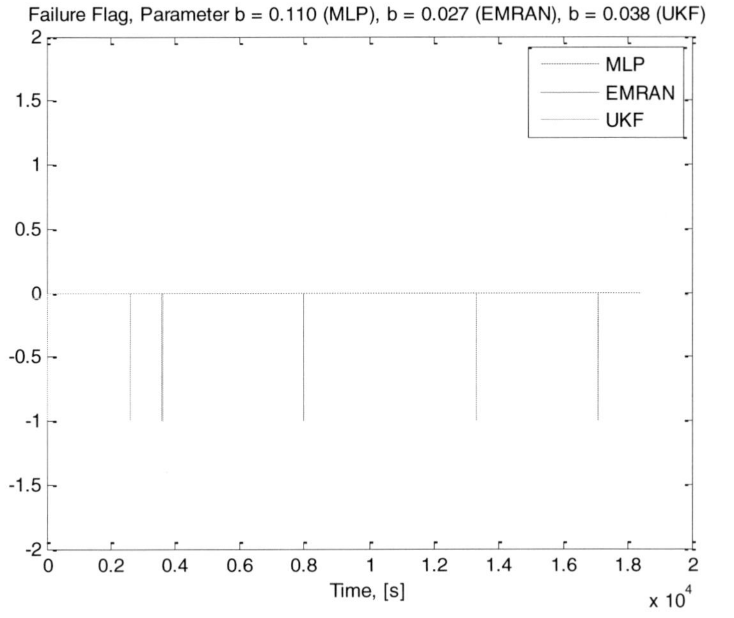

Figure 29. Tuning of FL parameter 'b'. Upper Bound of FL is not breached.

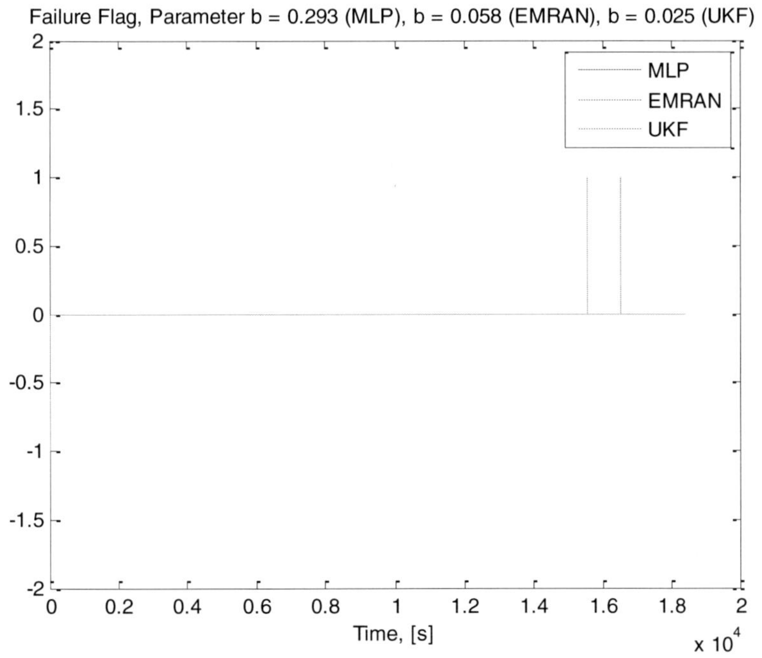

Figure 30. Tuning of FL parameter 'b'. Lower Bound of parameter 'b' not breached.

To assess the performance of the FL based SFDI scheme, a parametric study similar to the one performed with *CUSUM* filter was carried out by varying the amplitude of the faults as multiples of the FBU from 1 FBU to 40 FBUs for the different classes of considered faults. Tables 12 - 15 report the results for both Type 1 and Type 2 failures for a range of failure amplitudes from FBU = 1 to 40, in terms of the fault detection delay on data sets FDS 001 and FDS 005 respectively.

Table 12. Comparison of Detection Delays using CUSUM Filter and FL for Type 1 Failures (FDS 001)

	MLP ANN		EMRAN NN		UKF	
FBU	CUSUM	FL	CUSUM	FL	CUSUM	FL
1	---	---	---	0	---	---
2	---	---	69.34	0	---	---
3	53.26	---	39.4	0	---	0
4	30.72	0	28.58	0	---	0
5	21.8	0	22.88	0	138.5	0
10	9.22	0	10.96	0	82.52	0
40	2.06	0	2.60	0	23.32	0

Table 13. Comparison of Detection Delays using CUSUM Filter and FL for Type 1 Failures (FDS 005)

	MLP ANN		EMRAN NN		UKF	
FBU	CUSUM	FL	CUSUM	FL	CUSUM	FL
1	---	0	---	18.12	144.10	0
2	---	0	90.4	0	107.14	0
3	27.94	0	41.62	0	83.10	0
4	16.7	0	28.90	0	70.90	0
5	11.82	0	22.60	0	63.22	0
10	4.84	0	12.82	0	38.18	0
40	1.08	0	2.94	0	10.04	0

Table 14. Comparison of Detection Delays using CUSUM Filter and FL for Type 2 Failures (FDS 001)

	MLP ANN		EMRAN NN		UKF	
FBU	CUSUM	FL	CUSUM	FL	CUSUM	FL
1	---	---	---	---	---	---
2	---	---	149.32	---	---	---
3	175.38	---	106.12	---	---	---
4	107.22	---	87.28	---	---	---
5	90.64	---	75.76	---	---	---
10	55.10	---	49.80	---	122.72	---
40	23.18	---	23.96	---	65.94	---

Table 15. Comparison of Detection Delays using CUSUM Filter and FL for Type 2 Failures (FDS 005)

	MLP ANN		EMRAN NN		UKF	
FBU	CUSUM	FL	CUSUM	FL	CUSUM	FL
1	---	15.72	---	14.36	163.42	---
2	---	15.74	245.28	14.36	132.36	---
3	114.90	15.74	123.86	14.36	114.30	---
4	87.28	15.76	93.06	14.38	105.24	---
5	74.60	15.76	80.90	14.42	97.44	---
10	45.36	19.2	49.9	14.46	73.86	---
40	18.42	21.7	23.66	18.14	42.88	---

An analysis of the Tables 12 – 15 reveals that the failure detection performance of the *MLP ANN+CUSUM* filter combination is similar to that of the *EMRAN NN+CUSUM* filter, with the *EMRAN NN+CUSUM* combination performing slightly better in terms of being able to detect a smaller magnitude failure (2*FBU) earlier than the others, for both Type 1 and Type 2 failures. However, with increasing failure amplitudes the performance of the *MLP ANN+CUSUM* filter improves and is better than the *EMRAN NN+CUSUM* combination. It can also be seen that the average performance of the *UKF+CUSUM* filter combination is poorer than the other analytical redundancy methods, with slower detection times for both Type 1 and Type 2 failures.

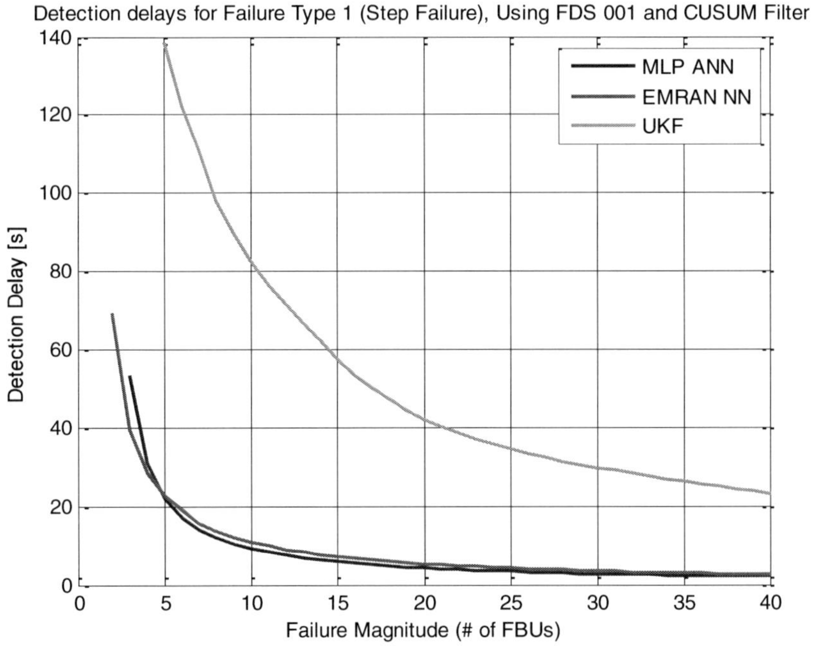

Figure 31. Comparison of failure detection performance of the CUSUM filter, for MLP ANN, EMRAN and UKF approaches - Type 1 failure – Step Bias; Flight Data Set 001.

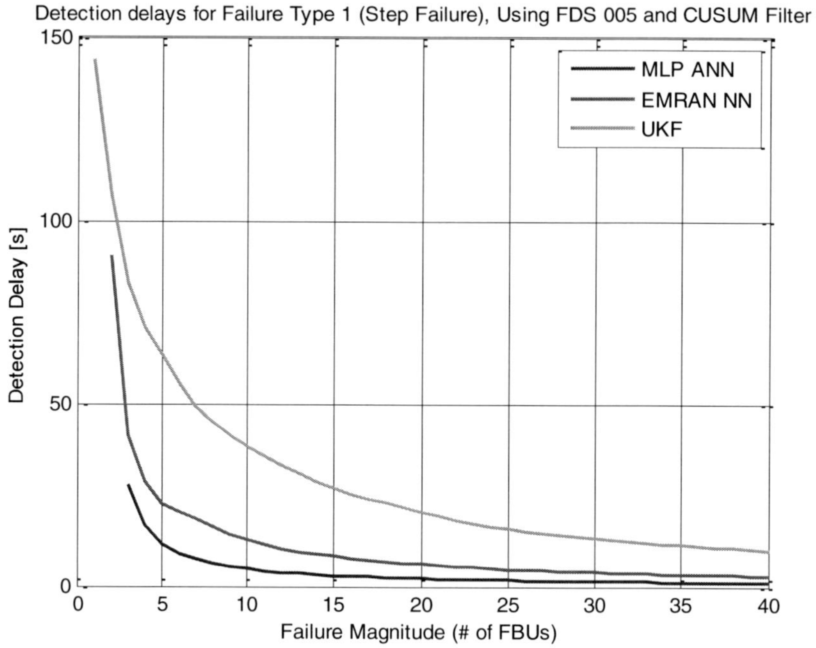

Figure 32. Comparison of failure detection performance of the CUSUM filter, for MLP ANN, EMRAN and UKF approaches - Type 1 failure – Step Bias; Flight Data Set 005

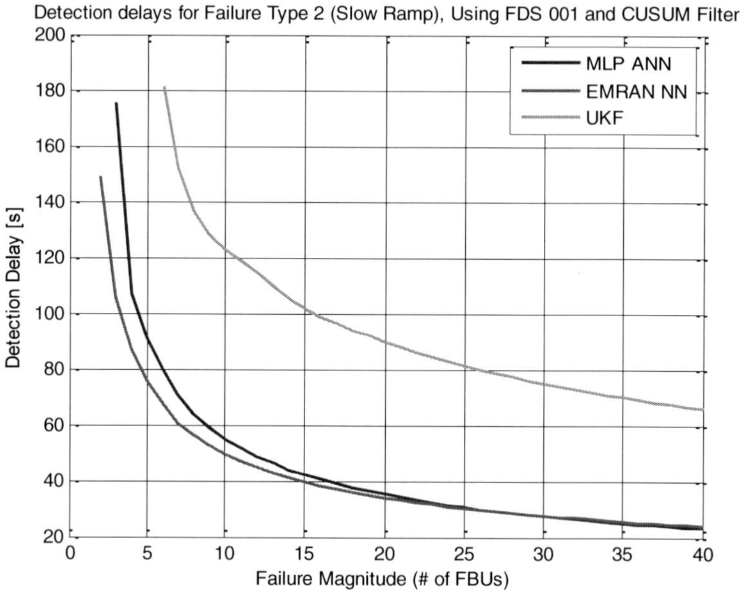

Figure 33. Comparison of failure detection performance of the CUSUM filter, for MLP ANN, EMRAN and UKF approaches - Type 2 failure – Slow Ramp; Flight Data Set 001.

This can be attributed to the higher fixed failure detection threshold, due to the fact that the UKF with FDS 003 has a higher estimation error standard deviation than FDS 001 or FDS 005. At the same time, as the tracking error standard deviation for FDS 003 using the MLP

ANN and EMRAN NN is lower than the same for FDS 001 and FDS 005, the failure detection thresholds are consequently lower, which indicates that the FDI performance is significantly influenced by the estimation performance of the analytical redundancy approaches – the higher the estimation error standard deviation, the poorer the FDI performance. The failure detection delays for both Type 1 and Type 2 failures on FDS 001 and FDS 005 with the fixed thresholds and CUSUM filter are shown in Figures 28 and 29.

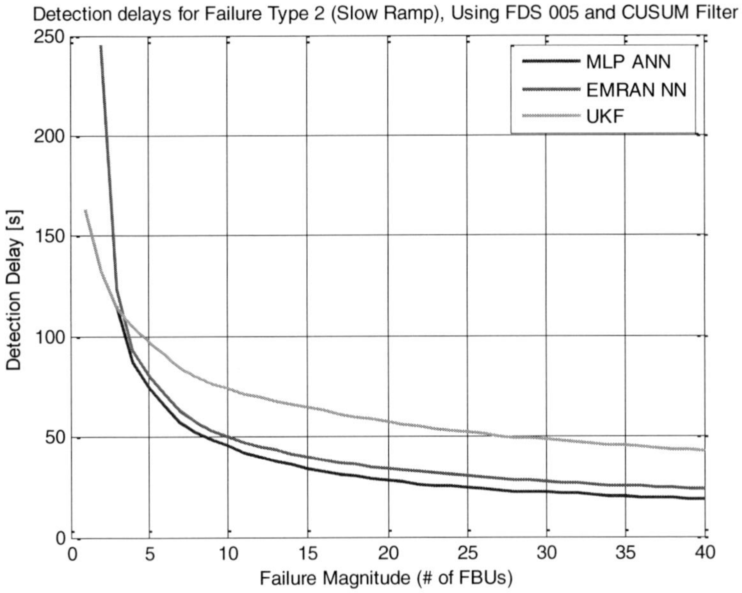

Figure 34. Comparison of failure detection performance of the CUSUM filter, for MLP ANN, EMRAN and UKF approaches - Type 2 failure – Slow Ramp; Flight Data Set 005.

The detection results from the implementation of the Floating Limiter and adaptive threshold based failure detection approach are also presented in Tables 12 – 15. In the implementation of the Floating Limiter based fault detection approach, an analysis of the Tables 12 – 15 reveals that for Type 1 failure, the average performance of the FL is superior to the performance of the CUSUM filter with fixed thresholds. As can be seen from the Table 12-15, with increasing magnitude, from 4*FBU and higher, the FL detects the failure instantaneously for Type 1 failure on data set FDS 001 for all analytical redundancy approaches.

The performance of the FL is better for the EMRAN NN approach, as it is able to detect smaller failures of magnitude 1-3*FBU that the CUSUM does not. This performance is consistent with the FDS 005 where the FL is able to detect failures of all magnitudes, with no delay, except for FDS 005, 1*FBU. For Type 2 failures, the performance of the FL is mixed, with the FL being able to detect slow moving failures only on FDS 005 for all magnitude of failures. For the FDS 001, the Floating Limiter fails to detect any of the slow developing failures irrespective of the magnitude of the failure.

Figures 30 and 31 show the failure detection performance of the FL for a Type 1 failure of magnitude 5*FBU, for data sets FDS 001 and FDS 005 respectively.

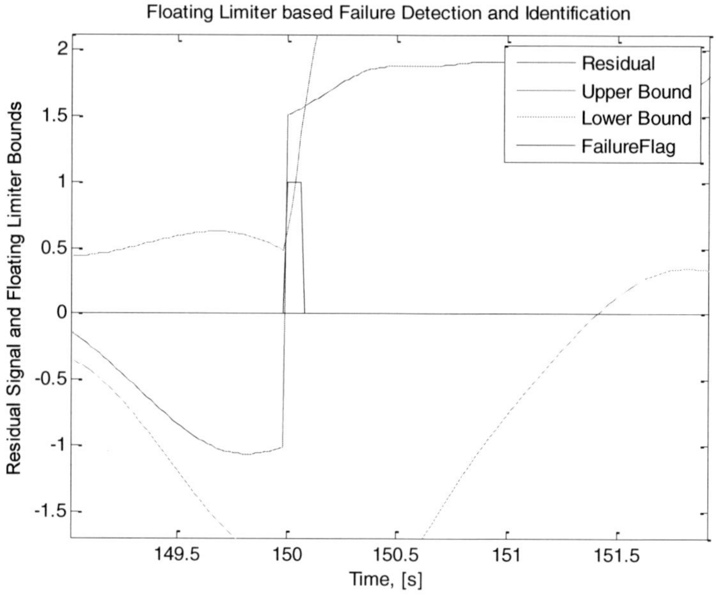

Figure 35. FL Performance, using Airspeed Estimate from MLP ANN. [SB Failure, FDS 001, 5*FBU].

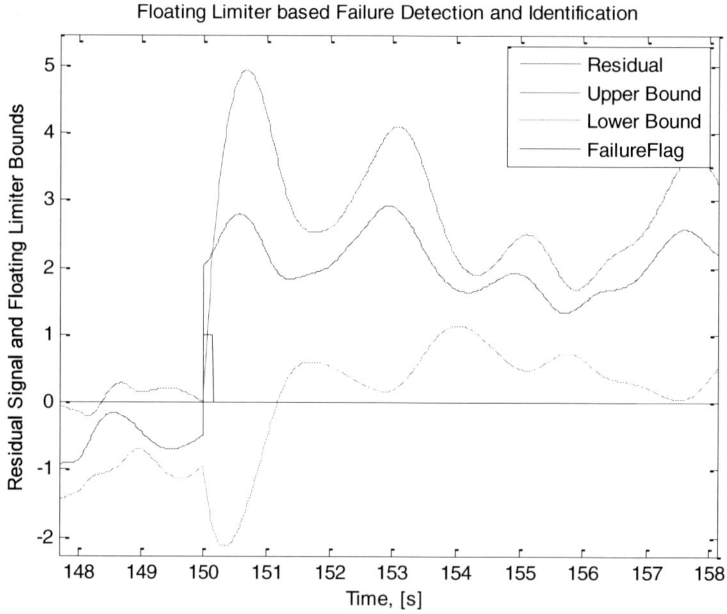

Figure 36. FL Performance, using Airspeed Estimate from MLP ANN. [SB Failure, FDS 005, 5*FBU].

"Model Based" Method - Identification of Model Parameters

Along with the "Model Free" methods, a concurrent study was performed using a "Model Based" approach. In this approach, the first step involved the identification of the 8 parameters W_i of the model in (40) using the approach described in Section IV.D. A segment of 400s (associated with 20,000 samples of data) was extracted from the flight data eliminating the take-off and landing segments of the flight. It should be noted that the model in (40) required the computation of the derivative of the angle of attack, whose signal is known to be affected by noise. It was verified that this derivative term has a small effect on the accuracy of the model estimation and, therefore, it was neglected.

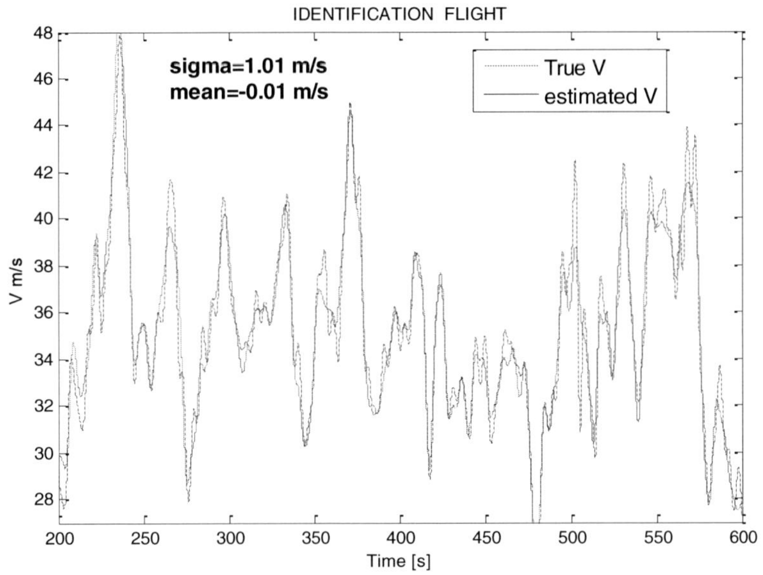

Figure 37. Airspeed velocity estimation for the identification flight. The mean value of the estimation error is $\eta = -0.01$ m/s and the standard deviation is $\sigma = 1.01$ m/s.

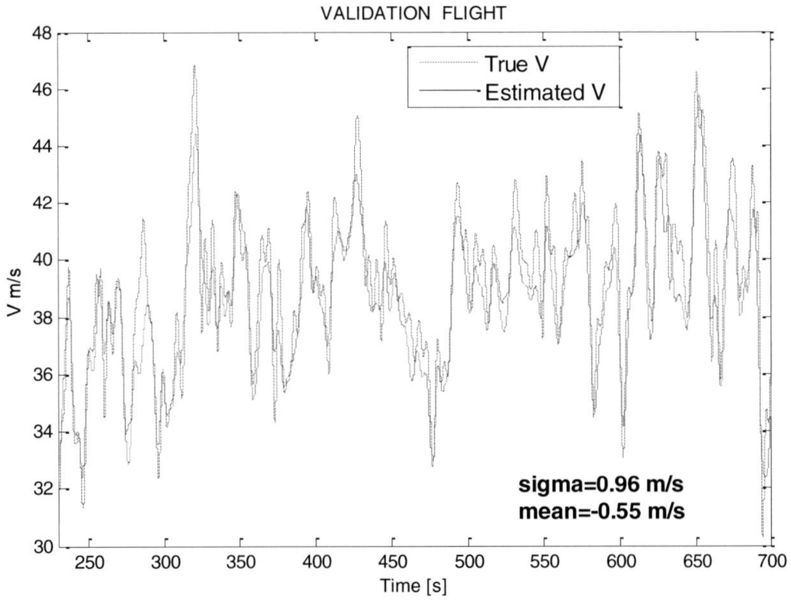

Figure 38. Airspeed velocity estimation for the validation flight. The mean value of the estimation error is $\eta = -0.55$ m/s and the standard deviation is $\sigma = 0.96$ m/s.

Once the model parameters were identified, the airspeed estimates were derived. The result of the estimation process is shown in Figure 39, plotted against the "True" airspeed derived from the Pitot sensor measurement. It can be observed that the estimation produces an almost unbiased estimation with a STD of about 1 m/s.

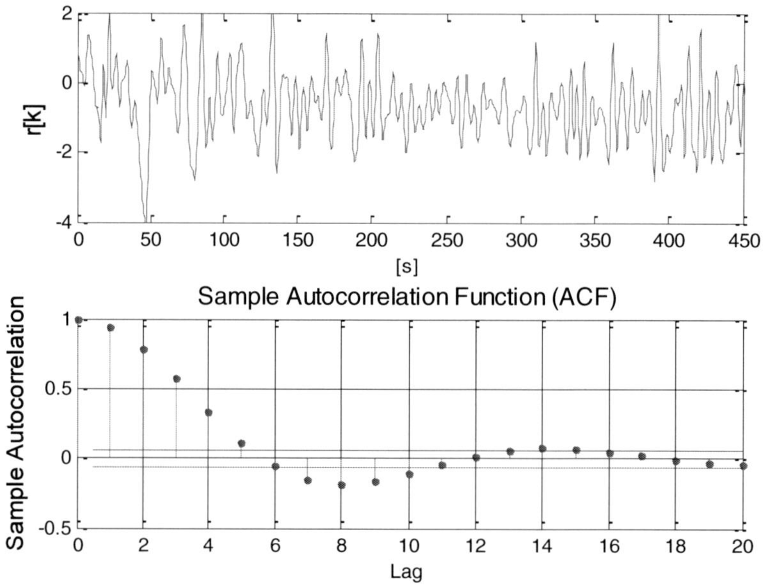

Figure 39. (a) The resampled residual r(k) and (b) The associated autocorrelation function.

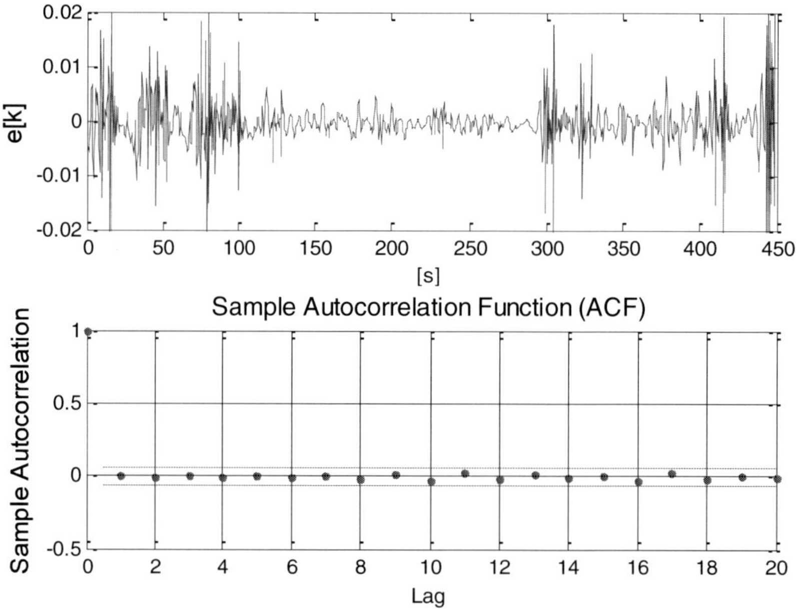

Figure 40. (a) The whitened residual $\varepsilon(k)$ and (b) The associated autocorrelation function.

The identified model was then validated with the data of another flight of similar length and the results are shown in Figure 40. It can be observed that for the validation flight a mean shift of about 0.5 m/s is introduced while the STD of the estimation error remains essentially unchanged. Considering the training and the validation performances it can be concluded that the proposed method provides an accurate estimation of the airspeed velocity for FDI purposes of the WVU-YF22 UAV.

Design of the Whitening Filter

Before designing the whitening filter, the frequency content of the "raw" residual $r(k) = V(k) - \hat{V}(k)$ in fault free condition was analyzed. The residual signal $r(k)$ has a band of about 0.05 Hz; therefore, a down sampling was first applied in order to reduce the number of the delays required by the FIR filter to produce a whitened signal. The down sampling was fixed so that the new sampling rate is T=0.8s. Figure 39(a) shows the resampled residual signal $r(k)$ computed from for the validation flight, and Figure 39(b) shows the corresponding autocorrelation function where a correlation length of about 20 samples is still evident (even after down sampling). The whitening of the resampled $r(k)$ was then carried out applying the approach outlined in Section III-D taking a FIR filter length of N_{FIR}=20 samples. Figure 40(a) shows the whitened residual $\varepsilon(k)$ at the output of the FIR filter while Figure 40(b) shows the corresponding autocorrelation function, where the almost whiteness of the signal $\varepsilon(k)$ is quite evident. Finally, Figure 41 shows the amplitude distribution for the whitened residual $\varepsilon(k)$ where it can be seen that distribution fits a Gaussian shape with sufficient accuracy and therefore, it can be assumed that the signal $\varepsilon(k)$ satisfies the requirement to be a Gaussian and white noise.

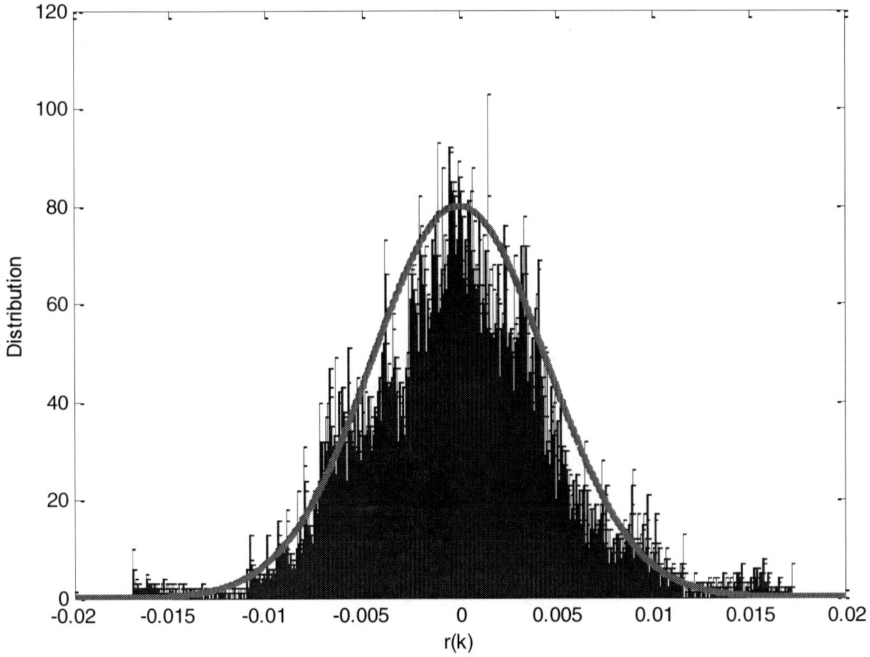

Figure 41. The experimental distribution function of the whitened residual $\varepsilon(k)$ and its Gaussian approximation.

Design of the GLRT Detector

The performance of the GLRT detector depends on some important parameters such as the length of the windows W_0 and W_1 and on the desired probability of false alarm α. While estimations based on very large time windows could produce reliable results, they could also lead to large detection delays; on the other hand, time windows that are too short could potentially detect faults very quickly, but also carry the potential of a high rate of false alarms.

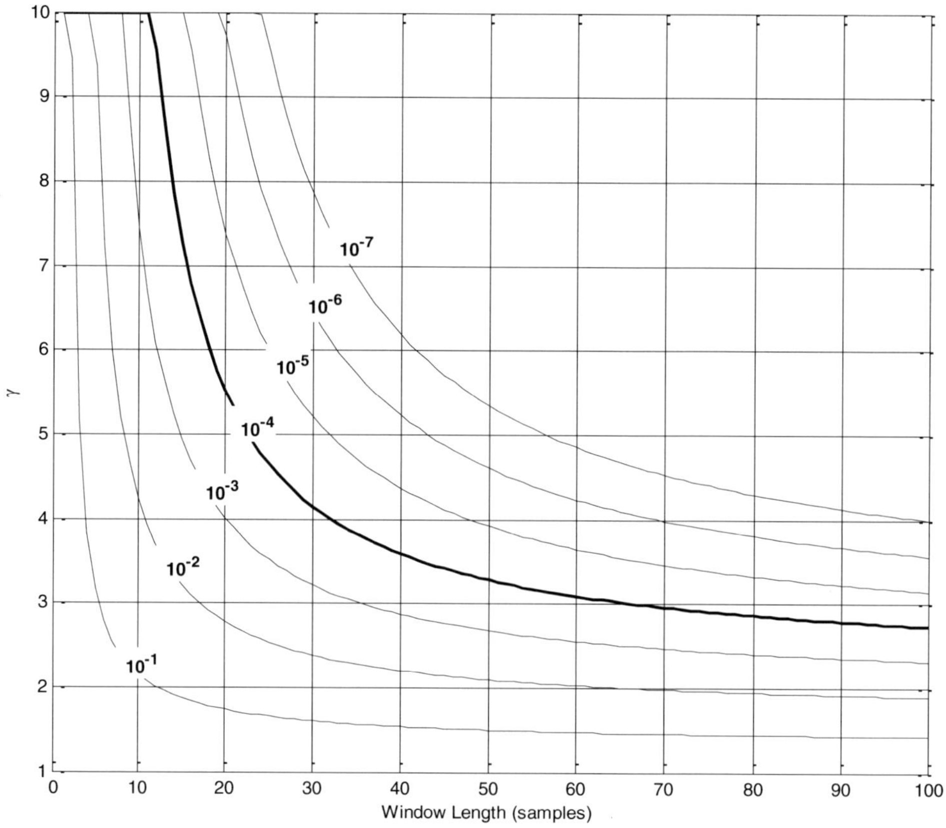

Figure 42. The detection threshold γ as function of length m of the moving window W_I for false alarm probabilities ranging from 10^{-1} to 10^{-7}. The length of the window W_0 is fixed and equal to n=100 samples.

In this study the estimation of the fault free variance $\hat{\sigma}_0^2$ under H_0 hypothesis was computed at the beginning of the flight when we can reasonably assume that the fault has not occurred yet. The initial time k_0 was assumed to be 50s while the window length of W_0 was fixed at n=100 samples in order to get a reliable estimation of $\hat{\sigma}_0^2$. As for the sliding window W_1 that is used for the online estimation of $\hat{\sigma}_1^2$ it is evident that its length has a significant effect on the detection performance as well as the selection of the detection threshold; therefore, the optimal window length W_1 was defined following a parametric study by varying its length (m samples) for different values of the desired probability of false alarm α. Figure 42 shows the value of the threshold γ for different lengths of the moving window W_1 for different probability of false alarms P_{FA}, assuming a fixed length for the window W_0: m=100. The curves in Figure 42 were derived from the F-distribution $F_{n-1,100-1}$.

From Figure 42 it can be observed that that for a fixed level of the probability of false alarm α, increasing the window length m results in a reduction of the detection threshold γ, while the reduction of the desired false alarm probability implies an obvious increase of the threshold γ. The above abacus is particularly useful to guide the designer in the selection of a suitable length for the window W_1 and of the associated value of γ. Within the scope of this chapter, the experiments were conducted with a fixed probability of false alarm at $P_{FA}=10^{-4}$ and a window W_1 of length m=30, resulting in a detection threshold of $\gamma = 4.16$.

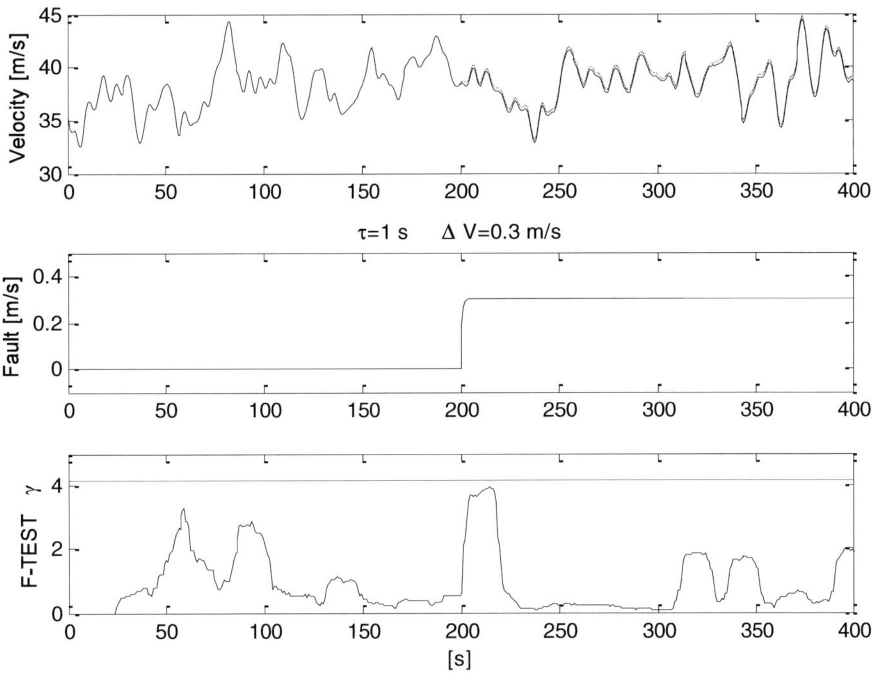

Figure 43. (Top) Velocity signal $V(k)$ and faulty signal $V_{faulty}(k)$ in red. Fault time is 200s. (Middle) The shape of the faulty signal (τ=1 s, ΔV=0.3 m/s). (Bottom) The result of the online F test and the detection threshed.

Sensor Fault Detection Results

Based on the detection threshold and window length settings described in the previous paragraph, the GLRT detector was implemented and an extensive simulation study was performed by injecting additive faults of variable amplitude and shape according to the fault model (41), with all simulated faults injected at time k_f =200s. The detector performs the F-test (50) repetitively at each sample time k deciding between the hypotheses H_0 or H_1.

One of the goals of this study was the determination of the "smallest" detectable faults by the GLRT. Figures 43-45 show the fault detection performance for faults of amplitudes ΔV ranging from 0.3 m/s to 3 m/s (0.6 – 6 FBU) with a time constant τ =1s.

It can be observed that for this (abrupt) fault the detector is not able to detect the fault with amplitude of 0.3 m/s (0.6 FBU), in fact, in this case the real time F-test output does not exceed the detection threshold; on the other hand, the detector detects, easily, faults of amplitude larger than 1 m/s (2 FBU).

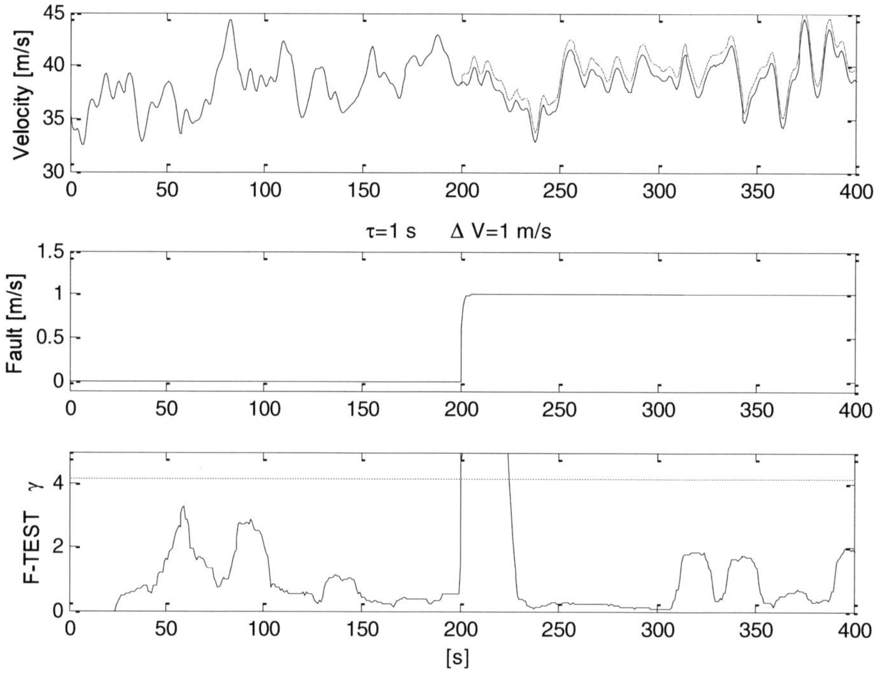

Figure 44. (Top) Velocity signal $V(k)$ and faulty signal $V_{faulty}(k)$ in red. Fault time is 200s. (Middle) The shape of the faulty signal (τ=1 s, ΔV=1 m/s). (Bottom) The result of the online F test and the detection threshed.

"Soft" failures were investigated by injecting additive faults having the same amplitudes of the abrupt failures but with significantly a larger time constant of τ =10s. The results of these simulations in case of "soft" failures are shown in Figures 46-48. As expected, it was observed that the detection performance degraded substantially. In fact, in this case the F-test is unable to detect "soft" faults with amplitude of 1 m/s while it is able to detect soft fault of higher amplitudes (3 m/s).

It can be observed from the time domain responses of the F-test detector that following the detection spike the signal quickly falls under the detection threshold even if the step fault persists. This can be attributed to the fact that the whitening filter is essentially a differentiator - that is $\varepsilon(k) = r(k) - r_{FIR}(k)$ (namely $\varepsilon(k)$ is the difference between $r(k)$ and its one step ahead prediction $r_{FIR}(k)$). As a consequence, as soon as the fault amplitude reaches an almost steady state value, its effect by default disappears from the $\varepsilon(k)$ residual signal. The duration of the detection spike is strictly related to the length of the moving window W_1 - the larger it is, the longer is the duration of the detection spike.

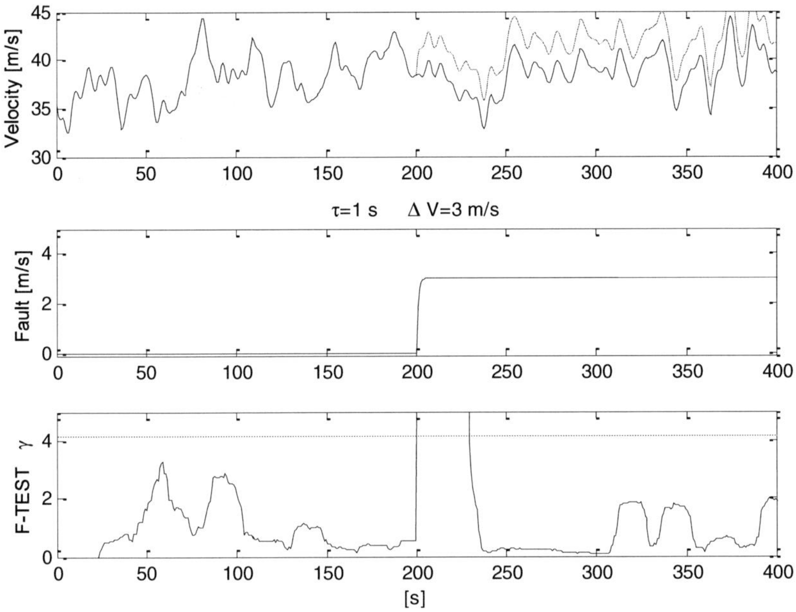

Figure 45. (Top) Velocity signal *V*(*k*) and faulty signal $V_{faulty}(k)$ in red. Fault time is 200s. (Middle) The shape of the faulty signal (τ=1 s, ΔV=3 m/s). (Bottom) The result of the online F test and the detection threshed.

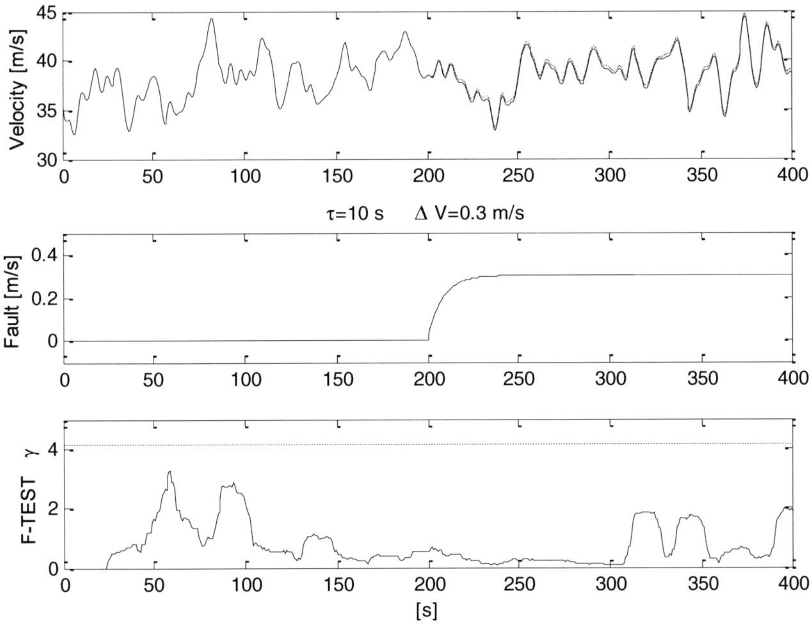

Figure 46. (Top) Velocity signal *V*(*k*) and faulty signal $V_{faulty}(k)$in red. Fault time is 200s. (Middle) The shape of the faulty signal(τ=10 s, ΔV=0.3 m/s). (Bottom) The result of the online F test and the detection threshed.

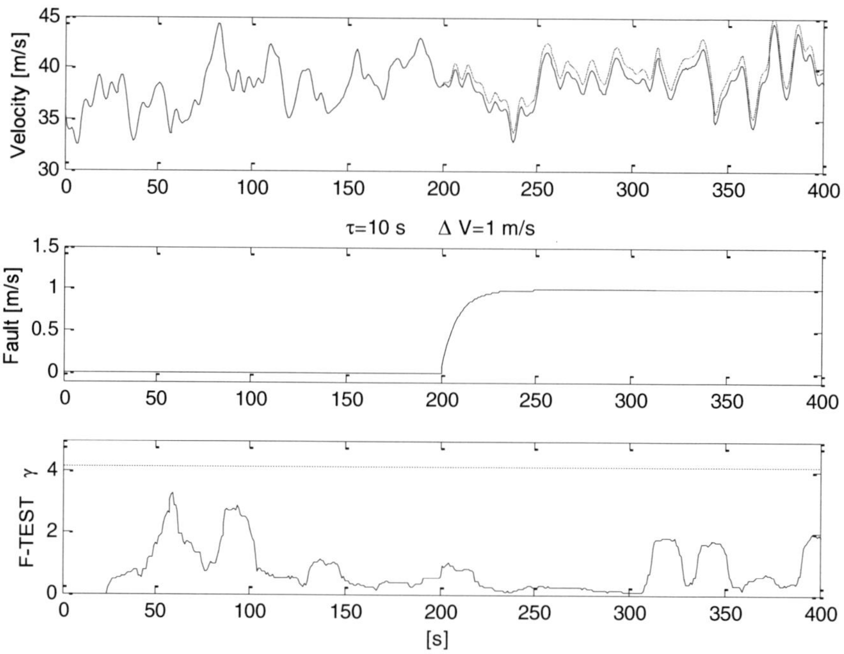

Figure 47. (Top) Velocity signal $V(k)$and faulty signal $V_{faulty}(k)$ in red. Fault time is 200s. (Middle) The shape of the faulty signal(τ=10 s, ΔV=1 m/s). (Bottom) The result of the online F test and the detection threshed.

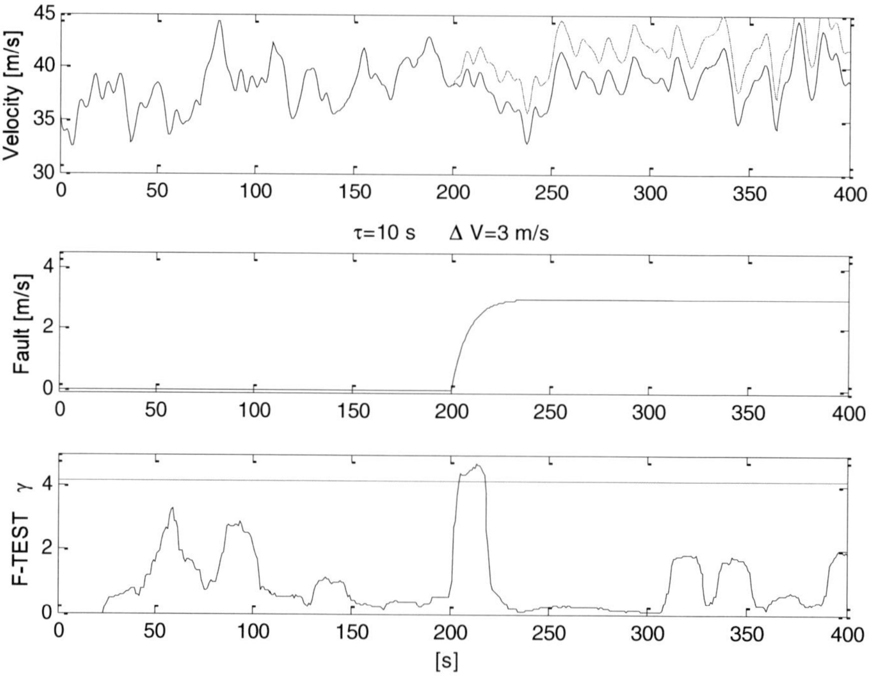

Figure 48. (Top) Velocity signal $V(k)$ and faulty signal $V_{faulty}(k)$ in red. Fault time is 200s. (Middle) The shape of the faulty signal(τ=10 s, ΔV=3 m/s). (Bottom) The result of the online F test and the detection threshed.

Another critical parameter that qualify the performance of a detector is the detection delay, that is the time that passes between the beginning of a faults and it detection (a fault is detected in case the F-test signal exceeds the threshold γ).

For the evaluation of the detection delay, a parametric study was conducted by varying the amplitude of the fault and its time constant. Table 16 reports the results of this study for faults beginning a time k_f =200 s. It can be observed that in case of hard failures (time constants τ <1 s) the detector is able to quickly detect faults, including faults of very small magnitudes (less than 1 m/s). On the other hand, "soft" failures are harder to detect; for instance, for a fault having a time constants τ = 10 s, it is only possible to detect faults of amplitudes larger than 3 m/s (with a detection time of 205.6 s). For a given value of the fault time constant, the detection time decreases with increasing fault amplitude. In Table 16, the empty boxes indicate that the corresponding faults were not detected by the GLRT. As a final consideration, it should be emphasized that the above performance are strictly dependent on the selection of the probability of false alarms, which was fixed at 10^{-4} in this study.

Table 16. Detection delay (the fault time is k_f=200s) expressed in seconds for different values of the fault amplitudes and of the fault time constant. An empty box indicates that for that fault the scheme was unable to detect the fault

		Fault Amplitude ΔV [m/s]													
		0.5	0.8	1	2	3	4	5	6	7	8	9	10	15	20
	0.1	---	0.6	0.8	0.8	0.0	0.0	0.0	0.0	0.0	0.0	0.0	0.0	0.0	0.0
Fault Time Const. τ[s]	0.5	---	1.6	0.8	0.8	0.0	0.0	0.0	0.0	0.0	0.0	0.0	0.0	0.0	0.0
	0.8	---	1.6	1.6	0.8	0.0	0.0	0.0	0.0	0.0	0.0	0.0	0.0	0.0	0.0
	1	---	1.6	1.6	0.8	0.0	0.0	0.0	0.0	0.0	0.0	0.0	0.0	0.0	0.0
	2	---	---	2.4	1.6	0.8	0.8	0.0	0.0	0.0	0.0	0.0	0.0	0.0	0.0
	5	---	---	---	---	2.4	1.6	0.8	0.8	0.8	0.0	0.0	0.0	0.0	0.0
	6	---	---	---	---	3.2	1.6	1.6	0.8	0.8	0.8	0.0	0.0	0.0	0.0
	7	---	---	---	---	4.0	1.6	1.6	1.6	0.8	0.8	0.8	0.8	0.0	0.0
	8	---	---	---	---	4.0	3.2	1.6	1.6	1.6	0.8	0.8	0.8	0.8	0.0
	10	---	---	---	---	5.6	4.8	1.6	1.6	1.6	1.6	0.8	0.8	0.8	0.0
	15	---	---	---	---	---	---	---	3.2	1.6	1.6	1.6	1.6	1.6	0.8
	20	---	---	---	---	---	---	---	---	4.8	3.2	1.6	1.6	1.6	1.6
	30	---	---	---	---	---	---	---	---	---	---	---	4.0	3.2	1.6
	40	---	---	---	---	---	---	---	---	---	---	---	---	---	2.4
	50	---	---	---	---	---	---	---	---	---	---	---	---	---	4.0

CONCLUSION

A Sensor Failure Detection Identification and Accommodation (SFDIA) approach was developed using data from a jet powered research UAV to address the specific problem of detecting and accommodating for common mode failures on the aircraft Air Data System. Within the framework of the SFDIA, three distinct Sensor Failure Accommodation (SFA) approaches based on analytical redundancy were designed and evaluated. The first approach was based on an Unscented Kalman Filter (UKF) formulation; the other approaches were based on a generic Multi-Layer Perceptron (MLP) ANN with a sigmoidal activation function, and an Extended Minimal Resource Allocating Network (EMRAN) ANN. During the training phase for the ANNs, the MLP ANN showed the best performance compared to the EMRAN ANN and the UKF in terms of the mean of the tracking error; however, the standard deviations in both NN based approaches were of the same order of magnitude, while the UKF approach did not perform as well. Based on the airspeed estimates from the analytical redundancy approaches using Flight Data Set 003 as the reference, two Failure Detection and Identification methods were evaluated. The first was based on the *CUSUM* filter and fixed failure detection thresholds and the second was based on a Floating Limiter with adaptive failure detection thresholds.

Two failures, a Sudden Bias and a Slow Ramp Bias were artificially injected on the flight data and the performance of the failure detection approaches evaluated. From the experiments and given the nature of the failures that were considered, the detection performance of the Floating Limiter with all the three analytical redundancy approaches was better than that of the *CUSUM* for Step Bias failures.

In case of Slow Ramp Bias failures, the *CUSUM* filter performed better for the NN base analytical redundancy approaches, but was unable to detect any failure, irrespective of the magnitude for the UKF.

A concurrent model based effort was also performed with a rigorous design of the fault detector that guarantees a-priori defined probabilistic performance of false alarm. Within this study, the airspeed estimator was designed following a standard LS estimation performed on a batch of flight data. Results from training and validation sets of flight data have shown that the proposed method is able to estimate the airspeed with mean bias of about 0.5 *m/s* and standard deviation of approximately 1 *m/s*. To facilitate the application of powerful and well known probabilistic tests to the problem of FDI, the raw residual was pre-whitened to satisfy the requirement of Gaussian white noise.

A detailed analysis was performed to set-up suitable parameters for the GLRT detector that, in this study, is a sequential F-test, designed to detect changes in the variance of the residual signal. In particular, the effect of the sliding window length in conjunction with the variation of the desired probability of false alarm was analyzed and allowed a rigorous computation of the detection threshold to be used by GLRT detector, to guarantee a probability of false alarms which was fixed at 10^{-4} in this study.

The results of the experiments with artificially injected hard and soft failures of different amplitudes and different time rise on the airspeed signal have shown that the scheme is able to promptly detect hard faults of amplitudes less than 1 *m/s*, while in case of faults having a arise time of about 30s.

ACKNOWLEDGMENTS

This work was partially supported by the NASA Aviation Safety Project, Grant # NNX10AI14G.

REFERENCES

[1] Williams-Hayes, P. S., "Flight Test Implementation of a Second Generation Intelligent Flight Control System," *NASA Technical Report TM-2005-213669*, 2005.

[2] Maine, T., Burken, J., Burcham, F. and Schaefer, P. "Design challenges encountered in a propulsion controlled aircraft flight test program," *30th AIAA/ASME/SAE/ASEE Joint Propulsion Conference*, Indianapolis, IN, Jun. 27-29, 1994.

[3] Edwards, C., Lombaerts, T., and Smaili, H., "Fault Tolerant Flight Control: a Benchmark Challenge," *Lecture Notes in Control and Information Sciences*, Vol. 399, DOI: 10.1007/978-3-642-11690-2_1, 2010.

[4] Goupil, P., "AIRBUS state of the Art and Practices on FDI and FTC in Flight Control System," *Control Engineering Practice*, Vol. 19, issue 6, pp. 524-539, 2011.

[5] Bureau d'Enquêtes et d'Analyses pour la sécurité de l'aviation civile, "Interim report no. 3: on the accident on 1 June 2009 to the Airbus A330-203 registered F-GZCP operated by Air France flight AF 447 Rio de Janeiro – Paris," Report, Jul. 29, 2011.

[6] J. Levine, X-31's loss. NASA X-Press, http://www.nasa.gov/centers/dryden/news/X-Press/stories/2004/013004/new_x31.html. 2004.

[7] McKenna, J.T., "Blocked Static Ports Eyed in Aeroperu 757 Crash," *Aviation Week and Space Technology*, Vol. 145, no. 20, pp. 76, 1996.

[8] Napolitano, M.R., An, Y., and Seanor, B., "A fault tolerant flight control system for sensor and actuator failures using neural networks," *Aircraft Design*, Vol. 3, issue 2, pp. 103-128, 2000.

[9] G. Campa, M.L. Fravolini, B. Seanor, M.R. Napolitano, D. Del Gobbo, G. Yu, S. Gururajan, "On-Line Learning Neural Networks for Sensor Validation for the Flight Control System of a B777 Research Scale Model," *International Journal of Robust and Nonlinear Control*, vol. 12, no. 11, pp. 987-1007, 2002.

[10] M.G. Perhinschi, M.R. Napolitano, G. Campa, M.L. Fravolini, B. Seanor, "Integration of Sensor and Actuator Failure Detection, Identification, and Accommodation Schemes Within Fault Tolerant Control Laws," *Control and Intelligent Systems*, vol. 35, no. 4, pp. 309-318, 2007.

[11] Heredia, G., Ollero, A., Mahtani, R., Bejar, M., Remuss, V., Musial, M., "Detection of Sensor Faults in Autonomous Helicopters," *Proceedings of the 2005 IEEE International Conference on Robotics and Automation*, 18-22 April 2005. doi: 10.1109/ROBOT.2005.1570444

[12] Samy, Ihab, Ian Postlethwaite, and Dawei Gu. "Neural network based sensor validation scheme demonstrated on an unmanned air vehicle (UAV) model," *47th IEEE Conference on Decision and Control*, 2008.

[13] Neppach, C., and Casdorph, V. (1995), "Sensor Failure Detection, Identification and Accommodation in a System Without Sensor Redundancy," *AIAA 33rd Aerospace Sciences Meeting and Exhibit*, Reno, NV, 9–12 January 1995.

[14] Major McLaren, S., "Velocity Estimate Following Air Data System Failure," Thesis. Air Force Institute of Technology, United States Air Force, 2009. Report No.: AFIT/GAE/ENY/08-M21.

[15] Samy, Ihab, Gu, Da-Wei, "Fault Detection and Flight Data Measurement: Demonstrated on Unmanned Air Vehicles using Neural Networks," *Lecture Notes in Control and Information Sciences*, Vol. 419, 2011.

[16] Fravolini, M. L., Pastorelli, M., Pagnottelli, S., Valigi, P., Gururajan, S., Chao, H., Napolitano, M.R., "Model-Based approaches for the Airspeed Estimation and Fault Monitoring of an Unmanned Aerial Vehicle," *2012 IEEE Workshop on Environmental, Energy, and Structural Monitoring Systems*, Perugia, Italy, September 2012.

[17] Napolitano, M.R. "Development of Formation Flight Control Algorithms Using 3 YF-22 Flying Models," AFOSR Final Report, AFOSR Grant F49620-01-1-0373, April 2005.

[18] Campa, G., Gu, Y., Seanor, B., Napolitano, M., Pollini, L., Fravolini, M.L., "Design And Flight Testing Of Nonlinear Formation Control Laws," *Control Engineering Practice*, vol.15, no. 9, pp.1077-1092, 2007.

[19] Gururajan, S., Gu, Y., Seanor, B., Prucz, J., Napolitano, M., "Evolution of the Flight Testing Program at West Virginia University in Support of Flight Control Research," *52nd Israel Annual Conference on Aerospace Sciences*, Tel Aviv, Israel, Feb 29 – Mar 1, 2012.

[20] Haykin, S., "*Neural Networks: A comprehensive Foundation*," McMillan College Publishing Company, Inc., 1994.

[21] Neural Network Toolbox 7– User's Guide. The Mathworks Inc., Natick, MA – 01760-2098.

[22] Lu Y., Sundararajan N., Saratchandran P., "Analysis of Minimal Radial Basis Function Network Algorithm for Real Time Identification of Nonlinear Dynamic Systems," *IEEE Proceedings on Control Theory and Application* 2000, vol.4, no 147, pp.476.

[23] Campa G., Fravolini M. L., Napolitano M. R., "A Library of Adaptive Neural Networks for Control Purposes," *IEEE International Symposium on Computer Aided Control System Design*, September 18-20, 2002 Glasgow, Scotland, UK.

[24] Fravolini M.L., Campa G., Napolitano M.R., La Cava M., "Comparison of Different Growing Radial Basis Functions Algorithms for Control Systems Applications," *Proceedings of the 2002 American Control Conference*, Anchorage AK, May 2002.

[25] Nguyen, D., and B. Widrow, "Improving the learning speed of 2-layer neural networks by choosing initial values of the adaptive weights," *Proceedings of the International Joint Conference on Neural Networks*, Vol. 3, 1990, pp. 21–26.

[26] Klein, V., and Morelli, E. A., *Aircraft System Identification: Theory and Practice*, American Institute of Aeronautics and Astronautics, Inc., Reston, VA, 2006.

[27] Kalman, R. E., and Bucy, R. S., "New Results in Linear Filtering and Prediction Theory," *Journal of Basic Engineering (Trans. of ASME)*, Vol. 83, 1961, pp. 95-108.

[28] Julier, S. and Uhlmann, J., "A New Extension of the Kalman Filtering to Non Linear Systems," *SPIE Proceedings Series*, 1997, Vol. 3068, pp. 182-193.

[29] Rhudy, M., Gu, Y., Gross, J., Gururajan, S., and Napolitano, M. R., "Sensitivity and Robustness Analysis of EKF and UKF Design Parameters for GPS/INS Sensor Fusion," *AIAA Journal of Aerospace Information Systems*, Vol. 10, No. 3, March 2013, pp. 131-143.

[30] Lewis, F. L., and Syrmos, V. L., *Optimal Control*, 2nd Ed., Wiley, NY, 1995.

[31] van der Merwe, R., Wan, E. and Julier, S., "Sigma-Point Kalman Filters for Nonlinear Estimation and Sensor Fusion- Applications to Integrated Navigation," *AIAA Guidance, Navigation and Control Conference*, Providence, RI, 2004.

[32] Napolitano, M.R., "Development of Formation Flight Control Algorithms Using 3 YF-22 Flying Models," AFOSR Final Report, AFOSR Grant F49620-01-1-0373, April 2005.

[33] Perhinschi, M.G., Napolitano, M.R., Campa, G., Seanor, B., Burken, J., Larson, R., "Design of safety monitor schemes for a fault tolerant flight control system," *IEEE Transactions on Aerospace and Electronic Systems*, Vol. 42, no.2, pp. 562- 571, 2006.

[34] Perhinschi M., Napolitano M. R., Seanor B., Burken J., Larson R., "An Adaptive Threshold Approach for Actuator failure Detection and Identification," *IEEE Transactions on Control Systems Technology*,Vol.14, No.3, May 2006.

[35] M. Basseville, I Nikiforov, (1993) "Detection of abrupt changes - theory and practice," Prentice-Hall. Englewood Cliffs, New York.

[36] Collinson, R.P.G. , "Fly-by-wire flight control," *Computing & Control Engineering Journal* , vol.10, no.4, pp.141-152, Aug. 1999. doi: 10.1049/cce:19990403

[37] Hornik, K., Stinchcombe, M., White, H., "Multilayer feedforward networks are universal approximators," *Neural Networks*, Volume 2, Issue 5, pp. 359-366, 1989.

[38] Frank P.M., "Fault Diagnosis in Dynamic Systems Using Analytical and Knowledge-Based Redundancy: A Survey and Some New Results," Automatica, vol. 26, no. 3, pp. 459-474, 1990.

[39] Hajiyev C.M., Caliskan F., "Fault detection in flight control systems via innovation sequence of Kalman filter," *Lecture Notes in Control and Information Sciences*, vol. 243, pp 63-74, 1999.

[40] Mammarella M., Campa G., Napolitano M., Fravolini, M.L., Perhinschi, M.G., Gu Y. "Machine Vision / GPS Integration Using EKF for the UAV Aerial Refueling Problem," *IEEE Transactions on Systems*, Man, and Cyber and Cybernetics, Part C: Applications and Reviews, vol. 38, no. 6, pp. 791-801, 2008.

[41] Frank P.M., Ding X., "Survey of robust residual generation and evaluation methods in observer-based fault detection systems," Journal of process control, vol.7, no.6, pp. 403–424 1997.

[42] Shields D.N., Ashton S.A., Daley A., "Design of nonlinear observers for detecting faults in hydraulic sub-sea pipelines," *Control Engineering Practice*, vol. 9, no. 3, pp. 297-311, 2001.

[43] Talebi H.A., Khorasani K., Tafazoli S., "A Recurrent Neural-Network-Based Sensor and Actuator Fault Detection and Isolation for Nonlinear Systems With Application to the Satellite's Attitude Control Subsystem," *IEEE Trans. on Neural Networks*, vol. 20, no.1, pp.45-60, 2009.

[44] Farrell J., Berger T., Appleby B.D., "Using learning techniques to accommodate unanticipated faults," *IEEE Control Systems*, vol. 13, no. 3, pp. 40-49, 1993.

[45] [45]Stevens, B., Lewis, F., "*Aircraft Control and Simulation*," John Wiley & Sons, NY, 1992.

[46] Campa G., Gu Y., Seanor B., Napolitano M.R., Pollini L., Fravolini M.L., "Design And Flight Testing Of Nonlinear Formation Control Laws," *Control Engineering Practice*, vol.15, no. 9, pp.1077-1092, 2007.

[47] Samara P.A., Fouskitakis G.N., Sakellariou J.S., Fassois S.D., "A Statistical Method for the Detection of Sensor Abrupt Faults, in Aircraft Control Systems," *IEEE Trans. Ctr. Sys Technology,* vol. 16. no. 4, pp. 798-799, 2008.

[48] Basseville M., Nikiforov I., "Detection of abrupt changes - theory and practice," Prentice-Hall. Englewood Cliffs, New York, 1999.

[49] Osadchy, M., Lindenbaum, M., Jacobs, D., "Whitening for Photometric Comparison of Smooth Surfaces under Varying Illumination," *Lecture Notes in Computer Science*, Vol. 3. pp. 217-228, 2004.

[50] Kay, S., "*Fundamentals of Statistical Signal Processing – Detection Theory*," Prentice Hall. 1998.

[51] Belcastro, C.M., Foster, J.V., "Aircraft Loss-of-Control Accident Analysis," *AIAA Guidance, Navigation, and Control Conference*, Toronto, Canada, Aug. 2-5, 2010.

[52] S. Hansen, M., Blanke, B., and Adrian, J., "Diagnosis of UAV Pitot Tube Defects Using Statistical Change Detection," In *proceedings of 7th Symposium on Intelligent Autonomous Vehicles*, 2010.

In: Fault Detection
Editors: F. P. G. Márquez and M. Papaelias
ISBN: 978-1-62808-999-8

Chapter 6

H_∞ Fault Detection and Diagnosis of a Class of Descriptor Systems under Feedback Control

Jun Xu*
Advanced Servo Technology, Western Digital, Singapore

Abstract

This chapter presents an enhanced approach to the H_∞ fault detection and diagnosis (FDD) problem for a class of descriptor systems. A new dynamic output feedback control structure via an elegant construction with adjustable parameters, which can simultaneously design the control input signal and the FDD signal, is proposed. In other words, the state and fault signal can be estimated at the same time with H_∞ performance tolerance. The design procedure is based on linear matrix inequality (LMI) techniques, which is numerically efficient. Some discussions are also given for the extension of this work.

Keywords: Fault detection and diagnosis, linear matrix inequalities, H_∞ performance, output feedback control, descriptor systems

AMS Subject Classification: 93B, 93C, 93D, 15A, 15B.

1. Introduction

The fault detection and diagnosis (FDD) problem has received considerable attention, as modern engineering systems are large-scale and sophisticated, and safety and reliability are of practical concerns. Many FDD approaches have been suggested in both frequency and time domains for both deterministic and stochastic systems (see, e.g., the survey papers [53, 51, 40, 38, 39, 9, 33, 18] and books [3, 30, 4, 34, 7, 19, 5, 15]. for a detailed classification and discussion). For example, Patton et al. [30] classified them into three types: (i) model-base (ii) knowledge-based (iii) empirical or signal processing approaches. In [49] Xu et al. simply divided them into two categories: data-based approach such as fuzzy logic

*E-mail addresses: jun.xu@wdc.com, jun.xu99@gmail.com

and neural network methods and statistical methods [5, 39, 52], and model-based approach such as high-gain observer, unknown input observer (UIO) [5, 6, 32, 44, 41, 23, 28], nonlinear adaptive estimator, sliding mode observer, extended Kalman filter, and parity space approach [19, 5, 31].

Among these approaches, the model-based fault detection approach is often used when the physical model is known. Particularly, the observer-based design technique, in which an observer plays the key role, is one of the most important and popular methods [45, 47, 49, 19, 5, 46]. Such technique falls into two categories: fault estimation and residual generation [17].

Although thousands of works have been presented in this area, the design of simultaneous controller and observer is very few. In [17], a residual generator is design in the H_∞ and H_- setting, where the controller information is embedded into FDD filter design. However, the design procedure is not LMI-based, unless the controller gain is fixed. Another inherited complexity lies in the threshold design. The results in [29, 36] are in the same category.

In this chapter, we propose a new approach based a descriptor model, which is mainly motivated by [50, 23, 16, 22] and partially by [14, 13, 11]. Descriptor systems are more general and natural than state-space systems in many cases, e.g., electric circuits, large-scale networks, economic systems, mechanical systems (e.g., robotics and aircraft) and other critical fields. Their properties have been extensively studied in the literature, e.g., [11, 42, 12, 14, 10, 26, 43]. In [50], a detailed stability analysis, controller and filter synthesis for descriptor systems are presented. As for the application in fault detection and/or diagnosis, few methods have been proposed. In [14, 10], the design is dedicated to fault filtering. In [23], an LMI-based H_∞ controller formulation is given for fault detection and isolation problems of LTI descriptor systems. The fault estimation signal is imbedded in the H_∞ controller design. In [16], an approach to the design of observers is addressed for simultaneous estimation of the state and unknown input for a class of nonlinear MIMO systems.

However, we propose here a different structure from [23, 16, 22, 14, 10, 26] to construct an augmented descriptor system, and then apply the expanded controlled input signal to design the so-called FDD-based extended dynamic output feedback controller. This controller can provide the controlled input signal and the FDD signal simultaneously. Meanwhile, it can also estimate the fault and state at the same time. Furthermore, the augmented system embeds several adjustable parameters to tune the system performance. As the design is based on linear matrix inequality (LMI) approach, it is efficient.

The remainder of this chapter is organized as follows. In Section 2, the system with unknown input is presented and the design objectives are discussed. The structure of the controller is also given. In Section 3, non-strict and strict LMI conditions are presented to design such controller. In Section 4, a numerical example is given to illustrate our design procedure. Conclusions are drawn in the last section with some possible extensions.

2. Problem Statement

Consider the following general linear descriptor system:

$$E\dot{x} = Ax + B_u u + B_d d + B_f f \tag{1}$$

$$y = Cx + D_u u + D_f f \tag{2}$$

where $x \in \mathcal{R}^n$ is the state vector, $u \in \mathcal{R}^m$ is the input vector, $d \in \mathcal{R}^l$ is the disturbance vector, $f \in \mathcal{R}^\ell$ is the fault vector and $y \in \mathcal{R}^r$ is the output vector. We assume that d and f are L_2 bounded. All matrices are properly dimensioned. $E \in \mathcal{R}^{n\times n}$ and $rank(E) = \imath \leq n$.

We define the following notations for future use.

- $\hat{x}$ and $\hat{f}$ are the estimation of x and f,respectively;
- the extended state $\xi = \begin{bmatrix} x \\ f - \hat{f} \end{bmatrix} \in \mathcal{R}^{n+\ell}$;
- the unknown input $w = \begin{bmatrix} d \\ f \end{bmatrix} \in \mathcal{R}^{l+\ell}$,
- the controlled input $\bar{u} = \begin{bmatrix} u \\ \hat{f} \end{bmatrix} \in \mathcal{R}^{m+\ell}$,
- some matrices

$$\bar{A} = \begin{bmatrix} A & \alpha B_f \\ 0 & -\beta I \end{bmatrix} \in \mathcal{R}^{(n+\ell)\times(n+\ell)}, \tag{3}$$

$$\bar{B}_{\bar{u}} = \begin{bmatrix} B_u & \alpha B_f \\ 0 & -\beta I \end{bmatrix} \in \mathcal{R}^{(n+\ell)\times(m+\ell)}, \tag{4}$$

$$\bar{E} = \begin{bmatrix} E & 0 \\ 0 & 0 \end{bmatrix} \in \mathcal{R}^{(n+\ell)\times(n+\ell)}, \tag{5}$$

$$\bar{B}_d = \begin{bmatrix} B_d & (1-\alpha)B_f \\ 0 & \beta I \end{bmatrix} \in \mathcal{R}^{(n+\ell)\times(l+\ell)}, \tag{6}$$

$$\bar{C} = [C\ D_f] \in \mathcal{R}^{r\times(n+\ell)}, \tag{7}$$

$$\bar{D} = [D_u\ D_f] \in \mathcal{R}^{r\times(m+\ell)}, \bar{H} = [H_1\ I] \in \mathcal{R}^{\ell\times(n+\ell)},$$

$$\tilde{A} = \begin{bmatrix} \bar{A} & 0 \\ 0 & -I \end{bmatrix} \in \mathcal{R}^{(n+\ell+r)\times(n+\ell+r)}, \tag{8}$$

$$\tilde{B}_{\bar{u}} = \begin{bmatrix} \bar{B}_{\bar{u}} \\ \bar{D} \end{bmatrix} \in \mathcal{R}^{(n+\ell+r)\times(m+\ell)}, \tag{9}$$

$$\tilde{E} = \begin{bmatrix} \bar{E} & 0 \\ 0 & 0 \end{bmatrix} \in \mathcal{R}^{(n+\ell+r)\times(n+\ell+r)}, \tag{10}$$

$$\tilde{B}_d = \begin{bmatrix} \bar{B}_d \\ 0 \end{bmatrix} \in \mathcal{R}^{(n+\ell+r)\times(l+\ell)}, \tag{11}$$

$$\tilde{C} = [\bar{C}\ I] \in \mathcal{R}^{r\times(n+\ell+r)}, \tag{12}$$

$$\tilde{H} = [\bar{H}\ 0] \in \mathcal{R}^{\ell\times(m+\ell+r)}, \tag{13}$$

where α,β are scalars and H_1 is a weight matrix, such that $\|H_1\| \le \theta$, $0 \le \theta < 1$.

Remark 1 *α and β are adjustable parameters that can be used to tune the system performance. As we can see from the examples later that these parameters affect the filter gain. However, for convenience, we may simply set $\alpha = \beta = 1$ for the first attempt during design procedure.*

Now the system (1)-(2) can be rewritten as

$$\bar{E}\dot{\xi} = \bar{A}\xi + \bar{B}_{\bar{u}}\bar{u} + \bar{B}_d w \tag{14}$$

$$y = \bar{C}\xi + \bar{D}\bar{u} \tag{15}$$

The signal to be estimated is

$$z = \bar{H}\xi \tag{16}$$

Remark 2 *When $H_1 \neq 0$, the estimated signal is influenced by the state x. If z includes the effect from $\bar{u}$, then there exists a $\bar{G} = [G_1\ G_2]$, such that $z = \bar{H}\xi + \bar{G}\bar{u}$.*

We consider the following FDD-based extended dynamic output feedback controller:

$$\begin{aligned} E_K\dot{\hat{\xi}} &= A_K\hat{\xi} + B_K y \\ \bar{u} &= C_K\hat{\xi} \end{aligned} \tag{17}$$

Hence the closed-loop system can be represented as

$$\begin{aligned} E_c\dot{\varsigma} &= A_c\varsigma + B_c w \\ z &= C_c\varsigma \end{aligned} \tag{18}$$

where

$$\varsigma = \begin{bmatrix} \xi \\ \hat{\xi} \end{bmatrix}, E_c = \begin{bmatrix} \bar{E} & 0 \\ 0 & E_K \end{bmatrix}, A_c = \begin{bmatrix} \bar{A} & \bar{B}_{\bar{u}}C_K \\ B_K\bar{C} & A_K + B_K\bar{D}C_K \end{bmatrix}$$

$$B_c = \begin{bmatrix} \bar{B}_d \\ 0 \end{bmatrix}, C_c = \begin{bmatrix} \bar{H} & \bar{G}C_K \end{bmatrix}$$

Our objective is to design a controller (17) such that

1. The closed-loop system (18) is admissible (stabilized);
2. The H_∞ norm of the closed-loop transfer function from w to z

$$G_c(s) = C_c(sE_c - A_c)^{-1}B_c \tag{19}$$

is less than γ.

Remark 3 *As stated in [22], the advantage of the simultaneous design of FDD filter and dynamic output feedback controller is that it does not consider control and FDD design as contrary objectives. However, unlike the approach in [22], wherein the filter is of order n, the method here gives order $n + \ell$.*

In our design, the extended state ξ has both state x and $f - \hat{f}$, and the "noise" w includes both disturbance d and fault f. The design objective (1) actually shows that $\hat{f}$ will converge to f in the case $w = 0$. In [23], without $f - \hat{f}$ embedded into extended state, the fault estimation may not converge. Meanwhile, as $\bar{u}$ includes $\hat{f}$, we can estimate the fault vector simultaneously. The objective (2) gives a kind of fault-tolerant properties, i.e., the estimated signal will be bounded with respect to w.

3. Design Procedure

From the discussion of last section, we can see that the FDD filter design problem actually has been converted into a dynamic output feedback controller design problem. There are some results related to this problem, e.g., [37, 50].

Lemma 1 *Given scalars $\gamma > 0$, α and β and the system (14)-(16). There exists a FDD-based extended dynamic output feedback controller (17) such that the closed-loop system (18) is admissible and the transfer function (19) satisfies $\|G_c\|_\infty < \gamma$ if and only if there exist matrices $X = \begin{bmatrix} X_1 & 0 \\ X_2 & X_3 \end{bmatrix}$ and $Y = \begin{bmatrix} Y_1 & 0 \\ Y_2 & Y_3 \end{bmatrix}$, $\Phi = \begin{bmatrix} \Phi_1 \\ \Phi_2 \end{bmatrix}$ and $\Psi = \begin{bmatrix} \Psi_{11} & \Psi_{12} \\ \Psi_{21} & \Psi_{22} \end{bmatrix}$ such that (20)-(24) ((23)-(24) in the next page) are satisfied.*

$$E^T X_1 = X_1^T E \tag{20}$$

$$EY_1 = Y_1^T E^T \tag{21}$$

$$\begin{bmatrix} X_1^T E & E^T \\ E & Y_1^T E^T \end{bmatrix} \geq 0 \tag{22}$$

$$\begin{bmatrix} A^T X_1 + X_1 A + \Phi_1 C + C^T \Phi_1^T & * & * & * & * \\ \alpha B_f^T X_1 - \beta X_2 + \Phi_2 C + D_f^T \Phi_1^T & \Xi_1 & * & * & * \\ B_d^T X_1 & 0 & -\gamma^2 I & * & * \\ (1-\alpha) B_f^T X_1 + \beta X_2 & \beta X_3 & 0 & -\gamma^2 I & * \\ H_1 & I & 0 & 0 & -I \end{bmatrix} < 0 \tag{23}$$

$$\begin{bmatrix} \Xi_2 & * & * & * & * \\ \Xi_3 & \beta(\Psi_{22} + \Psi_{22}^T - Y_3 - Y_3^T) & * & * & * \\ B_d^T & 0 & -\gamma^2 I & * & * \\ (1-\alpha) B_f^T & \beta I & 0 & -\gamma^2 I & * \\ \Xi_4 & Y_3 - G_1 \Psi_{12} - G_2 \Psi_{22} & 0 & 0 & -I \end{bmatrix} < 0 \tag{24}$$

where $\Xi_i = -\beta(X_3 + X_3^T) + \Phi_2 D_f + D_f^T \Phi_2^T$, $\Xi_2 = AY_1 + Y_1^T A^T - B_u \Psi_{11} - \Psi_{11}^T B_u + \alpha(B_f(Y_2 - \Psi_{21}) + (Y_2 - \Psi_{21})^T B_f^T)$ and $\Xi_3 = \beta(\Psi_{21} - Y_2) + \alpha(Y_3^T B_f^T - \Psi_{22}^T B_f^T) - \Psi_{12}^T B_u^T$ and $\Xi_4 = H_1 Y_1 + Y_2 - G_1 \Psi_{11} - G_2 \Psi_{21}$.

If (20)-(24) are feasible, we can design the controller with the following parameters:

$$\begin{aligned} A_K &= (Y^{-1} - X)^{-T}[(X^T\bar{B}_{\bar{u}} + \Phi\bar{D} + \bar{C}^T\bar{G})\Psi Y^{-1} - X^T\bar{A} \\ &\quad -\Phi\bar{C} - \gamma^{-2}X^T\bar{B}_d\bar{B}_d Y^{-1} - \bar{H}^T\bar{H} - \bar{A}^T Y^{-1}] \end{aligned} \tag{25}$$

$$E_K = \bar{E} \tag{26}$$

$$B_K = (Y^{-1} - X)^{-T}\Phi \tag{27}$$

$$C_K = -\Psi Y^{-1} \tag{28}$$

Proof: The proof procedure is similar to that of [50, Theorem 5.4]. Hence we omit the details.

Remark 4 *An optimization problem can be imposed for performance issue:*

$$\min \gamma^2, \ s.t. \ (20) - (24).$$

A necessary condition for inequalities (20)-(24) of Lemma 1 is that the system (14)-(15) is finite dynamic stablizable and finite dynamic detectable, i.e., there exists a constant matrix K such that $\{\bar{E}, \bar{A} + \bar{B}_u K\}$ *is stable and there exists a constant matrix L such that* $\{\bar{E}, \bar{A} + L\bar{C}\}$ *is stable.*

From (23)-(24), we can see that α *can affect the solvability while* β *cannot since it is actually a absorbable parameter. For example, we can redefine* X_3 *as* βX_3*, which will not change the result. However, as a scalar, it is useful to adjust the magnitude of the controller gain.*

Lemma 1 results in a non-strict LMI problem. A potential disadvantage is that the numerical problem may exist for handling the equations. A strict version LMI may be more computationally trackable. [37] provided such a solution for this problem.

To do so, we define a new variable $\tau = \bar{D}\bar{u}$. Hence the system (14)-(16) with the notations defined in (8)-(13) can be further expressed as

$$\tilde{E}\dot{\bar{\xi}} = \tilde{A}\bar{\xi} + \tilde{B}_{\bar{u}}\bar{u} + \tilde{B}_d w \tag{29}$$

$$y = \tilde{C}\bar{\xi} \tag{30}$$

$$z = \tilde{H}\bar{\xi} \tag{31}$$

where $\bar{\xi} = \begin{bmatrix} \xi \\ \tau \end{bmatrix} \in \mathcal{R}^{n+\ell+r}$.

The FDD-based extended dynamic output feedback controller is defined as follows:

$$\begin{aligned} \tilde{E}\dot{\hat{\bar{\xi}}} &= A_K\hat{\bar{\xi}} + B_K y \\ \bar{u} &= C_K\hat{\bar{\xi}} \end{aligned} \tag{32}$$

Now we can see that the disadvantage is that the order of controller is higher than that of (17), as the closed-loop system becomes

$$\begin{aligned} \tilde{E}_c\dot{\bar{\varsigma}} &= \tilde{A}_c\bar{\varsigma} + \tilde{B}_c w \\ z &= \tilde{C}_c\bar{\varsigma} \end{aligned} \tag{33}$$

where

$$\bar{\varsigma} = \begin{bmatrix} \bar{\xi} \\ \tilde{\bar{\xi}} \end{bmatrix}, \tilde{E}_c = \begin{bmatrix} \tilde{E} & 0 \\ 0 & \tilde{E} \end{bmatrix}, \tilde{A}_c = \begin{bmatrix} \tilde{A} & \tilde{B}_{\bar{u}} C_K \\ B_K \tilde{C} & A_K \end{bmatrix}$$

$$\tilde{B}_c = \begin{bmatrix} \tilde{B}_d \\ 0 \end{bmatrix}, C_c = \begin{bmatrix} \tilde{H} & 0 \end{bmatrix}$$

The result in [37] with strict LMI formulations can then be applied. For completeness, we restate the conditions as follows.

Lemma 2 *[37] Given scalars $\gamma > 0$, α and β and the system (14)-(16). There exists a FDD-based extended dynamic output feedback controller (32) such that the closed-loop system (33) is admissible and the transfer function $G_c(s) = \tilde{C}_c(s\tilde{E}_c - \tilde{A}_c)^{-1}\tilde{B}_c$ satisfies $\|\tilde{G}_c(s)\|_\infty < \gamma$ if and only if there exist symmetric matrices P, Q and matrices $R, S, K, L,$ such that the inequalities (34)-(36) are satisfied.*

$$\begin{bmatrix} \tilde{E}_r^T P \tilde{E}_r & \gamma I \\ \gamma I & \tilde{E}_l^T Q \tilde{E}_l \end{bmatrix} > 0 \tag{34}$$

$$\begin{bmatrix} \tilde{A}\Upsilon_P + \Upsilon_P^T \tilde{A}^T + \tilde{B}_{\bar{u}} K + K^T \tilde{B}_{\bar{u}}^T & * & * \\ \tilde{H}\Upsilon_P & -\gamma^2 I & * \\ \tilde{B}_d^T & 0 & -I \end{bmatrix} < 0 \tag{35}$$

$$\begin{bmatrix} \tilde{A}\Upsilon_Q + \Upsilon_Q^T \tilde{A}^T + L\tilde{C} + L^T \tilde{C}^T & * & * \\ \tilde{B}_d^T \Upsilon_Q & -\gamma^2 I & * \\ \tilde{H} & 0 & -I \end{bmatrix} < 0 \tag{36}$$

where E_r and E_l are of full column rank which is equal to rank of $\tilde{E}$, such that $\tilde{E} = E_l E_r^T$, V and U are of full column rank and composed of bases of $Null(\tilde{E})$ and $Null(\tilde{E}^T)$, respectively, $\Upsilon_P = P\tilde{E}^T + VSU^T$ and $\Upsilon_Q = Q\tilde{E} + URV^T$.

If (34)-(36) are feasible, the controller parameters can be obtained as follows:

$$\begin{aligned} A_K &= (\Upsilon_Q - \gamma^2 \Upsilon_P^{-1})^{-T}(\gamma^2 \tilde{A}^T \Upsilon_P^{-1} + \Upsilon_Q^T \tilde{A} + L\tilde{C} \\ &\quad + \Upsilon_Q^T \tilde{B}_{\bar{u}} K \Upsilon_P^{-1} + \tilde{H}^T \tilde{H} + \Upsilon_Q^T \tilde{B}_d \tilde{B}_d^T \Upsilon_P^{-1}) \end{aligned} \tag{37}$$

$$B_K = -(\Upsilon_Q - \gamma^2 \Upsilon_P^{-1})^{-T} L \tag{38}$$

$$C_K = K\Upsilon_P^{-1} \tag{39}$$

Remark 5 *Lemma 2 has a strong connection with Lemma 1. In fact, a non-strict necessary and sufficient condition based on Lemma 1 using the controller (32) is to solve the inequalities (35)-(36) and*

$$\begin{bmatrix} \tilde{E}^T & 0 \\ 0 & \tilde{E} \end{bmatrix} \begin{bmatrix} \Upsilon_Q & I \\ I & \Upsilon_P \end{bmatrix} = \begin{bmatrix} \Upsilon_Q^T & I \\ I & \Upsilon_P^T \end{bmatrix} \begin{bmatrix} \tilde{E} & 0 \\ 0 & \tilde{E}^T \end{bmatrix} \geq 0 \tag{40}$$

where K, L, Υ_P and Υ_Q are variables.

Due to the fact that $\tilde{E}^T U = 0$, $\tilde{E}V = 0$, and Q, P are symmetric, (40) is satisfied automatically, with the choice of Υ_P and Υ_Q in Lemma 2. Motivated by this connection, we can possibly design a lower order controller using (17) with strict LMI conditions.

Theorem 1 *Given scalars $\gamma > 0$, α and β and the system (14)-(16). There exists a FDD-based extended dynamic output feedback controller (17) with $E_K = \bar{E}$ such that the closed-loop system (18) is admissible and the transfer function (19) satisfies $\|G_c\|_\infty < \gamma$ if and only if there exist symmetric matrices P, Q and matrices R, S, Φ, Ψ, such that the inequalities (41)-(43) are satisfied.*

$$\begin{bmatrix} E_l^T Q E_l & I \\ I & E_r^T P E_r \end{bmatrix} > 0 \tag{41}$$

$$\begin{bmatrix} \bar{A}^T X + X^T \bar{A} + \Phi \bar{C} + \bar{C}^T \Phi^T & X^T \bar{B}_d & \bar{H}^T \\ \bar{B}_d^T X & -\gamma^2 I & 0 \\ \bar{H} & 0 & -I \end{bmatrix} < 0 \tag{42}$$

$$\begin{bmatrix} \bar{A} Y + Y^T \bar{A}^T - \bar{B}_{\bar{u}} \Psi - \Psi^T \bar{B}_{\bar{u}}^T & \bar{B}_d & Y^T \bar{H}^T - \Psi^T \bar{G}^T \\ \bar{B}_d^T & -\gamma^2 I & 0 \\ \bar{H} Y - \bar{G} \Psi & 0 & -I \end{bmatrix} < 0 \tag{43}$$

where E_r and E_l are of full column rank which is equal to rank of $\bar{E}$, such that $\bar{E} = E_l E_r^T$, V and U are of full column rank and composed of bases of $Null(\bar{E})$ and $Null(\bar{E}^T)$, respectively, $Y = P\bar{E}^T + VSU^T$ and $X = Q\bar{E} + URV^T$. If (41)-(43) are feasible, we can design the controller with the parameters (25)-(28).

Proof: "Sufficiency:" The only thing we need to prove is that Y is nonsingular, which can be done by the following result from [37]: $(P\bar{E}^T + VSU^T)^{-1} = \tilde{P}\bar{E} + U\tilde{S}V^T$, where $\tilde{P}$ is symmetric and $\tilde{S}$ is nonsingular such that $E_L^T \tilde{P} E_L = (E_R^T P E_R)^{-1}$, and $\tilde{S} = (U^T U)^{-1} S^{-1} (V^T V)^{-1}$ [1].

"Necessity:" Based on [50, Theorem 5.4], we can obtain (40), where $\tilde{E}$, Υ_Q and Υ_P are replaced by $\bar{E}$, X and Y, respectively, and (42)-(43) for some X, Y, Φ and Ψ. The structure of $X = Q\bar{E} + URV^T$ and $Y = P\bar{E}^T + VSU^T$ can be proved by noting the fact that there exist nonsingular matrices M and N, such that $ME_cN = \begin{bmatrix} I_\imath & 0 \\ 0 & 0 \end{bmatrix}$ and $M\bar{A}_C N = \begin{bmatrix} A_1 & 0 \\ 0 & I_{2n+2\ell-\imath} \end{bmatrix}$, where A_1 is stable matrix and $\imath$ is the rank of E_c. (41) can be seen via the connection between Y^{-1} and X and the fact that $E_l^T Q E_l$ and $E_r^T P E_r$ are nonsingular. Thus we complete the proof.

Remark 6 *In the case that the output is also disturbed by noise d_y, i.e., $y = Cx + D_u u + D_d d_y + D_f f$, we can set $f_s = [f^T \; d_y^T]^T$ and make some some changes correspondingly.*

4. A Numerical Example

Now we consider an example with the following parameters:

$$A = \begin{bmatrix} 10 & 1 \\ 1 & -10 \end{bmatrix}, B_u = \begin{bmatrix} 1 \\ 1.5 \end{bmatrix}, B_d = \begin{bmatrix} 0 \\ 0.01 \end{bmatrix},$$

[1] Without loss of generalities, we can assume that S is nonsingular [50, 37].

$$B_f = \begin{bmatrix} 1 & 0 \\ 0.5 & 0 \end{bmatrix}, C = \begin{bmatrix} 0.01 & 1 \\ 0 & 0.02 \end{bmatrix}, D_u = \begin{bmatrix} 0 \\ 1 \end{bmatrix},$$

$$D_f = \begin{bmatrix} 0 & 0.5 \\ 0 & 0 \end{bmatrix}, H_1 = 0, \bar{G} = 0, E = I$$

We use Matlab toolboxes YALMIP [24] and SeduMi [35] as the LMI solver. Based on Lemma 1, we can obtain the following observer gain with optimal $\gamma = 1$:

$$A_K = \begin{bmatrix} 2.0229e+7 & -1.31468e+7 & -2.33872e+13 & 3.89434e+13 \\ 185590 & -120614 & -2.70378e+11 & 4.57588e+11 \\ -3.67358e-5 & 2.3885e-5 & -12.3742 & 31.5721 \\ -9.15882e-5 & 5.95494e-5 & -50.0283 & 103.241 \end{bmatrix}$$

$$B_K = \begin{bmatrix} -3207.26 & 2.08976e+6 \\ -39.2801 & 19174 \\ -4.75342e-9 & -3.79481e-6 \\ -1.23729e-8 & -9.46098e-6 \end{bmatrix}$$

$$C_K = \begin{bmatrix} -9.68043 & 6.27282 & -3.65297e-5 & 4.31771e-5 \\ 1.75052 & -0.606098 & 3.80392e+10 & -4.49613e+10 \\ -0.637528 & 0.248769 & -1.51168e+10 & 2.48354e+10 \end{bmatrix}$$

We assume the fault is $f = \begin{cases} [0\ 1]^T, & if\ 2s \le t \le 3s \\ [0\ 0]^T, & otherwise \end{cases}$ and the noise d is generated by Matlab routine "rand(1)" (uniform distribution on the unit interval). Figures 1 shows the simulation results with the initial condition $x = [0.1\ \ -0.1]^T$ and $\hat{x} = [0\ 0]^T$. However, we note that the magnitude of this controller is too high. In order to reduce the magnitude, we may impose some additional constraints on the variables X, Y, Φ, Ψ, tune α, β and allow γ to be a little bigger value. For instance, the magnitude of Y and Y^{-1} may not be large based on the observation on the expressions (25)-(28). The following parameters are obtained when $\gamma = 1.04$:

$$A_K = \begin{bmatrix} 6076430 & -3978310 & 421221 & -484805 \\ 121347 & -79478.2 & 9134 & -10501.7 \\ 1024.17 & -708.662 & 1019.72 & -1209.19 \\ 1000.42 & -689.203 & 876.225 & -1047.62 \end{bmatrix}$$

$$B_K = \begin{bmatrix} 13975.8 & 630774 \\ 300.578 & 12598 \\ 40.0899 & 106.368 \\ 36.1609 & 103.898 \end{bmatrix}$$

$$C_K = \begin{bmatrix} -9.63326 & 6.2651 & -5.38089e-015 & 5.32124e-015 \\ 0.0328759 & 0.0112816 & 391.092 & -386.584 \\ 0.00401275 & 0.00131936 & -68.96 & 77.1325 \end{bmatrix}$$

Figures 2 shows the simulation result under the low gain controller. From the simulation results, we can see that the estimated fault approaches the actual fault only after the state is approaching steady range.

In this example, due to the small B_d, the effect of disturbance with small magnitude noise is not significant. Now we change the noise to $10 * rand(1)$ and $100 * rand(1)$, respectively, and still use the low gain controller. The simulation results are shown in Figures 3-4 correspondingly. All these simulation results indicate the good fault estimation performance of the designed controller/filter.

Conclusion

In this chapter, we have discussed the fault detection and diagnosis problem for a class of descriptor systems. One of the main contributions lies in the novel structure to stabilize the system and estimate the fault simultaneously. Once the structure is decided, we can apply some existing results to solve the corresponding controller/filter gain based on non-strict/strict LMI conditions. We have also briefly discussed the connection between this woek and some existing results. The simulation example shows the efficiency and feasibility of this approach.

However, as a preliminary work, we still leave some questions unsolved. For example, a systematical way should be established to adjust the parameters α, β, X, Y such that the controller gain are harmonious, since descriptor systems are numerically sensitive generally. Furthermore, we shall also pay attention to the effectiveness of fault estimation, such as the bound of the state estimated error and the energy. These issues can be converted into a multiple objective problem. A practical solution is to adjust the structure of $\bar{H}$ and $\bar{G}$. If only sensor fault is considered, it is possible to construct fault-tolerant controller (17) by replacing y as $y - D_f\hat{f}$. Another interesting problem is the adaptive threshold. Although the H_∞ bound can be used a guideline for the threshold design, it may be rather loose in the sense of up-bound.

Our results can be also extended to many cases, and namely few of them are as follows.

- *The descriptor systems with structured uncertainties and/or Lipschitz-like nonlinearities or polytopic uncertainties, i.e.,*

$$\begin{array}{l} E\dot{x} = (A + \Delta A)x + (B_u + \Delta B_u)u + \theta(x, u) + B_d d + B_f f, \\ \|\theta(\hat{x}, u) - \theta(x, u)\| \leq |\ \Theta(\hat{x} - x)\|,\ \forall(\hat{x}, u), (x, u) \in \mathcal{R}^{n+m} \end{array} \tag{44}$$

or

$$\begin{array}{l} E\dot{x} = \sum_i^N \rho_i(A_i x + B_{ui} u + B_{di} d + B_{fi} f), \\ \sum_i^N \rho_i = 1,\ \rho_i \geq 0 \end{array} \tag{45}$$

For example, we can use the results from [1, 43] to deal with structured uncertainties or apply the skills from [13, 11] to deal with nonlinearities or employ the techniques from [2] to handle the polytopic uncertainties. For the polytopic uncertainties, the dilated matrix techniques are useful [48] to relax the design conversation with multiple matrix inequalities constraints.

- *The systems with different operating frequencies of the noise and the fault.* When the difference is significant, we may consider the generalized KYP lemma [21, 20, 25]

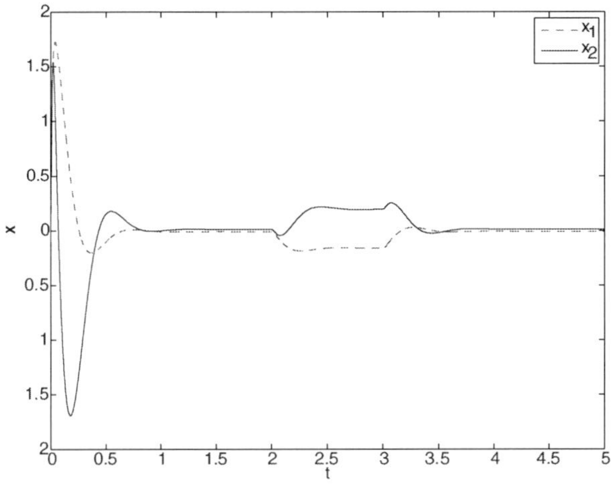

The actual state

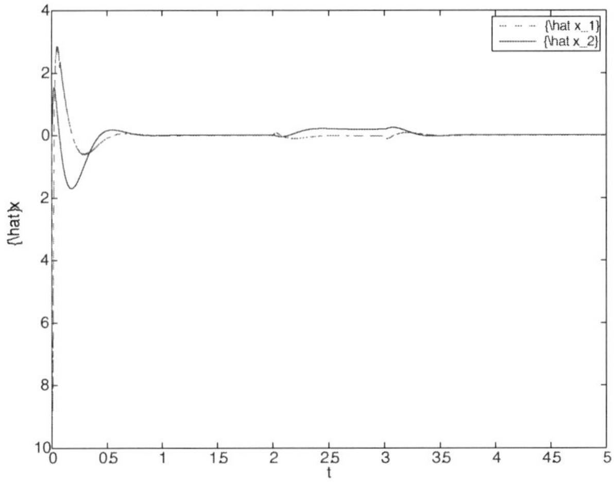

The estimated state

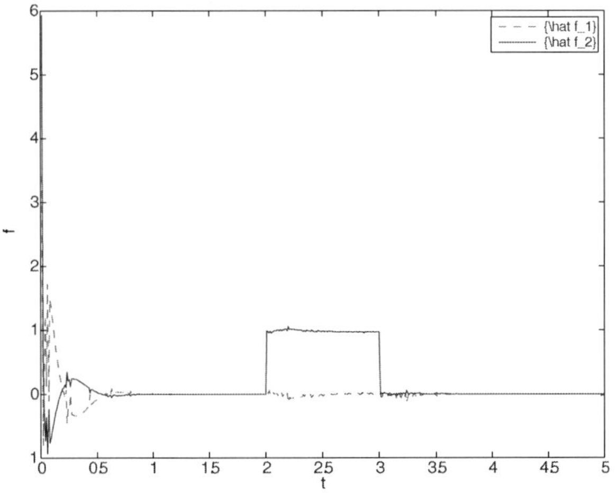

The estimated fault

Figure 1. The simulation result with high magnitude gain.

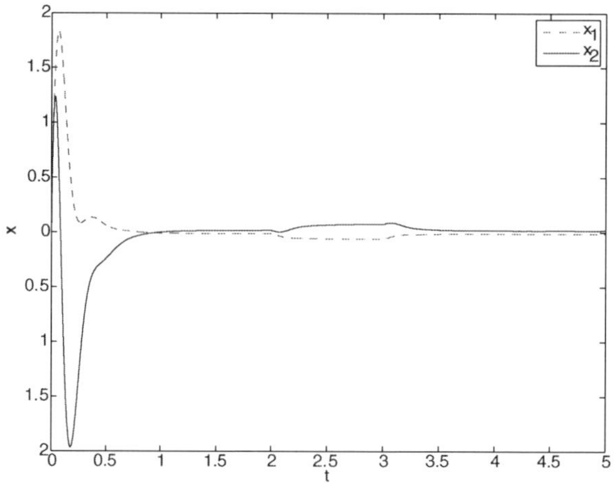

The actual state

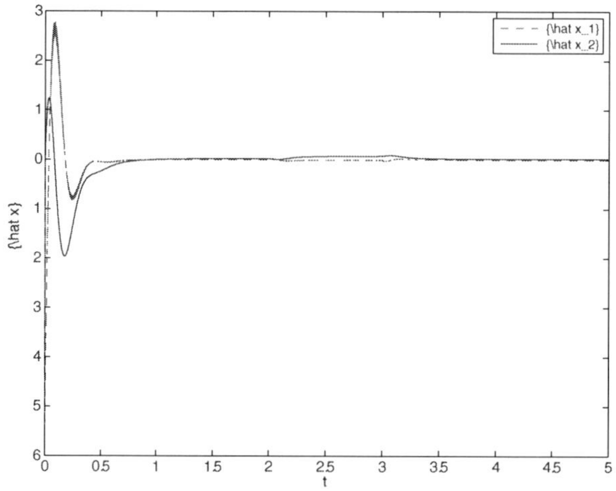

The estimated state

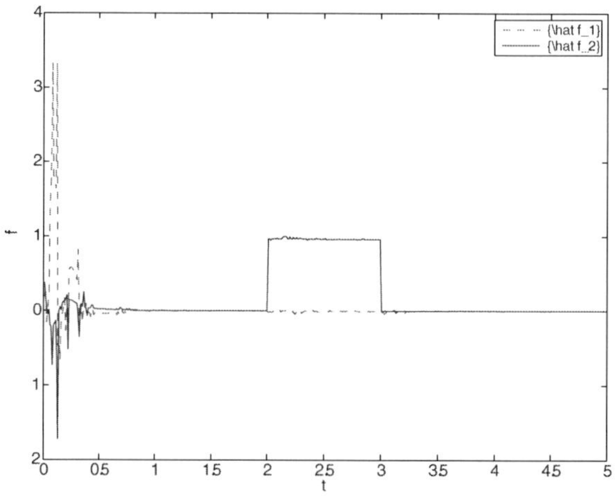

The estimated fault

Figure 2. The simulation result with low magnitude gain.

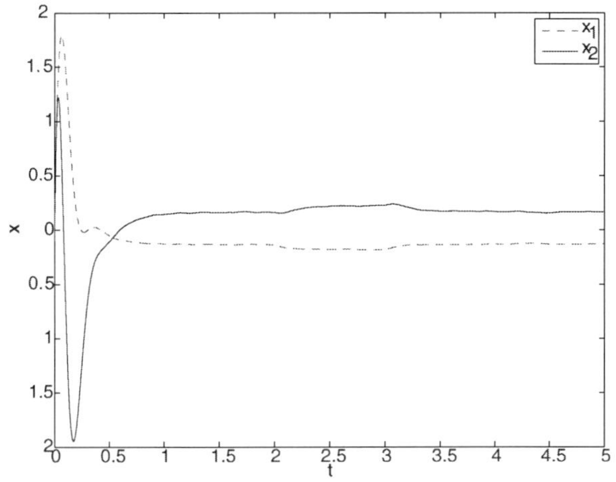

The actual state

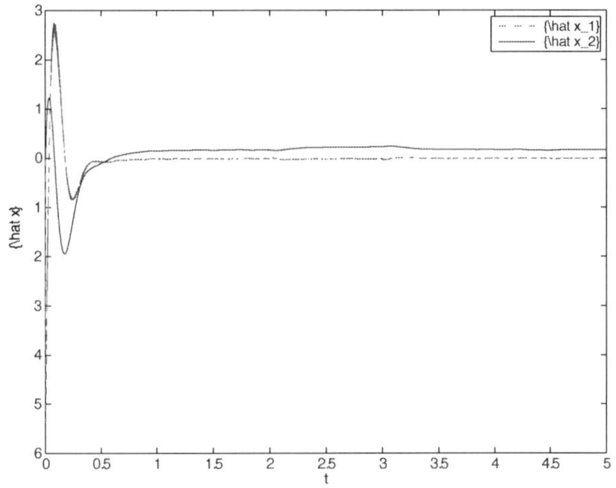

The estimated state

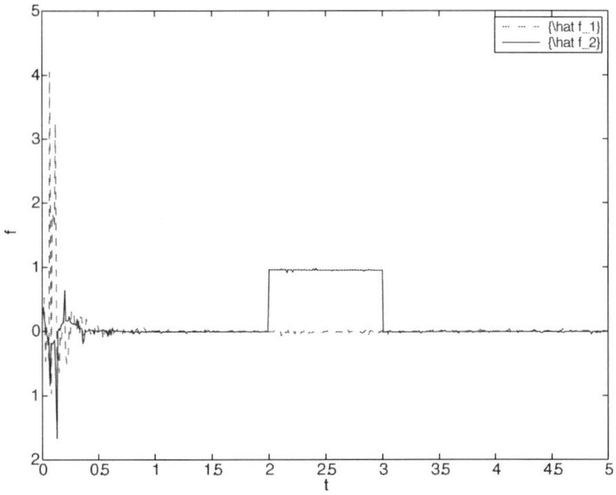

The estimated fault

Figure 3. The simulation result with the noise=$10 * rand(1)$.

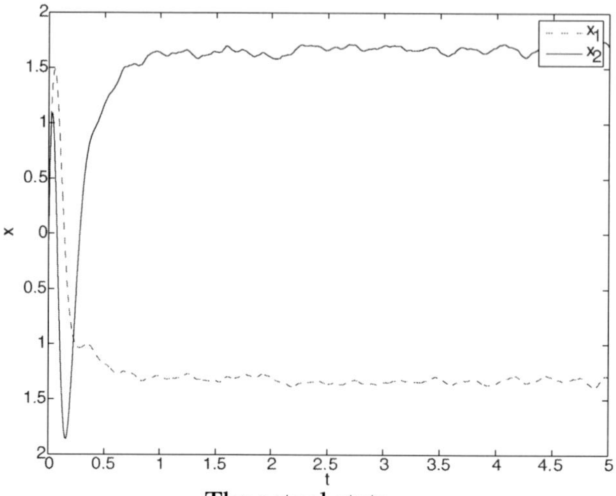

The actual state

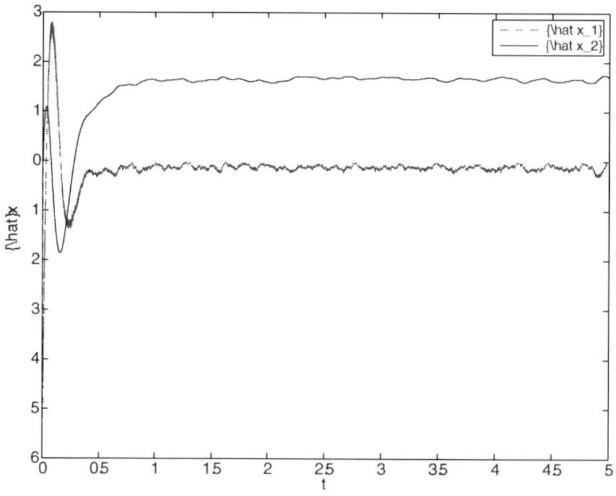

The estimated state

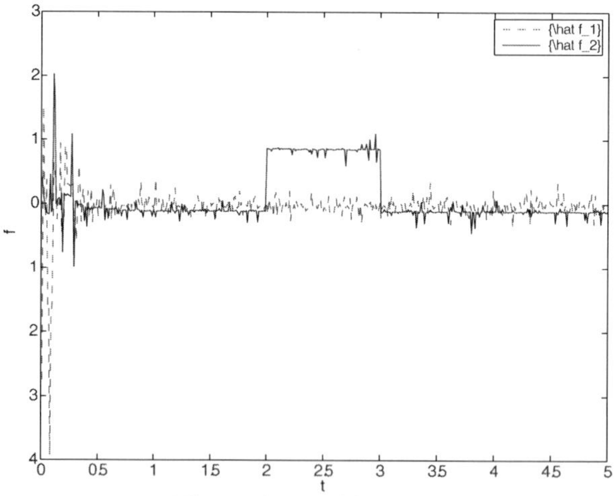

The estimated fault

Figure 4. The simulation result with the noise=$100 * rand(1)$.

or the weighted approach [8]. The approach stated in [27] can be applied when a closed-loop dissipativity condition is imposed.

- *Discrete-time descriptor systems*, i.e.,

$$Ex(k+1) = Ax(k) + B_u u(k) + B_d d(k) + B_f f(k) \tag{46}$$

In fact, we can apply the similar techniques of continuous-time case to handle it [50]. The difference is the matrix transformation when dealing with the multiplication of matrix variables. Some elegant matrix constructive methods may be required to convert the design procedure into an LMI problem.

Our continuing work will attempt to solve these questions. And collaboration is also welcome.

References

[1] W. Assawinchaichote, S. K. Nguang, and P. Shi. H_∞ output feedback control design for uncertain fuzzy singularly perturbed systems: an LMI approach. *Automatica*, 40:2147–2152, 2004.

[2] G. I. Bara. Robust analysis and control of parameter-dependent uncertain descriptor systems. *Systems and Control Letters*, 60:356364, 2011.

[3] M. Blanke, M. Kinaert, J. Lunze, and M. Staroswiecki. *Diagnosis and fault-tolerant control*. Springer, Berlin Heidelberg, 2003.

[4] F. Caccavale and L. Villani, editors. *Fault diagnosis and fault torlerance for mechatronic systems: recent advances*. Springer, Berlin Heidelberg, 2003.

[5] J. Chen and R. J. Patton. *Robust model-based fault diagnosis for dynamic systems*. Massachusetts: Kluwer Academic Publishers, 1999.

[6] W. Chen and M. Saif. Unknown input observer design for a class of nonlinear systems: an LMI approach. In *Proc. American Control Conference*, pages 834–838, Mineapolis, Minnesota, USA, Jun 2006.

[7] L. H. Chiang, R. D. Braatz, and E. Russell. *Fault detection and diagnosis in industrial systems*. Advanced textbooks in control and signal processing. Springer, 2001.

[8] Y. Feng, M.Yagoubi, and P. Chevre. H_∞ control with unstable and nonproper weights for descriptor systems. *Automatica*, 2012.

[9] W. G. Fenton, T. M. McGinnity, and L. P. Maguire. Fault diagnosis of electronic systems using intelligent techniques: A review. *IEEE Trans. on systems, man, and cybernetics part c: applications and reviews*, 31(3):269–281, 2001.

[10] Wang H. Gao, Z.W. and T. Cai. A robust fault detection filtering for stochastic distribution systems via descriptor estimator and parametric gain design. *IET Proc. - Control Theory Appli.*, 1(5):1286–1293, 2007.

[11] Z. W. Gao and S. X. Ding. Actuator fault robust estimation and fault-tolerant control for a class of nonlinear descriptor systems. *Automatica*, 43:913–920, 2007.

[12] Z. W. Gao and S. X. Ding. Fault estimation and fault-tolerant control for descriptor systems via pmid observer design. *IET Proc. - Control Theory Appli.*, 1(5):1208–1218, 2007.

[13] Z. W. Gao and D. W. C. Ho. State/noise estimation for descriptor systems with application to senser fault diagnosis. *IEEE Trans. on Signal Processing*, 54(4):1316–1326, 2006.

[14] Z. W. Gao and H. Wang. Descriptor observer approaches for multivarible systems with measurement noises and application in fault detection and diagnosis. *Systems and Control Letters*, 55:304–313, 2006.

[15] J. Gertler. *Fault detection and diagnosis in engineering systems*. Marcel Dekker. Inc, New Year, USA, 1998.

[16] Q. P. Ha and H. Trinh. State and input simultaneous estimation for a class of nonlinear systems. *Automatica*, 40:1779–1785, 2004.

[17] D. Henry and A. Zolghadri. Norm-based design of robust FDI schemes for uncertain systems under feedback control: Comparison of two approaches. *Control Engineering Practice*, 14:1081–1097, 2006.

[18] R. Isermann. Model-based fault-detection and diagnosis - status and applications. *Annual reviews in control*, 29:71–85, 2005.

[19] R. Isermann. *Fault-diagnosis systems: an introduction from fault detection to fault tolerance*. Springer, 2006.

[20] T. Iwasaki and S. Hara. Generalized KYP lemma: unified frequency domain inequalities with design applications. *IEEE Trans. on Automatic Control*, 50(1):41–59, 2005.

[21] T. Iwasaki and S. Hara. Feedback control synthesis of multiple frequency domain specification via generalized KYP lemma. *International Journal of Robust and Nonlinear Control*, 17:415–434, 2007.

[22] S. Kilsgaard, M. L. Rank, H. H. Niemann, and J. Stoustrup. Simultaneous design of controller and fault detector. In *Proc. Conference on Decision and Control*, pages 628–629, Kobe, Japan, Dec 1996.

[23] D. Koenig, S. Mammar, and B. Marx. H_∞ fault detection and isolation for descriptor systems: A matrix inequality approach. In *Proc. American Control Conference*, pages 1080–1081, Anchorage, AK, May 2002.

[24] J. Löfberg. YALMIP : A toolbox for modeling and optimization in MATLAB. In *the CACSD Conference*, Taipei, Taiwan, 2004.

[25] K. Y. Lum, J. Xu, L. Xie, and A. P. Loh. A novel H_∞ UIO design for actuator FDI in finite frequency domain. *International journal of control, automation and systems*, 9(3):601–605, 2011.

[26] B. Marx, D. Koenig, and D. Georges. Robust fault diagnosis for linear descriptor systems using proportional integral observers. In *Proc. Conference on Decision and Control*, pages 457–462, Maui, Hawaii USA, Dec 2003.

[27] I. Masubuchi. Output feedback controller synthesis for descriptor systems satisfying closed-loop dissipativity. *Automatica*, 43:339–345, 2007.

[28] S. Mondal, G. Chakraborty, and K.Bhattacharyya. Robust unknown input observer for nonlinear systems and its application to fault detection and isolation. *J. Dyn. Sys., Meas., Control*, 130(4):044503, 2008.

[29] H. Nakamura, T. Suzuki, and S. Okuma. Simultaneous design of controller and fault detector and its application to motor drive control system. In *IEEE int. conf. control applications*, pages 82–87, Mexico City, Mexico, Sep 2001.

[30] R. J. Patton, P. M. Frank, and R. N. Clark, editors. *Issues of fault diagnosis for dynamic systems*. Springer-Verlag, London, U.K., 2000.

[31] C. D. Persis and A. Isidori. A geometric approach to nonlinear fault detection and isolation. *IEEE Trans. on Automatic Control*, 46(6):853–865, 2001.

[32] A. M. Pertew, H. J. Marquez, and Q. Zhao. Design of unknown input observers for lipschitz nonlinear systems. In *Proc. American Control Conference*, pages 4198–4203, Portland, USA, Jun 2005.

[33] S. J. Qin. Statistical process monitoring: basics and beyond. *Journal of chemometrics*, 17:480C502, 2003.

[34] S. Simani, C. Fantuzzi, and R. J. Patton. *Model-based fault diagnosis in dynamic systems using identification tecniques*. Springer, London, 2003.

[35] J. F. Sturm, O. Romanko, and I. Pólik. Sedumi: http: // sedumi.mcmaster.ca/. User manual, McMaster University, 2006.

[36] M. L. Tyler and M. Morari. Optimal and robust design of integrated control and diagnostic modules. In *Proc. American Control Conference*, pages 2060–2064, Maryland, USA, Jun 1994.

[37] E. Uezato and M. Ikeda. Strict LMI conditions for stability, robust stabilization, and H_∞ control of descriptor systems. In *Proc. Conference on Decision and Control*, pages 4092–4097, Phoenix, Ariona USA, Dec 1999.

[38] V. Venkatasubramanian, R. Rengaswamy, and S.N. Kavuri. A review of process fault detection and diagnosis part II: Qualitative models and search strategies. *Computers and Chemical Engineering*, 27:313–326, 2003.

[39] V. Venkatasubramanian, R. Rengaswamy, S.N. Kavuri, and K. Yin. A review of process fault detection and diagnosis part III: Process history based methods. *Computers and Chemical Engineering*, 27:327–346, 2003.

[40] V. Venkatasubramanian, R. Rengaswamy, K. Yin, and S.N. Kavuri. A review of process fault detection and diagnosis part I: Quantitative model-based methods. *Computers and Chemical Engineering*, 27:293–311, 2003.

[41] D. Wan and K. Y. Lum. Adaptive unknown input observer approach for aircraft actuator fault detection and isolation. *International Journal of Adaptive Control and Signal Processing*, 21:31–48, 2007.

[42] H. Wang, C. Yung, and F. Chang. *H_∞ control for nonlinear descriptor system*, volume 326 of *LNCIS*. Springer-Verlag, London.

[43] Y. Wang and Q. Zhang. Dynamic output feedback robust H_∞ control for uncertain descriptor systems. *Control and Decision*, 17(6):948–951, 2002.

[44] K. Watanabe and D. M. Himmelblau. Instrument fault detection in systems with uncertainties. *International Journal of System Science*, 12(2):137–158, 1982.

[45] J. Xu, K. Y. Lum, and A. P. Loh. A gain-varying uio approach with adaptive threshold for FDI of nonlinear f16 systems. *Journal of control theory and application*, 8(3), May 2010.

[46] J. Xu, K. Y. Lum, A. P. Loh, and L. Xie. A SOS-based approach to residual generators for discrete-time polynomial nonlinear systems. In *Conference on Decision and Control*, pages 372–377, New Orleans, LA, USA, Dec 2007.

[47] J. Xu, K. Y. Lum, A. P. Loh, and L. Xie. Two improved approaches to fault detection with unknown input. In *IFAC World Congress*, pages 10136–10141, Seoul, South Korea, Jul 2008.

[48] J. Xu and L. Xie. Dilated LMI characterization and a new stability criterion for polytopic uncertain systems. In *IEEE World Congress on Intelligent Control and Automation*, pages 243–247, Dalian, China, Jun 2006.

[49] J. Xu, L. Xie, and K. Y. Lum. A nonlinear uio approach to FDI of disturbed nonlinear systems with sos techniques. *ASME Journal of Dynamic Systems, Measurement and Control*, 134(4):041005, 2012.

[50] S. Xu and J. Lam. *Robust control and filtering of singular systems*, volume 332 of *LNCIS*. Springer, 2006.

[51] P. Zhang and S. X. Ding. Multiobjective design of robust fault detection systems. In *the 6th IFAC symp. on fault detection, supervision and satety of technical processes*, pages 1453–1458, Beijing, China, Aug 2006.

[52] Y. Zhang, L. Guo, and H. Wang. Filter-based fault detection and diagnosis using outout pdfs for stochastic systems with time delays. *International Journal of Adaptive Control and Signal Processing*, 20:175–194, 2006.

[53] Y. M. Zhang and J. Jiang. Bibliographical review on reconfigureable fault-tolerant cotnrols systems (survey paper). In *the 5th IFAC symp. on fault detection, supervision and satety of technical processes*, pages 265–276, Washington, D. C., USA, Jun 2003.

In: Fault Detection
Editors: F. P. G. Márquez and M. Papaelias
ISBN: 978-1-62808-999-8

Chapter 7

INTELLIGENT DATA COMPRESSION, DIAGNOSTICS AND PROGNOSTICS USING AN EVOLUTIONARY-BASED CLUSTERING ALGORITHM FOR INDUSTRIAL MACHINES

***Jun Chen*[*1], *Michael Gallimore*[1], *Chris Bingham*[1], *Mahdi Mahfouf*[2] *and Yu Zhang*[1]**

[1]School of Engineering, University of Lincoln, Lincoln, UK
[2]Department of Automatic Control and Systems Engineering, University of Sheffield, Sheffield, UK

ABSTRACT

Intelligent signal processing methods play an important role in identifying 'novelty' in machine behaviour, thereby facilitating greater operational availability and early detection of emerging faults. Complex industrial systems often capture a high volume of data, with high dimensionality, and therefore clustering techniques have emerged as primary candidates for machine fault detection and classification due to their pattern-matching and data compression capabilities. To address well-known 'sensitivity' problems associated with classical clustering techniques, however, a real-coded Genetic Algorithm (GA) is now considered that is integrated into a traditional *K*-means clustering methodology. The resultant hybridisation provides an enhanced search algorithm by incorporating the local search capabilities rendered by hill-climbing optimisation and the global search ability provided by a GA. Performance of the proposed algorithm is compared with other clustering techniques using two artificial data sets, and a benchmark iris classification problem. Results show that, in all cases, the proposed algorithm outperforms its counterparts in terms of global search capability and robustness. Moreover, the scalability of the proposed algorithm to higher-dimensional problems featuring a greater number of data points, is also considered, and validated using an application to provide a reduced data-representation of measurements taken from industrial gas turbine units, and for fault detection based on rundown vibration signatures.

---

[*] E-mail: juchen@lincoln.ac.uk.

Keywords: hybridised clustering algorithm, genetic algorithms, *K*-means clustering algorithm, data compression, industrial machines, diagnostics and prognostics, and rundown vibration signature

INTRODUCTION

Clustering techniques have attracted significant attention for machine fault detection systems as a result of two key characteristics: i) retrieving representatives or prototypes from anfractuous information through a degree of abstraction, and ii) the ability to readily classify normal and abnormal attributes. Representing data by fewer prototypes necessarily loses fine detail, but achieves simplification and improved interpretability. This characteristic can be employed to compress a large data set into a few representative points that can be more readily handled. Furthermore, the ability to classify different failure modes can have significant commercial impact by rapidly identifying parts that need replacement prior to failure, thereby reducing operational down-time and providing input to flexible maintenance strategies.

With increasing dimensionality of recorded data, many existing fault classification techniques, such as those based on traditional *K*-means for instance, may only offer suboptimal solutions that are more likely to be trapped in regions of local optima, with consequential effects such as, a) less separation of data groupings, that will degrade diagnostic performance, and b) the 'compressed data' being less representative, which further hinders accurate classification. Here, then, to address such 'sensitivity' problems, the integration of a real-coded Genetic Algorithm (GA) into a *K*-means clustering method, is considered, which is also computationally efficient compared to other hybridised clustering techniques due to its underpinning encoding scheme. Section II therefore provides an account of background research relating to hybridised clustering algorithms, along with associated issues attributed to early implementations; Section III describes the proposed evolutionary-based clustering algorithm, and Section IV provides a comparison of performance with other clustering methodologies. Two artificial data sets and a benchmark iris problem are used; results show that, in all cases, the proposed algorithm outperforms its counterparts in terms of global search capability and robustness. Further, two additional case studies are given in Section V, where the proposed approach is used to provide a compressed representation of field data sets from industrial gas turbines, and for fault detection based on rundown vibration signatures.

BACKGROUND AND PREVIOUS RESEARCH WORK

Significant effort to hybridize GAs with conventional partition-based clustering algorithms has taken place over recent decades (Bezdek, et al., 1994; Hall, et al., 1999; Krishna, et al., 1999; Bandyopadhyay, et al., 2002; Sheng, et al., 2006): the rationale being that most optimization techniques used in partition-based clustering are inherently based on 'hill-climbing' that is known to be sensitive to initial conditions and leads to convergence towards local optima. One of the earliest GA-based clustering implementations was given in

Bezdek, et al. (1994) by hybridizing a GA with Fuzzy *C*-Means (FCM) (Bezdek, 1981), in which a binary coded chromosome was adopted to represent cluster centres so that a GA could be used to iteratively search for optimal fuzzy partitions. Hall, et al. (1999) improved the efficacy of such techniques by coding the centres with a binary Gray code, whereby any two adjacent numbers are different by only one 'bit'. Both contributions considered that coding the centres in the chromosome to be more efficient than coding the membership matrix. Similar progress has been made in hybridizing GAs with *K*-means clustering. However, as an alternative to coding centres in the chromosome, Murthy, et al. (1996) proposed to code the cluster identity number that is assigned to each data point. Hence, the length of the chromosome is the same as the number of data points—ultimately making the algorithm's computational performance sensitive to the use of large data sets. Krishna, et al. (1999) proposed to code the hard membership matrix in a binary form so that a GA framework can be applied to find the optimal hard membership matrix that minimises the 'within-cluster variance'. Bandyopahyay, et al. (2002) subsequently acknowledged the contribution by Bezdek (1994) and noticed that the clustering problem under a GA framework could be reformulated as a real-valued optimization problem. Hence, a real-valued GA was adopted for their work, and cluster centres were encoded in the chromosome with floating-point values. More recently, Sheng, et al. (2006) incorporated GAs into *K*-medoids clustering, where variable length individuals that encode different number of medoids (clusters) are used for evolution with a modified Davies-Bouldin index as a measure of the fitness of the corresponding partitionings. As a result, the algorithm 'evolves' appropriate partitionings without any *a priori* assumption about the number of clusters present in the datasets. Nevertheless, despite these achievements, several related problems still remain to be addressed:

1. Although binary-coded GAs have been widely adopted, applying them to a continuous search space can result in 'Hamming cliffs': a difficulty associated with particular strings (such as 0111 and 1000) where a transition to a neighbouring solution (in real space) requires the alteration of many bits (Deb, 2001, p. 110).
2. Binary-coded GA-based clustering can suffer from problems due to imprecision, especially when they are used for real-valued optimisation problems. More precision requires the use of longer strings that in-turn increases the search space. It also leads to the requirement for larger population sizes in order to provide an effective search.
3. Both Murthy et al.'s (1996) and Krishna et al.'s work (1999) cannot effectively accommodate large data sets since the length of the chromosome increases as the number of the data points increases.
4. Although Bandyopahyay, et al. (2002) adopted a real-valued GA, a so-called Naive crossover (single-point crossover) is used, similar to the crossover operators used in binary-coded GAs. However, as discussed in Deb (2001, p. 112), this crossover operator does not have an adequate search power for real-valued cases. Hence, it surrenders its search responsibility to mutation operators, which are not considered ideal or generally effective.

In view of the above issues, here, a novel clustering technique, termed G3Kmeans (Chen, 2009, 2012), is proposed, which uses a real-coded GA, namely G3PCX (Deb, et al., 2002), as its optimisation method. G3PCX is then incorporated into *K*-means clustering, which only encodes cluster centres in the chromosome. Hence, the length of the chromosome relies only on the number of the cluster centres. The search power of G3Kmeans is significantly enhanced by combining the Parent-Centric Recombination (PCX) operator with hill-climbing. Such a combination takes full advantage of the global search capability attributed to the PCX recombination operator, and local search ability rendered by hill-climbing. As a result, G3Kmeans is more robust to initial conditions compared to other, more conventional partition-based clustering techniques, and is more efficient than the previously reported techniques discussed above.

THE PROPOSED EVOLUTIONARY BASED CLUSTERING ALGORITHM

Introduction to G3PCX

Here, G3PCX (Deb, et al., 2002) is formed from an abbreviation of Generalised Generation Gap (G3) and Parent-Centric Recombination (PCX), and is considered a computationally efficient genetic algorithm, especially developed for real parameter optimisation. Design of the G3 model emanates from the realisation that a population alteration model also plays a vital role in a real-valued optimisation process, and should be considered different from standard binary-based genetic algorithms. The G3 model employs the following four steps:

Step 1: From the population P, select the best parent and '$\mu - 1$' other parents randomly, where μ is set to 3, for instance, in (Deb, et al., 2002).
Step 2: Generate λ offspring from the chosen μ parents using the PCX, where λ is set to2 according to (Deb, et al., 2002).
Step 3: Choose two parents at random from the population P.
Step 4: From a combined subpopulation of the two chosen parents and λ-created offspring, choose the 'best' two solutions and replace the two parents (in step 3) with these.

Regarding variation operators, Deb (2001, p. 110-112) considers that a binary coded GA or a real-valued GA with simple naive crossover is no longer sufficient for real-parameter optimisation. Hence, a new crossover (recombination) based on the parent-centric principle was proposed. The underlying concept is based on the assumption that good solutions are most likely to appear in regions that are close to their parents, which have already satisfied the 'fitness test' in the selection operator. The PCX operator first calculates the mean vector $\vec{g}$ of the chosen μ parents so that for each offspring, one parent $X^{(P)}$ is chosen with equal probability. The direction vector $\overrightarrow{d^{(P)}} = X^{(P)} - \vec{g}$ is then calculated. Subsequently, from each of the other $(\mu - 1)$ parents, perpendicular distances D_i to the vector $\overrightarrow{d^{(P)}}$ are calculated along with their mean, $\overline{D}$. The offspring is then calculated from:

$$X_{offspring} = X^{(P)} + \omega_\zeta \cdot \overrightarrow{d^{(P)}} + \sum_{i=1,i\neq p}^{\mu} \omega_\eta \cdot \overline{D} \cdot \vec{e}^{(i)} \tag{1}$$

where, $\vec{e}^{(i)}$ are the $(\mu - 1)$ orthonormal bases that span the subspace perpendicular to the vector $\overrightarrow{d^{(P)}}$. The parameters ω_ζ and ω_η are zero-mean normally distributed variables with variance σ_ζ^2 and σ_η^2, respectively. Figure 1 indicates the underlying philosophy behind the use of PCX and shows the density of the potential solutions produced by three parents in a 2-dimensional decision variable space (represented by red circles at $[1.2\ 1.1], [1.2\ 1.25], [1.1\ 1.1]$).

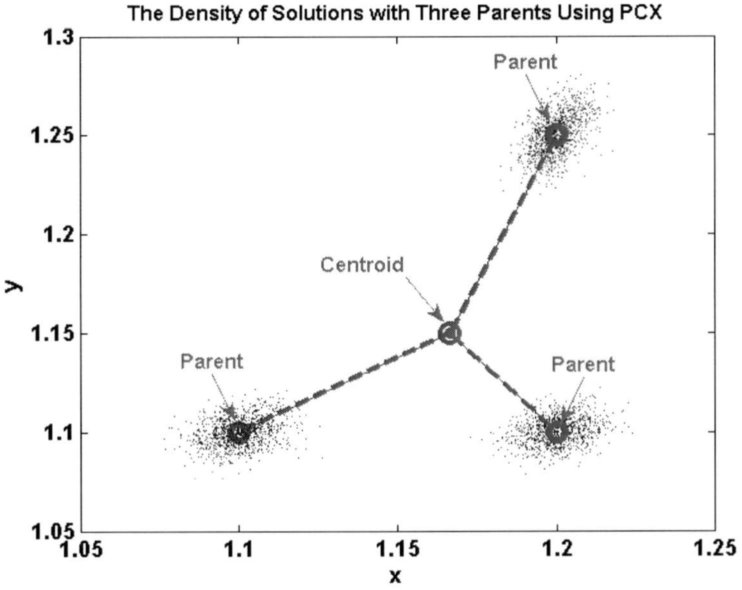

Figure 1. Solution density with three parents using PCX.

Description of G3Kmeans

G3PCX is now hybridised with *K*-means to address two well-documented problems associated traditional *K*-means, viz. sensitivity to initial conditions, and convergence to local optima. The resulting hybridised clustering technique is termed G3Kmeans, and is described by the following procedure:

Step 1: Initialisation: the randomly generated '*k*' cluster centres are encoded in each chromosome in a concatenated form. *P* initial chromosomes are generated from the initial population.

Step 2: Assigning data points: Each data point is assigned to one cluster with centre, C_i calculated using:

$$X_m \in C_i : if \begin{cases} \|X_m - C_i\| < \|X_m - C_l\| \\ m = 1,2,\ldots,N; i,l = 1,2,\ldots,k; l \neq i \end{cases} \tag{2}$$

where, is the Euclidean norm, X_m is the *m* th data point, C_i is the *i* th cluster centre, k is the pre-specified number of cluster centres, and *N* is the number of data samples. After the assignment, cluster centres encoded in the chromosome are updated by calculating the mean value of each cluster.

Step 3: Fitness computation: the fitness of each individual is calculated using (3):

$$\varpi(C_1, C_2, \dots, C_k) = \sum_{l=1}^{k} \sum_{X_m \in C_l} \|X_m - C_l\|^2 \; l = 1, \dots, k \tag{3}$$

where, ϖ is a within-cluster-distance metric to be optimised (minimised) and $C_1, C_2, \dots, C_3$ are k cluster centres.

Step 4: Parent-Centric Crossover (PCX): Generate λ offspring from the μ parents using the PCX recombination, as described in (1).

Step 5: Fitness Calculation: the cluster centres and fitness values of the offspring are updated and re-calculated, as in Steps 2 and 3.

Step 6: Parents to be replaced: choose two parents at random from the population P.

Step 7: Replacement: from the combined subpopulation of two chosen parents and λ created offspring, choose the 'best' two solutions and replace the chosen two parents (Step 6) with the new solutions.

Step 8: Iteration: Step 2 onwards is repeated for a pre-determined number of generations or until the standard deviation of the fitness values of the last five iterations becomes less than some threshold (*stable*), and the final solution is that with the smallest fitness value at the end of the execution.

Of note in the following case studies, is that all user-specified parameters are set to those suggested by Deb, *et al.* (2002) unless otherwise stated. Hence, $\sigma_\zeta^2 = \sigma_\eta^2 = 0.1, P = 100, \lambda = 2, \mu = 3, stable = 0.001$.

Case Studies

Two synthetically generated datasets along with a benchmark iris dataset are now employed to investigate the performance of the proposed G3Kmeans algorithm. The results are compared to those from FCM (Bezdek, 1981), *K*-means (Hartian, 1975, 1979), Subtractive Clustering (Chiu, 1994), GA-clustering (Murthy, et al., 1996) and KGA (Bandyopadhyay, et al., 2002).

Case Study 1: Data Set 1

The first test problem consists of 4 randomly generated Gaussian clusters about a nominal cluster centre[1] (dataset 1). Each cluster contains 100 data points, as shown in Figure 2. Figure 3 shows the identified cluster centres obtained by G3Kmeans, and Figure 4 the

[1] They are considered 'nominal centres' as the objective value of the clusters generated around these centres may not represent the minimum (see Table 1 for more details).

evolution curve of G3Kmeans[2]. Although the evolution has taken 11 generations to complete, notably it takes only 5 generations to reach the minimum objective value.

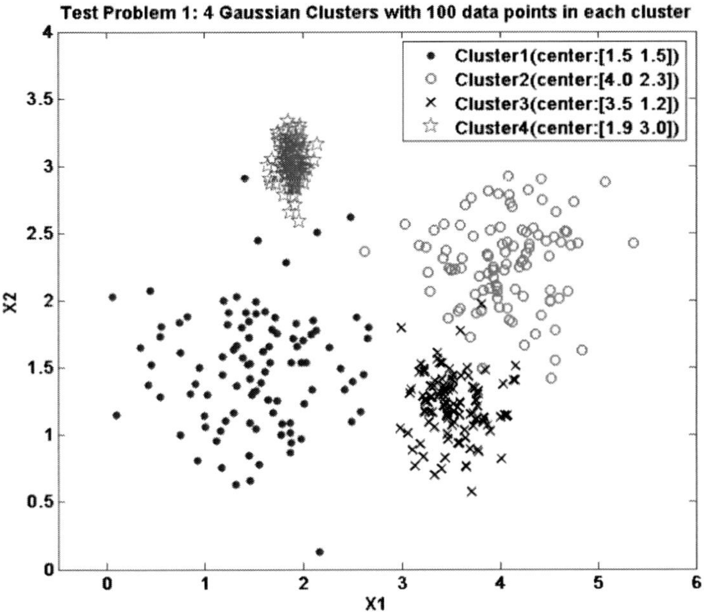

Figure 2. Four Gaussian clusters with 100 data points per clusters.

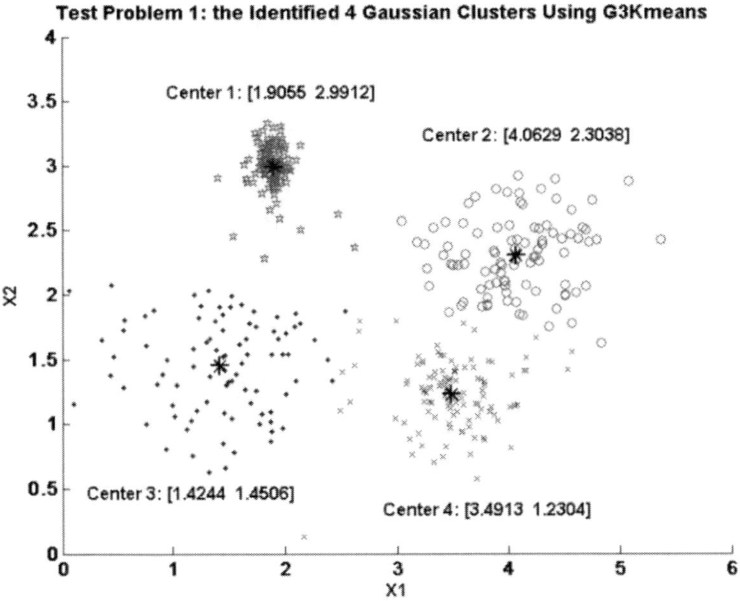

Figure 3. The four identified clusters of the first test problem using G3Kmeans.

[2] The objective value is calculated using normalised data and this is fixed for the following problems unless otherwise stated.

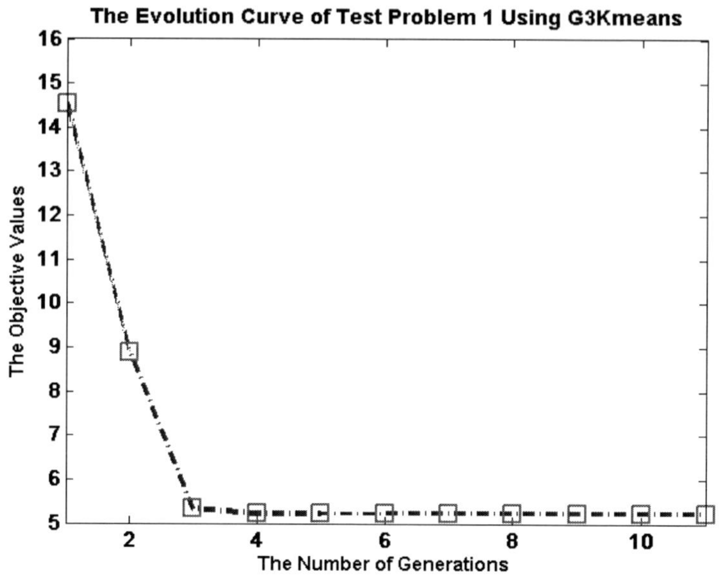

Figure 4. Evolution curve of the first test problem using G3Kmeans.

To provide a quantitative comparison of the performance of the various clustering algorithms, objective values are calculated using (3) and are used as a measure of the algorithms' efficacy. The objective value of the original clusters is also calculated using their nominal centres as a baseline in order to determine how close the resulting solutions are.

Table 1 provides the average values of 20 independent executions of the algorithms. The standard deviation of the results is also used as an indication of robustness to different initial conditions.

Table 1. Performance Comparison

Methods	Maximum	Minimum	Mean	Standard Deviation	Time (second)
Original Clusters*	5.2880	5.2880	5.2880	0	-
FCM	5.2533	5.2522	5.2522	5.1640e-005	0.0271
K-means	5.2284	5.2284	5.2284	0	0.0090
Subtractive Clustering	7.6686	7.6686	7.6686	0	0.0706
G3Kmeans	5.2284	5.2284	5.2284	0	0.7283

It can be seen that both *K*-means and G3Kmeans approach the global optimal partition (nominal centres) and outperform the other techniques in this instance. Notably the standard deviation of the *K*-means trials is zero, indicating that for this problem there are no local optima. The FCM-based algorithm also consistently approaches the optimal solution. However, the small variations of the objective values associated with FCM indicate that even for this basic problem, different initial conditions inevitably lead to different solutions. The radius of Subtractive Clustering is set to its default value, i.e. 0.5, and the results consistently show a lack of solution accuracy. This is a commonly reported feature, and hence, Subtractive Clustering is often used as an a-priori method of estimating the number of clusters as a precursor to applying other techniques. For completeness, Figure 5 shows the distribution of the nominal centres, and the identified cluster centres, using the tested methods.

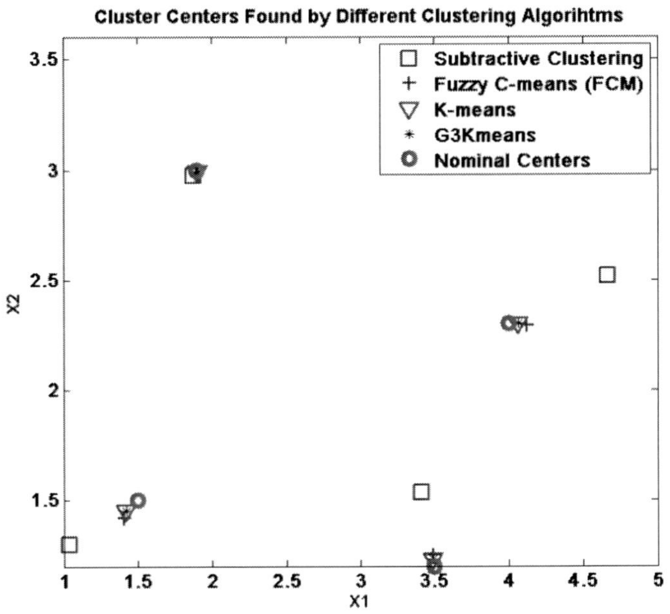

Figure 5. The distribution of identified cluster centres for dataset 1.

Case Study 2: Dataset 2

To investigate the relative robustness properties of the algorithms, random white noise is used to contaminate the data. Specifically, similar Gaussian clusters as those used in dataset1 are combined with 200 randomly distributed points of 'noise data'. The difficulties of this problem lie in the fact that the classes are not well separated and the search space presents many local optima due to the presence of noise. Figure 6 shows the resulting data distribution associated with the new dataset. Applying each algorithm, it is notable that G3Kmeans takes 11 generations to finish, but has converged to the minimum objective value within 6 generations. Again, for comparison purposes, the results are summarised in Table 2, which shows the solution means from 20 independent runs.

Table 2. Performance Comparison

Methods	Max.	Min.	Mean	Standard Deviation	Time (second)
Original Clusters	13.8697	13.8697	13.8697	0	-
FCM	13.1898	13.1895	13.1897	1.038e-004	0.0269
K-means	18.3056	13.0382	13.5173	1.5881	0.0130
Subtractive Clustering	16.2954	16.2954	16.2954	0	0.2000
G3Kmeans	13.0382	13.0382	13.0382	0	0.8008

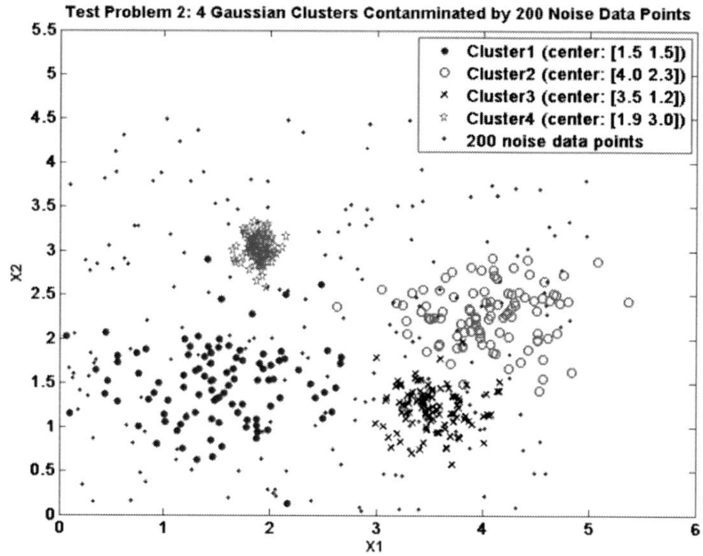

Figure 6. 4 Gaussian clusters contaminated by 200 ‘noise’ data points.

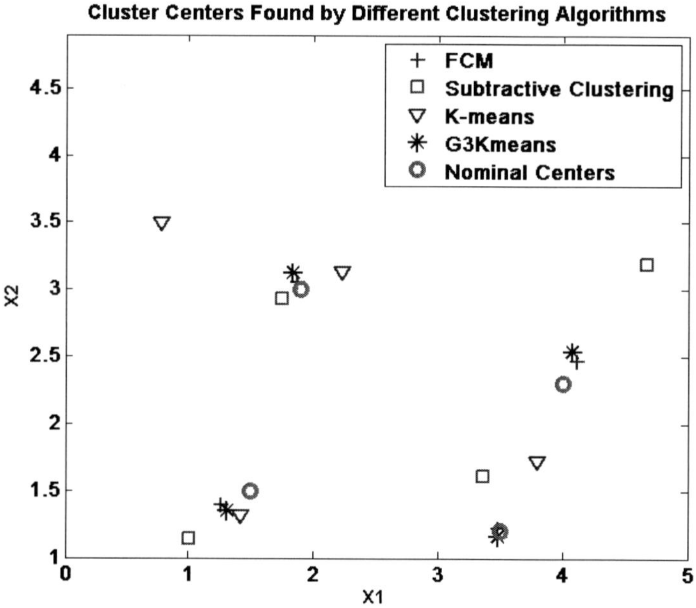

Figure 7. The distribution of the identified cluster centres for artificial data set 2.

It is clear from the results in Table 2 that both FCM and *K*-means are now sensitive to initial conditions, as a consequence of the presence of noise. Moreover, since the problem now contains several local optima, a non-zero standard deviation from *K*-means occurs in this case. Also of note is that *K*-means misclassified clusters in 2 out of the 20 runs. Figure 7 shows the distribution of the identified cluster centres from the various techniques, along with the nominal centres. It can be seen that the results generated by G3Kmeans are very similar to those of the first data set, indicating that G3Kmeans provides a degree of robustness to the presence of noise. The results from Table 2 and Figure 7 also indicated that the solution centres found by *K*-means can be remote from the nominal centres.

Case Study 3: Benchmark Problem-Iris data

The performance of G3Kmeans is now compared to alternative GA-based *K*-means techniques, e.g. GA-Clustering (Murthy, et al., 1996) and KGA (Bandyopadhyay, et al., 2002). The Iris dataset (Fisher, 1936) is used for the investigation, which consists of 150 patterns belonging to three categories of Iris. Each pattern is described by four real-valued features (in centimetres) viz. sepal length, sepal width, petal length and petal width. Each category consists of 50 patterns. Difficulties in classifying the Iris data are due to two overlapping classes that present many local optima. On applying the algorithms, G3Kmeans takes 11 generations to terminate, although convergence took place after 5 generations. Table 3 summarises the performance of all techniques[3] after 20 independent runs.

Table 3. Performance Comparison Using the Iris Data

Methods	Max.	Min.	Mean	Standard Deviation	Time (second)
FCM	79.4566	79.4516	79.4557	1.6000e-3	0.0252
K-means	142.9149	79.0031	95.0244	27.4598	0.0052
Subtractive Clustering	84.6800	84.6800	84.6800	0	0.0114
GA-clustering	139.7782	124.1274	135.4048	-	-
KGA	97.1008	97.1008	97.1008	0	-
G3Kmeans	79.0031	79.0031	79.0031	0	0.4623

The results from GA-clustering and KGA have been taken from Bandyopadhyay,*et al.* (2002) for consistency. In this case, *K*-means misclassified clusters 4 times in the 20 runs. A high standard deviation is also associated with *K*-means, again indicating that the data consists of many local optima and confirms that *K*-means is sensitive to such circumstances. The results of G3Kmeans are shown to be far superior to those of GA-clustering and KGA, as expected from the reasons discussed in Section II.

Specifically, GA-clustering is unable to provide meaningful clusters within 1000 iterations (Bandyopadhyay, et al., 2002). This is mainly due to the coding scheme that unnecessarily increases the search space and thereby requires more computational effort to converge. Figure 8 shows the identified Iris classes and their centres on two-dimensional planes of two features.

[3] Unlike the results presented in Tables 1 and 2, the objective values shown in Table 3 are calculated using the original data so that the results produced by G3kmeans can be compared directly with GA-clustering and KGA.

INDUSTRIAL APPLICATION

To investigate the scalability of G3Kmeans to high dimensional problems, and its efficacy for providing compressed representations of large datasets and machine fault diagnostics, two sets of sensor measurements from sub-15MW industrial gas turbines, are now used to form 2 case studies.

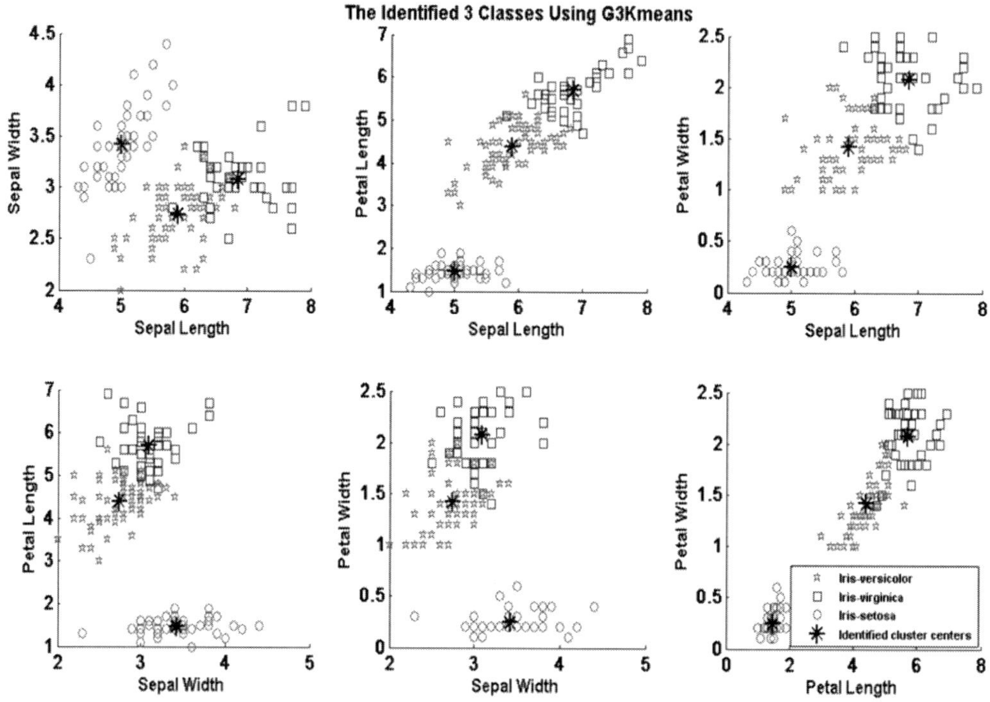

Figure 8. Three Iris classes from G3Kmeans.

Case Study 4: Sub-15MW Industry Gas Turbines

Here the data is collected from four groups of sensors viz. bearing vibration, shaft speed, bearing temperature, and lubrication oil temperature, sited on a twin shaft industrial gas turbine, as depicted in Figure 9. The datasets consist of up to 20 feature variables. Measurements were recorded during March 2011 and consist of 420928 data samples that were considered to reflect 'normal' operation. A further 49074 data samples were taken during December 2011 when an emerging fault was noted. The sampling frequency is 1Hz.

The technique developed by Zhang, et al. (2012), specifically, the Principal Component Analysis (PCA) based Y-distance Indexes, is initially used to distinguish between normal and abnormal operational characteristics. However, it is notable that in general, the method is very restrictive for high numbers of data points due to the reliance on traditional PCA, where large data sets require a high storage capacity to hold the PCA-matrix. In this case then, G3Kmeans is applied for the purpose of providing a reduced representation of the large data set whilst still preserving 'sufficient' information contained in the original data.

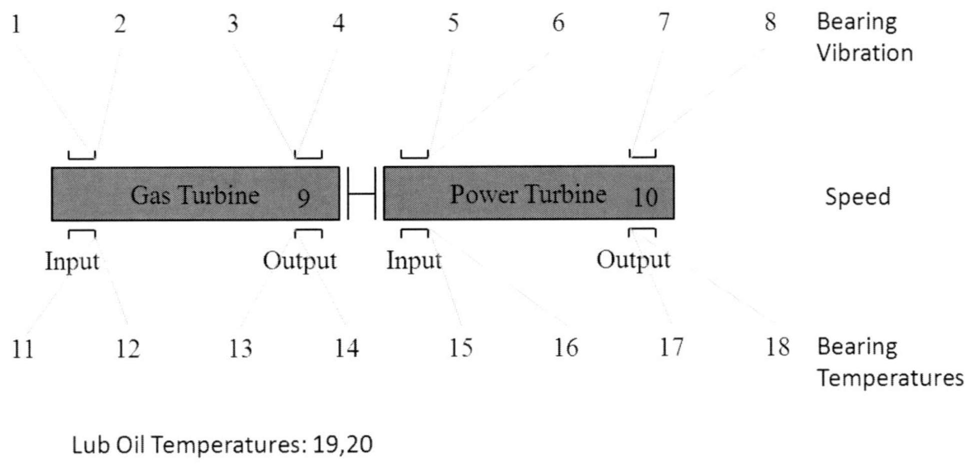

Figure 9. The location of 20 sensors on a twin shaft industrial gas turbine.

This has been achieved by considering clustered centres as representing the original data. Due to restrictions on space and visualisation, Figures 10 and 11 only give two examples of the reduced representation based on two feature variables.

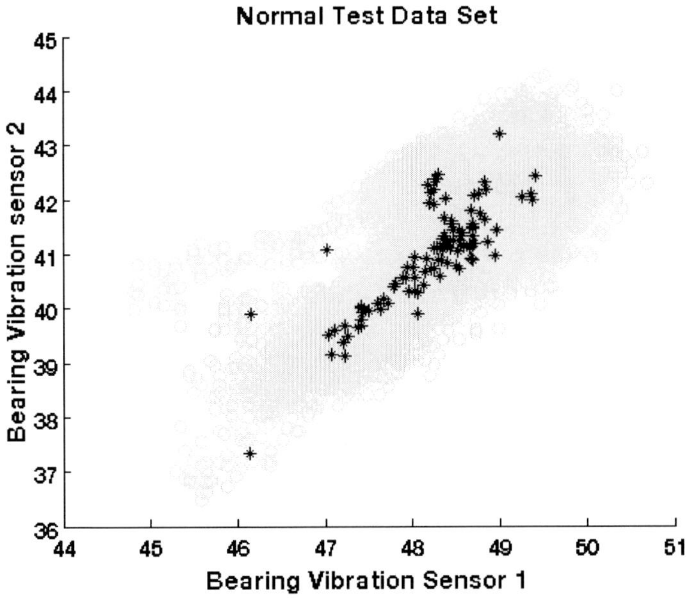

Figure 10. Reduced presentation of the 42098 point 'normal operation' test data by 100 data points using G3Kmeans (yellow circles are the original data; black asterisks are the reduced representation).

It can be seen from Figure 10 that for normal operating conditions, sensor measurements, and resulting clusters are well bounded. By contrast, Figure 11shows results from the alternative operational scenario where measurement data is spread across a much wider region—with some well outside of what would be considered 'normal operation'.

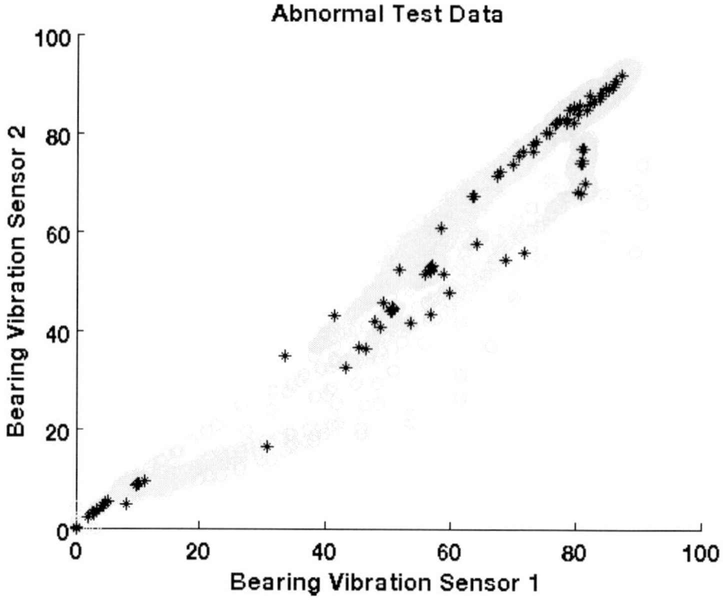

Figure 11. Reduced presentation of the 49047 point 'abnormal operation' test data by 100 data points using G3Kmeans (yellow circles are the original data; black asterisks are the reduced representation).

It is therefore crucial that a data reduction algorithm can capture not only the 'normal' characteristics but also be capable of providing cluster centres that are sensitive to the 'abnormal' variations. Visual inspection of Figure 11shows that all the important features of the data are captured by G3Kmeans in this case (the compression ratios in each case are, respectively, 420.98:1 and 490.47:1), and it can therefore be used to facilitate significant reductions of required storage capacity.

Case Study 5: Fault Diagnosis from Rundown Vibration Signatures

Vibration signatures taken from rundowns of gas turbine units are considered information-rich for determining the health of the underlying component parts. This has previously been reported by Gallimore, et al. (2013), where, during a rundown, the engine will normally pass through at least one critical rotor speed or natural frequency. A typical rundown vibration signature is shown in Figure 12—in this case the measurement data has been interpolated using a shape-preserving algorithm (Matlab® Curve Fitting Toolbox) to provide values with a resolution of 1rpm. To enable a consistent reference datum between the rundown series, only data between the speeds of 6000 rpm and 1000rpm are considered for comparison. The objective is to group the run-downs in order to identify those that show 'novel' characteristics and thereby act as an early warning of emerging fault conditions.

Given the high volume of data and the need to perform real-time similarity searches, dimensionality reduction techniques are often used as a pre-processing step for subsequent search (or novelty detection) techniques. Recently, Gallimore, et al. (2013) proposed to use Piecewise Aggregate Approximation (PAA) (Keogh, et al., 2000) as the dimensionality reduction technique to approximate vibration signatures by segmenting the sequences into equal-length sections and recording the mean values of each section. The horizontal bars shown in Figure 12 are the mean values of 10x chosen equal-length sections[4]in this case, thereby reducing the rundown signature to a point in 10-dimensional hyperspace.

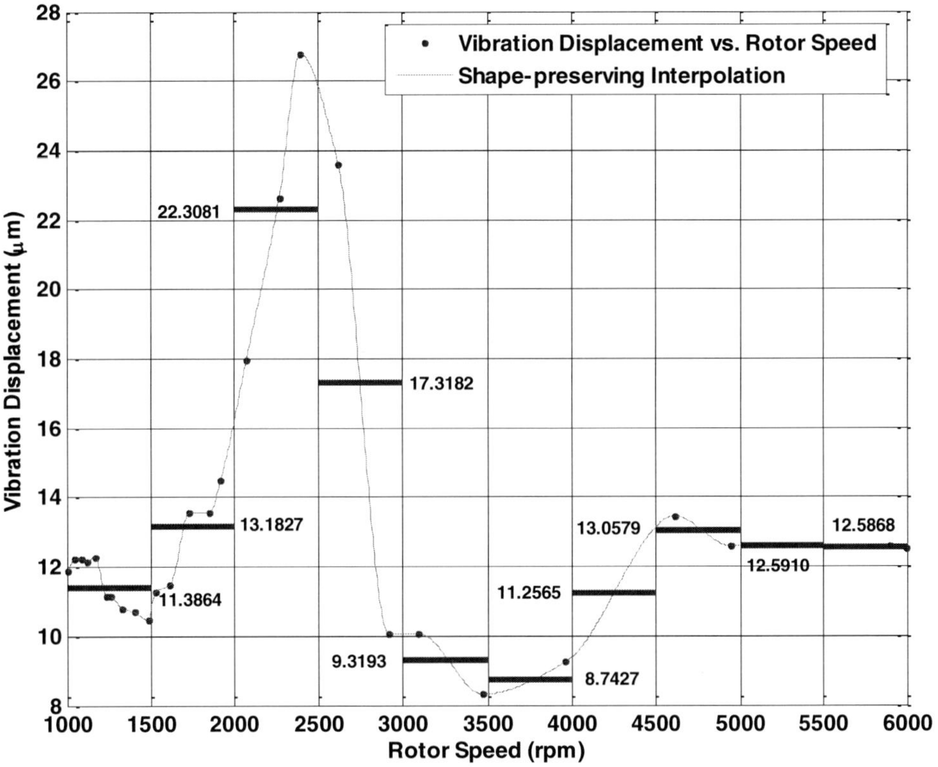

Figure 12. Example rundown vibration signature.

A series of 54 rundown characteristics are used for classification. The series include what are considered to be 52 'normal' rundowns and 2 rundowns that are associated with known emerging fault conditions. Figure 13 provides a 3D plot of the raw data of the 54 run-downs. The challenge is to identify signatures that correspond to emerging fault conditions i.e. signatures 39 and 43in this case.

As has been previously demonstrated, local clustering algorithms such as *K*-means can very easily misclassify conditions in circumstances such as these, and thereby signal unnecessary false alarms to the operator. By way of example, *K*-means regularly misclassifies signatures 38, 40, 45 and 49 as abnormal in this case.

This is because *K*-means is sensitive to initial conditions when multiple local optima are present, and therefore also sensitive to only a portion of the signature for pattern matching—see Figure 14, for example. This leads to classifications that are only locally optimal, and therefore necessitates that use global clustering algorithms, such as G3K means, for addressing problems of this complexity.

[4] This number applies to the rest of the chapter.

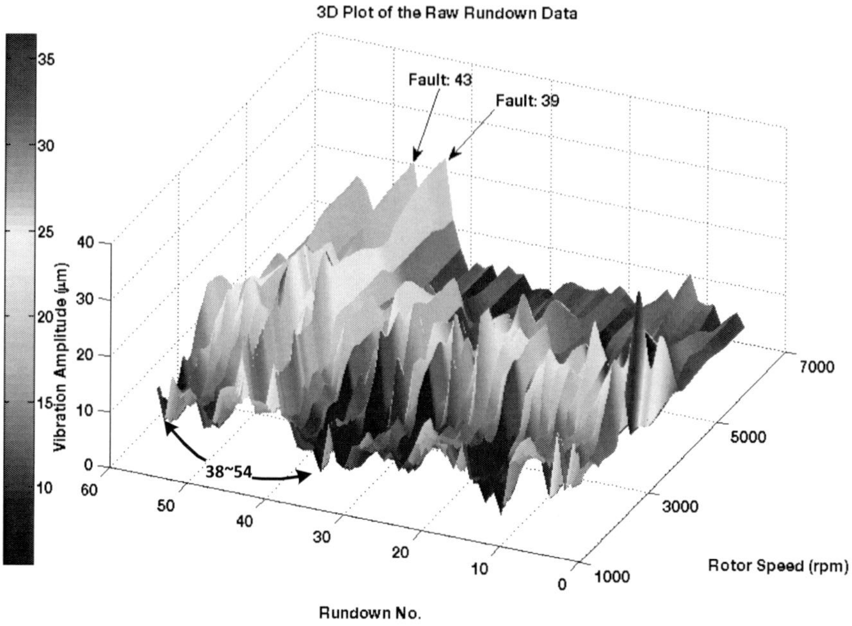

Figure 13. The 3D plot of raw data from 54 run-downs.

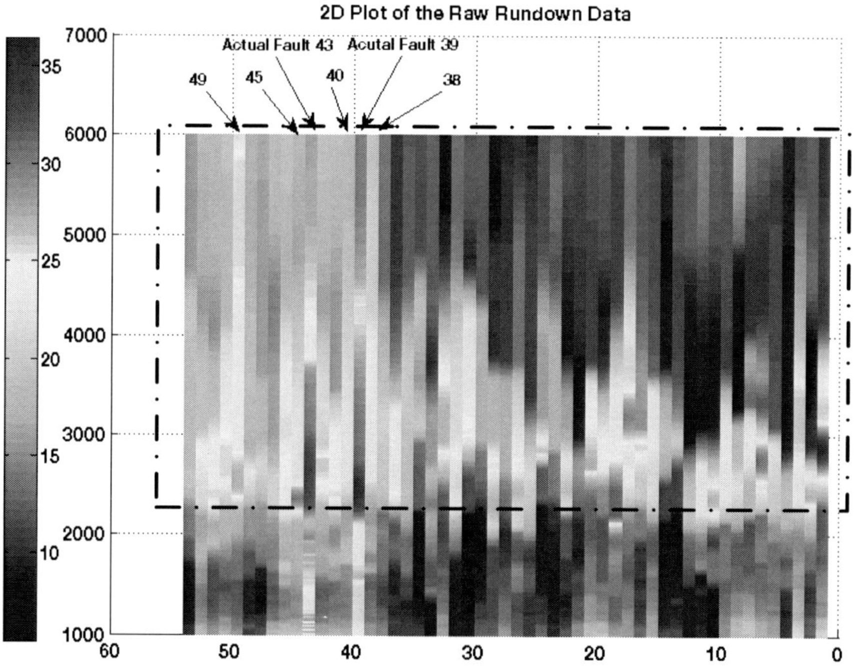

Figure 14. A fraction of rundown vibration signatures used by *K*-means.

A vector representation of the 54 signatures is obtained using PAA, and are then used as input to G3Kmeans to investigate the performance of the algorithm for detecting 'novelty'. A nominal 4 clusters are used in this case according to Gallimore, et al. (2013).To provide a comparison of performance, *K*-means clustering is used as a benchmark, and two metrics are employed. The first measures the density of identified clusters, as given in (3), and should be as small as possible to make coherent decisions on health classification. The second measures the separation between the clusters—defined as the sum of the distance between any two identified cluster centres. A higher index in this case indicates better clustering performance and will reduce the scope for misclassification. The results presented in Table 4 summarise the maximum, minimum, mean and standard deviation of the objective values obtained by the two clustering techniques, along with calculation of the two metrics. 20 independent runs of each algorithm are used. Notably, G3Kmeans required a total of 10 generations although it converged in 8.

Table 4. Comparisons of the Objective Values, Density and Separation, for G3Kmeans and *K*-Means

Methods	Maximum (Obj.)	Minimum (Obj.)	Mean (Obj.)	Standard Deviation	Density (Mean)	Separation (Mean)
G3Kmeans	8.9541	8.9541	8.9541	0	8.9541	167.4564
K-means	18.5550	11.1193	14.4565	2.0177	14.4565	114.5850

From Table 4, it can be seen that G3Kmeans consistently produces the lowest objective values with no standard deviation, whilst the results of *K*-means was highly dependant on initial conditions. Of particular note is that in only 1 out of the 20 runs did *K*-means correctly classify the 2 rundowns associated with the known fault conditions (corresponding to the minimum objective value shown in Table 4). G3Kmeans also produces denser and more separable clusters by comparison—Figure 15 shows the identified clusters using the 6th and 7th segments, for instance.

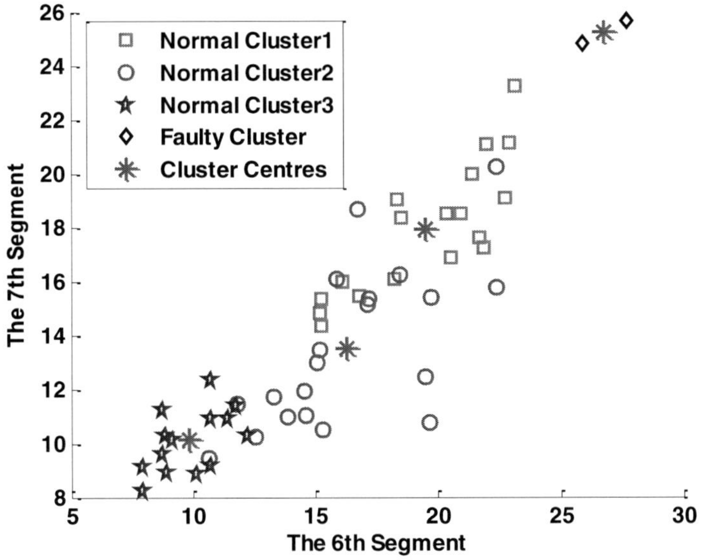

Figure 15. Clustering of the vibration signatures.

Table 5 shows the classified rundowns using G3Kmeans.

Table 5. The Classified Rundowns Using G3Kmeans

Normal Cluster1	3, 17, 31, 38, 40, 41, 42, 44, 45, 46, 47, 48, 49, 50, 51, 52, 53, 54
Normal Cluster2	1, 6, 7, 9, 15, 16, 18, 19, 20, 23, 24, 26, 28, 29, 30, 32, 34, 35, 36, 37
Normal Cluster3	2, 4, 5, 8, 10, 11, 12, 13, 14, 21, 22, 25, 27, 33
Faulty Cluster	39, 43

Conclusion

An evolutionary-based clustering algorithm, G3Kmeans, is proposed. Performance comparisons using a number of case studies have demonstrated that the proposed algorithm is more robust than other, more traditional clustering techniques. Specifically, it is shown to be robust to initial conditions, and although global optimal partitions are not guaranteed, G3Kmeans can approach very closely to such partitions, even for high-dimensional problems. The method is also more computationally efficient compared to other evolutionary based clustering algorithms—in a specific case study concerning the classification of iris data, GA-clustering required more than 1000 iterations to converge, whilst G3Kmeans took only 11. The performance improvement is mainly attributed to the combined local and global search operators adopted in G3Kmeans, which are specially designed for the real-valued optimisation. The encoding scheme of the proposed method also ensures a more reasonable search space when compared to GA-clustering algorithms, for instance. Moreover, results show that G3Kmeans can be effective for data reduction whilst 'preserving' important features contained in the original data set, and be used as an effective means of novelty detection. At present it is notable that the current generation of G3Kmeans only uses within-cluster distance as the objective function. Other objectives are also being investigated for inclusion in order to provide better separation between clusters and promote the generation of clusters with higher density.

References

Bandyopadhyay, S.; Maulik, U. (2002). An Evolutionary Technique Based on K-Means Algorithm for Optimal Clustering in $\mathbb{R}^N$, *Information Sciences*, vol. 146, 221-237.

Bezdek, J. C. Pattern Recognition with Fuzzy Objective Function Algorithms, MA: Kluwer Academic Publishers Norwell, 1981.

Bezdek, J. C.; Hathaway, R. J. (1994).Optimization of Fuzzy Clustering Criteria Using Genetic Algorithms, *The 1st IEEE Conference on Evolutionary Computation*, vol. 2(2), 589-594.

Chen, J. Biologically Inspired Optimisation Algorithms for Transparent Knowledge Extraction Allied to Engineering Materials Processing, PhD. Thesis, *University of Sheffield*, 2009.

Chen, J.; Mahfouf, M.; Bingham, C.; Zhang, Y.; Yang Z. J.; Gallimore, M. (2012). An Evolutionary Based Algorithm Applied to Data Compression for Industrial Systems, *Advances in Intelligent Data Analysis XI*, LNCS, vol. 7619, 103-113.

Chiu, S. L. (1994). Fuzzy Model Identification Based on Cluster Estimation, *J. of Intelligent & Fuzzy Systems*, vol. 2 (3).

Deb, K. Multi-Objective Optimization using Evolutionary Algorithms; Wiley: Chichester, U.K., 2001.

Deb, K.; Anand, A.; Joshi, D. A.(2002).Computationally Efficient Evolutionary Algorithm for Real-Parameter Optimization, Evolutionary Computation, vol. 10 (4), MIT Press, 371-395.

Fisher, R. A. (1936). The Use of Multiple Measurements in Taxonomic Problems, *Ann. Eugen.*, vol. 7(2), 179-188.

Gallimore, M.; Riley, M.; Bingham, C.; Yang, Z. J.; Zhang, Yu.(2013). Intelliget Diagnositics and Prognostics for Industrial Machines Using an Optimization Approach, 4th International Conference on Integrity, Reliability and Failure, Funchal.

Hall, L. O.;Özyurt, I. B.;Bezdek, J. C. (1999). Clustering with a Genetically Optimized Approach", *IEEE Transactions on Evolutionary Computation*, vol. 3(2), 103-112.

Hartigan, J. A. Clustering Algorithms, New York: Wiley, 1975.

Hartigan, J. A.; Wong, M. A. (1979). A k-means Clustering Algorithm, *Journal of the Royal Statistical Society*, Series C (Applied Statistics), vol. 28(1), 100-108.

Keogh, E.; Chakrabarti, K.; Pazzani, M.; Mehrotra, S. (2000). Dimensionality reduction for fast similarity search in large time series databases, *Journal of Knowledge and Information Systems.*

Keogh, E.; Chakrabarti, K.; Mehrotra, S.; Pazzani, M. (2001). Locally Adaptive Dimensionality Reduction for Indexing Large Time Series Databases, *ACM SIGMOD*, Santa Barbara, California, USA.

Krishna, K.; Murty, M. Narasimha. (1999). Genetic K-Means Algorithm, IEEE Transactions on Systems, *Man, and Cybernetics-Part B: Cybernetics*, vol. 29(3), 433-439.

Murthy, C. A.; Chowdhury, N. (1996).In Search of Optimal Clusters Using Genetic Algorithms, *Pattern Recognition Letters*, vol. 17, 825-832.

Sheng, W.; Liu, X. (2006).A Genetic K-medoids Clustering Algorithm, *Journal of Heuristics*, vol. 12, 447-466.

Zhang, Y.; Bingham, C. M.; Yang, Z.; Gallimore, M.; Ling, W. K. (2012).Sensor Fault Detection for Industrial Gas Turbine System by Using Principal Component Analysis Based Y-distance Indexes, proceeding on 8th IEEE, *IET International Symposium on Communication Systems, Networks and Digital Signal Processing*.

In: Fault Detection
Editors: F. P. G. Márquez and M. Papaelias
ISBN: 978-1-62808-999-8

Chapter 8

MODEL-BASED FAULT DIAGNOSIS FOR INDUSTRIAL MOBILE ROBOTS

Andrea Monteriù* and Sauro Longhi
Dipartimento di Ingegneria dell'Informazione
Università Politecnica delle Marche, Ancona, Italy

Abstract

This chapter describes how a suitable design of a fault diagnosis system can be successfully applied on industrial mobile robots for increasing their reliability and safety, improving their operability and maximize their performance for different tasks and different environments.Structural analysis is applied to the nonlinear model of an industrial mobile robot for residual generation. An adaptive/thresholding decision algorithm is described for developing the residual evaluation module. This includes a threshold test based on the minimum and maximum values of the generated residuals to decide if a fault has occurred or not. Moving thresholds are updated using a sliding window of a fixed number of sequential sample readings. With this solution, probability density functions of residuals are unknown and do not need to be estimated. Experimental validation of the proposed scheme is performed on a differential drive industrial mobile robot, and it is presented in the last part of the chapter.

Keywords: fault detection, fault isolation, fault diagnosis, model-based techniques, sensor fusion, industrial robot, mobile robot

1. Introduction

There is an increased interest in mobile robotics across all industries. This is mainly due to the ability of one mobile robot to service several locations and perform a greatly expanded range of tasks which offers a great appeal for specialized applications. Thank to this, mobile robots are proliferating, especially in industrial applications, where they facilitate the connection between manufacturing cells, transporting materials from donation areas and taking these raw materials to a work cell. In the industrial space, mobile robots are redefining the playing field for Autonomous Guided Vehicles (AGVs) in that modern mobile

*E-mail address: a.monteriu@univpm.it

platforms are capable of operating in areas without requiring alterations or investment into existing infrastructure ([34]). Mobile robots are finally overcoming a historical impediment of AGVs consisting on their inability to dynamically reroute themselves. This is due to the fact that mobile robots can be now designed and outfitted with advanced sensory and enhanced intelligence systems ([35, 23, 1]). On the other hand, this increases the complexity of the industrial mobile robots, and thus the chances for malfunctions.

In industrial mobile robots, the complexity is permanently growing due to the continuos requirements of higher positioning accuracy and degree of automation ([39]). In order to realize lean manufacturing, mobile robots must be highly intelligent and able to autonomously deliver parts from any random origin to any random destination. Thus, several kinds of sensors need to be integrated together to ensure a more reliable guidance control system ([36, 8]). Data fusion is required to process the navigation information collected from heterogeneous navigation sensors ([6, 3]). Among information coming from the multisensor system, some can be erroneous or faulty. The reason must be either a sensor fault or a temporary inadequacy of the sensor in the industrial work environment of the robotic system ([29]). The existence of a sensor fault can cause undesired reactions and consequences to technical part of the robotic system, and thus aggravating the impact for man and for the industrial environment ([30]). Therefore, in the design and operation of an industrial mobile robot, increased attention has to be paid to reliability and safety. In order to maintain a high level of safety, performance and reliability, it is important that sensor faults are detected promptly and that the source and severity of each fault is diagnosed. The backbone of modern fault diagnosis is the *model-based approach*, where the mathematical model of the supervised system is used as a reference and contains the analytical redundancy ([32, 25, 21, 15, 28]). Making use of dynamic models of the system under consideration provides the most powerful approach, allowing to determine the time, size and cause of even small faults during all phases of dynamic system operations. Design and development of model-based fault diagnosis systems for industrial mobile robots is the topic of this chapter. The industrial mobile robots, in this case, are supposed to be equipped with a sensor suite that includes two incremental optical encoders mounted on motors measuring motor rotation, a gyroscope measuring the robot orientation, and a laser scanner mounted at the front of the robot measuring the robot distances with respect to the industrial environment.

Given an industrial mobile robot (or any system), faults may be overcome by using robust "model-based fault diagnosis" defined as *"... the determination of faults of a system from the comparison of available system measurements with a priori information represented by the mathematical model of the system through generation of residual quantities and their analysis"* ([9]). Residuals are zero when no faults occur and non-zero otherwise, detecting the presence of a fault. Residual quantities differ for each fault, allowing detection of the specific occurred fault. This chapter describes the development of model-based fault diagnosis systems, which comprises both residual generation and residual evaluation, applying them to industrial mobile robots. Additive and abrupt sensor faults are considered here, describing changes in the system states interpreted as sensor faults of the industrial mobile robot.

The structural analysis ([27, 22]) is chosen to generate the residual signals, applying it on the nonlinear model of the industrial mobile robot. The residual generator module generates the specific residuals to detect sensor faults following structural analysis of the mobile robot nonlinear model, and determining existing redundancies. Structural analysis depends not on analytical relations, but on relations between a variable and a constraint, allowing exploration of fundamental system properties using structure graphs. This is essentially a way to indicate which equations (or constraints) in a system (industrial mobile robot in this case) are needed to find a solution for its variables. If there are more equations than needed, excess equations may be used to check the validity of observations. If such excess equation is not valid, the unmatched constraint called residual indicates the presence of a fault in the system ([5]).The residual evaluation module reduces to the problem of detecting a change in the mean of a random sequence ([31]). This chapter describes a feasible solution for developing the residual evaluation module of diagnosis system for industrial mobile robots. An adaptive/thresholding decision algorithm is described which includes a threshold test based on the minimum and maximum values of the generated residuals to decide if a fault has occurred or not. Moving thresholds are updated using a sliding window of a fixed number of sequential sample readings (the window size may vary). With this solution, probability density functions of residuals are unknown and do not need to be estimated. Experimental validation of the proposed scheme is performed on a differential drive mobile robot, and it is presented in the last part of this chapter.

Fault diagnosis techniques, that permits to monitor online the navigation system for sensor faults, can contribute significantly to increase the reliability of industrial mobile robots. Quick detection and, possibly, isolation of sensor faults can prevent serious damages and possible irreparable consequences. This chapter describes how a suitable design of a fault diagnosis system can be successfully applied on industrial mobile robots for increasing their reliability and safety, improving their operability and maximize their performance for different tasks and different environments. The diagnostic solutions presented in this chapter can be easily applied and extended to a wide class of industrial mobile robots.

2. Industrial Robot Model

The major advantage of the model-based approach is that no additional hardware components are required in order to realize a fault detection and isolation algorithm ([16]). In fact, a model-based Fault Detection and Isolation (FDI) algorithm can be implemented via software on a process control computer. This advantage is paid in terms to model the system to be monitoring as better as possible. This model is the key tool to find the redundancy of the system, and thus generate the residual signals. Model-based fault diagnosis methods rest on the model of the system behavior. This model have to describe the nominal as well as the faulty behavior of the considered dynamic system.In this section, the model assumptions and simplifications of an autonomous industrial mobile robot are argued and, finally, the robot non-linear model is developed.

The considered industrial mobile robot is equipped with a suitable navigation sensor equipment, widely used and adopted in industrial mobile robots, which includes odometric, gyroscopic and laser range measures. These measurements are used in the developed model-based fault diagnosis algorithm to detect and, possibly, isolate sensor faults, which can cause a wrong localization of the industrial mobile robot.

2.1. The Robot Navigation System and the Sensor Equipment

Autonomous navigation is one of the most challenging competences required of an industrial mobile robot. For this reason, this topic has received great research attention ([12, 18, 19, 38, 14, 37, 33]) in the past decade and, as a result, significant advances have been made on this front.

Any autonomous industrial mobile robot is equipped with a navigation system which permits to obtain its relative localization ([20]). In the considered study case, the following sensors are integrated and used by the navigation module: two incremental optical encoders in the rear wheels, an optical gyroscope (HITACHI mod. HOFG-1), and a laser range finder (SICK LMS mod. 200). These sensors are used for the on-line estimation of the industrial mobile robot pose and for detecting unknown obstacles in the environment, that is assumed *a priori* known. This sensor equipment is a basic sensor suite but is quite common for autonomous mobile robots used in industrial environments.

2.1.1. Odometric Measures

Assume the kinematic model of an industrial robot as an unicycle-like mobile robot with two driving wheels, mounted on the left and right sides of the robot, with their common axis passing through the center of the robot (see Figure 1).

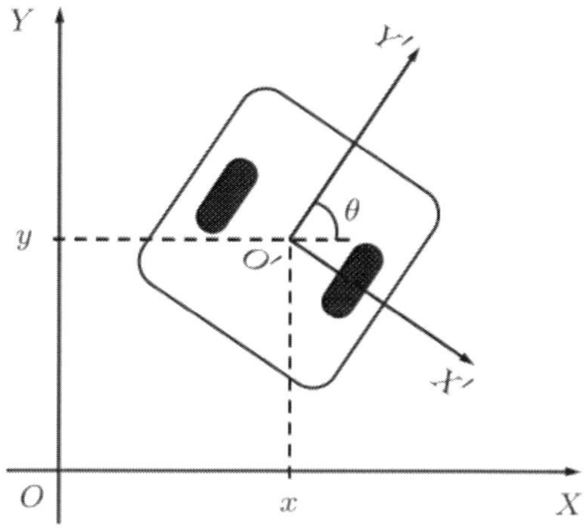

Figure 1. The scheme of the unicycle robot.

Localization of this mobile robot in a two-dimensional space requires the knowledge of coordinates x and y of the midpoint between the two driving wheels and of the angle θ between the main axis of the robot and the X-direction of the considered inertial coordinate system (O, X, Y). The kinematic model of the unicycle robot is described by the following equations

$$\dot{x}(t) = \nu(t)\cos\theta(t) \tag{1}$$
$$\dot{y}(t) = \nu(t)\sin\theta(t) \tag{2}$$
$$\dot{\theta}(t) = \omega(t) \tag{3}$$

where $\nu(t)$ and $\omega(t)$ are, respectively, the displacement velocity and angular velocity of the robot which are expressed by

$$\nu(t) = \frac{D}{4}\left(\omega_l\left(t\right) + \omega_r\left(t\right)\right) \tag{4}$$
$$\dot{\theta}(t) = \frac{D}{2L}\left(\omega_r\left(t\right) - \omega_l\left(t\right)\right) \tag{5}$$

where $\omega_r(t)$ and $\omega_l(t)$ are the angular velocities of the right and left wheels, respectively, D is the wheel diameter and L is the distance between the wheels.

2.1.2. Fiber Optic Gyroscope Measures

The operative principle of a Fiber Optic Gyroscope (FOG) is based on the Sagnac effect [7, 10, 24, 40]. The FOG readings show the angular velocities of the robot and are denoted by $\omega_g(\cdot) \equiv \omega(\cdot)$.

2.1.3. Laser Scanner Measures

The distance readings by the Laser Measurement System (LMS) are related to the indoor environment model and to the configuration of the industrial mobile robot. Let denote with l the distance between the center of the laser scanner and the origin O' of the coordinate system $(O',\ X',\ Y')$ fixed to the mobile robot, as reported in Figure 2.

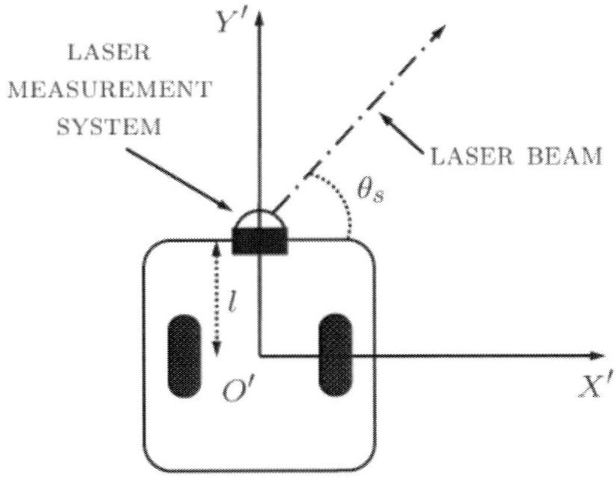

Figure 2. The scheme of the unicycle robot.

At the sampling time t_k, the position x_s, y_s and orientation θ_s of the center of the laser scanner, referred to the inertial coordinate system $(O,\ X,\ Y)$, have the following form

$$x_s(t_k) = x(t_k) + l\ \cos\theta(t_k) \tag{6}$$

$$y_s(t_k) = y(t_k) + l\ \sin\theta(t_k) \tag{7}$$

$$\theta_s(t_k) = \theta(t_k) \tag{8}$$

The walls and the obstacles in an indoor environment are represented by a proper set of planes orthogonal to the plane XY of the inertial coordinate system. Each plane P is represented by the triplet P_r, P_n and P_ν, where P_r is the normal distance of the plane from the origin O, P_n is the angle between the normal line to the plane and the X-direction and P_ν is a binary variable, $P_\nu \in \{-1,\ 1\}$, which defines the face of the plane reflecting the laser beam. In such a notation, the expectation of the laser reading $d(t_k)$, relative to the present distance of the center of the laser scanner from the plane P, has the following expression (see Figure 3)

$$d(t_k) = P_\nu \frac{P_r - x_s(t_k)\cos(P_n) - y_s(t_k)\sin(P_n)}{\cos(\theta_d)} \tag{9}$$

where $\theta_d = P_n - \theta^*$ with $\theta^* \in [\theta_0,\ \theta_1]$ given by

$$\theta^* = \theta + \theta_s - \frac{\pi}{2} \tag{10}$$

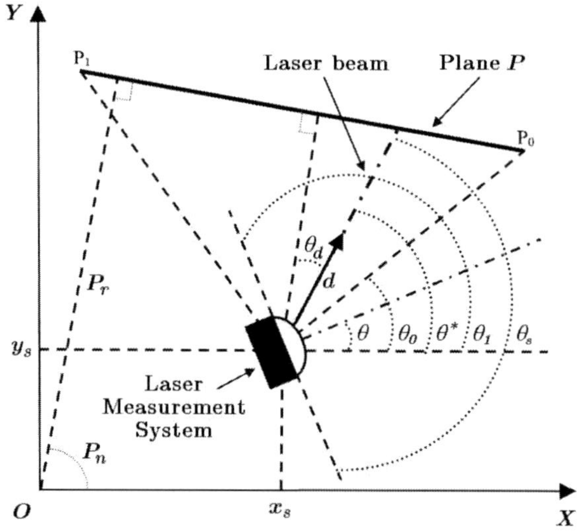

Figure 3. Laser scanner measure.

3. Sensor Fault Diagnosis System

Any industrial mobile robot, as all artificial systems controlled by the use of sensors, is vulnerable to faults such as defects in sensors or failures, which can cause undesired reactions and consequences as damage to technical parts of devices, to the user or to the environment. For this reason, fault detection is an important and a key issue to consider [4] and, when possible, fault isolation must be achieved.

In this section a sensor fault diagnosis system is proposed for introducing in the industrial robotic system a set of functions for increasing the navigation reliability of the autonomous industrial mobile robot. The proposed sensor fault detection and isolation system will be integrated with the navigation system. The diagnostic system concerns of all fault detection and isolation steps. Firstly, a residual generator module is developed to generate the specific residuals to detect sensor faults following structural analysis ([5]) of the mobile robot nonlinear model, and determining existing redundancies. Secondly, a residual evaluation module ([31]) permits to isolate sensor faults when possible, showing which sensor is faulty.

In the following subsection, the structural analysis is applied for residual signal generation to the nonlinear model of an industrial mobile robot.

3.1. Structural Analysis: The Residual Generator Module

The residual generator module [13, 11] is able to generate the residuals exploiting the main ideas of the structural analysis [13, 27], which is applied on the non-linear model of the autonomous industrial mobile robot.

Making use of the structural analysis ([5]), the model of the industrial mobile robot is considered as a set of constraints (in the following (11)–(19))

$$\mathcal{C}=\{c_1, c_2, \cdots, c_9\}$$

which are applied to a set of variables $\mathcal{Z}=\mathcal{X}\cup\mathcal{K}$, where $\mathcal{X}$ denotes the subset of the unknown variables while $\mathcal{K}$ denotes the subset of the known ones: sensor measurements, variables with known values (constants, parameters), and reference variables. The set of unknown variables is

$$\mathcal{X}=\left\{x, y, \theta, \dot{x}, \dot{y}, \dot{\theta}, \nu\right\}$$

and the set of known variables is

$$\mathcal{K}=\{\omega, \omega_l, \omega_r, d\}$$

The set of constraints for the industrial mobile robot is

$$c_1 : \quad \dot{x} = \nu \cos\theta \tag{11}$$

$$c_2 : \quad \dot{y} = \nu \sin\theta \tag{12}$$

$$c_3 : \quad \dot{\theta} = \omega \tag{13}$$

$$c_4 : \quad \omega_l = \frac{2}{D}\left(\nu - \frac{L}{2}\omega\right) \tag{14}$$

$$c_5 : \quad \omega_r = \frac{2}{D}\left(\nu + \frac{L}{2}\omega\right) \tag{15}$$

$$c_6: \quad d = P_\nu \frac{P_r - x\cos(P_n) - y\sin(P_n)}{\cos\left(P_n - \theta - \theta_s + \frac{\pi}{2}\right)} \tag{16}$$

$$c_7: \quad \dot{x} = \frac{dx}{dt} \tag{17}$$

$$c_8: \quad \dot{y} = \frac{dy}{dt} \tag{18}$$

$$c_9: \quad \dot{\theta} = \frac{d\theta}{dt} \tag{19}$$

The structure is described by the following binary relation

$$S : \mathcal{C} \times \mathcal{Z} \rightarrow \{0, 1\}$$

$$(c_i, z_j) \rightarrow \begin{cases} S(c_i, z_j) = 1 \text{ iff } c_i \text{ applies to } z_j \\ S(c_i, z_j) = 0 \text{ otherwise} \end{cases}$$

Thus, the structure of the considered system can be represented by the incidence matrix illustrated in Table 1.

Table 1. Incidence matrix

	KNOWN				UNKNOWN						
↗	ω	ω_l	ω_r	d	x	y	θ	$\dot{x}$	$\dot{y}$	$\dot{\theta}$	ν
c_1							1	1			1
c_2							1		1		1
c_3	1									1	
c_4	1	1									1
c_5	1		1								1
c_6				1	1	1	1				
c_7					1			1			
c_8						1			1		
c_9							1			1	

The matching algorithm identifies the over-determined parts of the system [11]. The result is lists of the matched and unmatched constraints. The matched constraints result

$$\mathcal{M} = \{c_1, c_2, c_3, c_5, c_7, c_8, c_9\},$$

while the unmatched constraints are

$$\mathcal{U} = \{c_4, c_6\}.$$

Each of the unmatched constraints give a parity equation. In this way, the detection equations are derived by back tracing the matching of the unknown variables involves in the unmatched constraints until only known variables are part of the expression. The resulting parity equations are

$$c_4(\omega, \nu, \omega_l) = 0 \tag{20}$$

$$c_6(x, y, \theta, d) = 0 \tag{21}$$

and back tracing unknown variables to known variables results

$$c_4 \left[\omega, c_5 \left(\omega, \omega_r\right), \omega_l\right] = 0 \tag{22}$$

$$c_6 \left[c_7 \left(c_1 \left(c_9 \left(c_3 \left(\omega\right)\right), c_5 \left(\omega, \omega_r\right)\right)\right), c_8 \left(c_2 \left(c_9 \left(c_3 \left(\omega\right)\right), c_5 \left(\omega, \omega_r\right)\right)\right), c_9 \left(c_3 \left(\omega\right)\right), d\right] = 0 \tag{23}$$

Thus, the following residuals are obtained

$$r_\alpha = f_\alpha \left(\omega, \omega_l, \omega_r\right) \tag{24}$$

$$r_\beta = f_\beta \left(\omega, \omega_r, d\right) \tag{25}$$

3.2. Change Detection Algorithm: The Decision Module

The role of the decision module is to determine whether the residuals differ significantly from zero and, from pattern of zero and non-zero residuals, to decide which are the most likely fault effects, and in turn, which component/s could be the origin of a fault. Residual evaluation consists to detect a change in the mean of a normally distributed random sequence, which can be achieved by sequential change detection algorithms like the cumulative sum algorithm ([2]), particle filter-based approach ([26]), and others ([17]).

A novel algorithm has been developed based on an adaptive threshold logic approach, and is presented in this subsection. This approach has been experimentally validated on an industrial mobile robot, it has shown good performance in terms of fault detectability and isolability, with low computational efforts. This algorithm is described in the following.

Let $\{r_i (kT_s) : k \in [((j-1) \cdot n + 1), j \cdot n]\}$ be an observed sequence of the i-th residual $r_i (kT_s)$ in a generic j-th sliding window of size n (number of readings in the sliding window). Before the unknown change time $k_0 T_s$, the mean value θ of the residual is constant and equal to θ_0. After a change the parameter is equal to θ_1. The hypotheses are:

$$\mathcal{H}_0 : \theta = \theta_0 \text{ for } ((j-1) \cdot n + 1) \le k \le j \cdot n$$
$$\mathcal{H}_1 : \theta = \theta_0 \text{ for } ((j-1) \cdot n + 1) \le k \le k_0 - 1 \text{ and } \theta = \theta_1 \text{ for } k_0 \le k \le j \cdot n$$

The on-line problem is to detect the occurrence of the change as soon as possible. The *decision test* is based on the following:

$$\text{if } r_i (kT_s) < H_{i,j} \text{ or } r_i (kT_s) > h_{i,j} \text{ accept } \mathcal{H}_0 \text{ and set } g_i (kT_s) = 0$$
$$\text{if } r_i (kT_s) \ge H_{i,j} \text{ accept } \mathcal{H}_1 \text{ and set } g_i (kT_s) = +1$$
$$\text{if } r_i (kT_s) \le h_{i,j} \text{ accept } \mathcal{H}_1 \text{ and set } g_i (kT_s) = -1$$

where $g_i (kT_s)$ is the decision function of the i-th residual, $H_{i,j}$ and $h_{i,j}$ are upper and lower thresholds, respectively (to detect deviations from θ_0 in both directions). For each specific residual, $H_{i,j}$ and $h_{i,j}$ should be properly chosen.

In the study case of the considered industrial mobile robot, because of noise, it is not possible to set constant values for $H_{i,j}$ and $h_{i,j}$ for the residuals r_α and r_β. In order to set threshold values, an ad-hoc algorithm has been applied to the i-th residual $r_i (kT_s)$ with $i = \alpha, \beta$. This proposed algorithm is as follows:

1. Let $j = 1$. In the first sliding window $j = 1$ of size n, consider that the system operates in a fault free working mode, that is, $\mathcal{H}_0$ is accepted. Determine the absolute minimum $m_{i,j}$ and absolute maximum $M_{i,j}$ values of the observed sequence of the i-th residual $r_i(kT_s)$. Set upper and lower thresholds for the next sliding window as follows:

$$H_{i,j+1} = M_{i,j} + \Delta_{i,j}^{+} \tag{26}$$

$$h_{i,j+1} = m_{i,j} - \Delta_{i,j}^{-} \tag{27}$$

   with $\Delta_{i,j}^{+}$ and $\Delta_{i,j}^{-}$ constant values resulting from statistical investigations of faultless operation residuals.

2. Let $j = j + 1$. In the sliding window j of size n at each sample time $k \in [(j-1) \cdot n + 1, j \cdot n]$, the *decision test* runs on-line and the decision function $g_i(kT_s)$ of the i-th residual is updated:

   if $\mathcal{H}_0$ is accepted, set $g_i(kT_s) = 0$
   if $\mathcal{H}_1$ is accepted, set $g_i(kT_s) = +1$ or set $g_i(kT_s) = -1$

   If $\mathcal{H}_0$ is accepted, determine the absolute minimum $m_{i,j}$ and absolute maximum $M_{i,j}$ of the i-th residual $r_i(kT_s)$ in the considered sliding window. Set the upper and lower thresholds for the next sliding window as in (26) and (27), respectively. Otherwise, if $\mathcal{H}_1$ is accepted, also only in one sample time $k \in [(j-1) \cdot n + 1, j \cdot n]$, then $H_{i,j+1}$ and $h_{i,j+1}$ keep the predetermined values, i.e. $H_{i,j+1} = H_{i,j}$, $h_{i,j+1} = h_{i,j}$.

3. Go to step 2).

Figure 4 shows how the novel algorithm works. Supposing that no faults are occurred in j-th window, absolute minimum value $m_{i,j}$ and absolute maximum value $M_{i,j}$ of the residual r_i are updated in this j-th window. These values will be used in the next $(j+1)$-th window. In the $(j+1)$-th window, lower and upper thresholds are updated, and specifically $H_{i,j+1} = M_{i,j} + \Delta_{i,j}^{+}$ and $h_{i,j+1} = m_{i,j} - \Delta_{i,j}^{-}$, where $\Delta_{i,j}^{-}$ and $\Delta_{i,j}^{+}$ are updated using the minimum and maximum values of the residual in the previous window (j-th window). If the residual crosses or the lower threshold or the upper threshold, a fault alarm will be issued, as detailed in Figure 4.

Note that, in the j-th sliding window, if $\mathcal{H}_0$ is accepted, then upper and lower thresholds are based on maximum and minimum values of $r_i(kT_s)$ determined in the previous (($j-$ 1)-th) sliding window, otherwise they keep the previous values. This allows for updated thresholds in every new sliding window. However, if a fault occurs in a sliding window, the residual changes its mean and its maximum value, but the positive and negative thresholds do not change. Hence, the thresholds are independent from the occurrence of the faults, permitting to detect them.

Table 2 shows the $\Delta_{i,j}^{+}$ and $\Delta_{i,j}^{-}$ values obtained by the experimental trials performed on the considered industrial mobile robot.

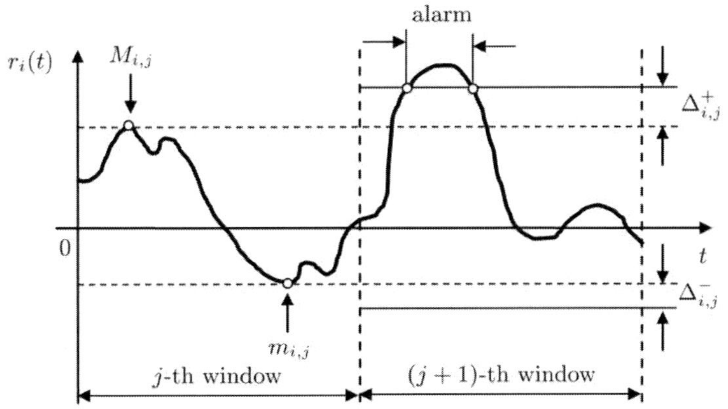

Figure 4. Application of the novel algorithm.

Table 2. Delta values for residual r_α and r_β

RESIDUAL	$\Delta^+_{i,j}$	$\Delta^-_{i,j}$
$r_\alpha(kT_s)$	$2\,\lvert M_{\alpha,j}\rvert$	$2\,\lvert m_{\alpha,j}\rvert$
$r_\beta(kT_s)$	$\lvert M_{\beta,j}\rvert + \frac{\lvert m_{\beta,j}\rvert}{2}$	$\lvert M_{\beta,j}\rvert + \frac{\lvert m_{\beta,j}\rvert}{2}$

4. Experimental Results

The simulation results have confirmed the developed analysis. The same results are also obtained by experimental tests performed on the an industrial mobile robot where the implementation of the designed fault diagnosis system has been performed online on the PC used by the navigation module. The PC installed also manages the sensor readings.

The experiments were performed in the indoor environment considering only a single fault for each trial. Additive faults have been considered, simulated and injected. The sensor fault has been generated by adding random signals to the sensor readings. The results obtained are satisfactory. For all the experiments which were performed, the false alarm and the missed-detection were zero.

The designed sensor fault diagnosis system is able to detect every single fault of the considered sensor equipment. Moreover, the faults on the laser scanner and the left encoder can be isolated. Re-running the matching algorithm, it is possible to introduce another couple of residuals on the sensor fault diagnosis system which allow to isolate faults on the laser scanner and the right encoder. The situation and the results are perfectly symmetric to these shown in this section, and they are not reported for sake of brevity. The integration of this system with the navigation module introduces new capabilities for localizing the mobile base when a single sensor fault occurs. Therefore the designed sensor fault diagnosis system increases the navigation reliability of the considered industrial mobile robot. The faults signatures are resumed in the Table 3.

Table 3. Effects of the faults on the residuals

↗	f_{gyro}	$f_{enc,l}$	$f_{enc,r}$	f_{laser}
r_α	1	1	1	0
r_β	1	0	1	1

As a result, the sensor system is able to detect all sensor faults, since at least one residual is really sensitive to each of sensor fault. Adding new sensors is possible to improve the performances of the proposed sensor fault diagnosis system.

A significant experimental activity has been developed to analyze the reliability of the proposed sensor fault diagnosis system for different environment conditions and mobile tasks. Sensor faults have been simulated, introducing a random signal on the measure of the relative sensor. Following, a series of significant experimental results are presented which show the fault detectability in each sensor fault case.

4.1. Right Encoder Faults

Faults on right encoder affect residual r_α and r_β (see Table 3). In this experiment, the sensor fault has been created starting at 55 s. Experimental results are given in Figure 5. It is seen that residuals r_α and r_β are really sensitive to faults of right encoder. The change detector, specifically designed for residual r_α, is able to issue an alarm at $55, 1$ s, and that one designed for residual r_β, issues an alarm at $55, 3$ s.

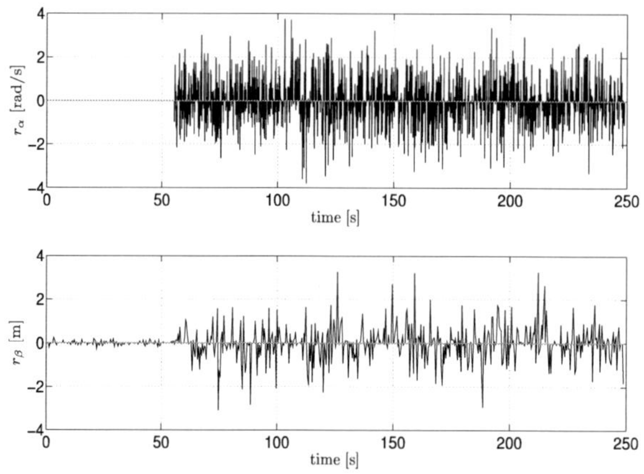

Figure 5. Right encoder fault: residuals r_α and r_β.

4.2. Left Encoder Faults

If a fault occurs on left encoder, this affects only the residual r_α, as shown in Table 3. Figure 6 shows the experimental results, where only the residual r_α reacts to the fault occurrence,

r_β is insensitive to it. The left encoder fault has been simulated starting at $77,6$ s, and the change detector relative to the residual r_α has been able to detect it after only $0,1$ s.

Note that the scale of the graph of residual r_β has been intentionally changed, to show the magnitude of the residual r_β in faultless situation. The residual is really close to zero.

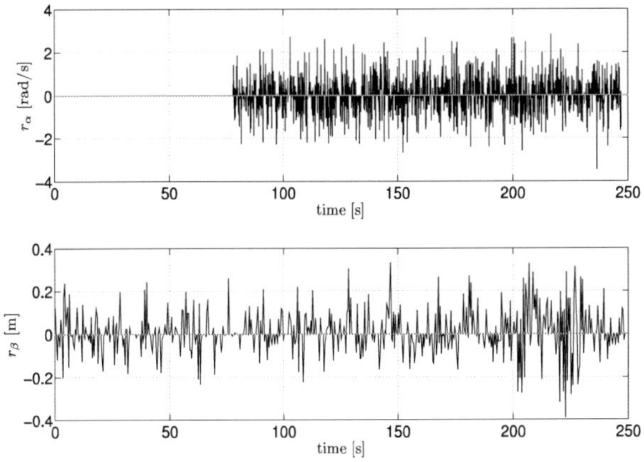

Figure 6. Left encoder fault: residuals r_α and r_β.

4.3. Gyroscope Faults

Table 3 shows that gyroscope faults affect both residuals, and this is also validated by experimental results shown in Figure 7. In this case, the gyroscope fault has been generated starting at $160,1$ s and the change detector of the residual r_α issues an alarm after $0,2$ s, and that one of the residual r_β after $0,1$ s.

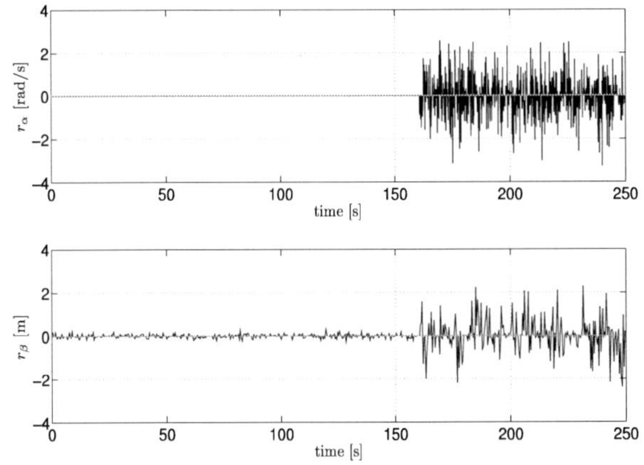

Figure 7. Gyroscope fault: residuals r_α and r_β.

4.4. Laser Range Finder Faults

Faults on laser range finder act only on residual r_β, as specified in Table 3. The sensor fault has been generated by adding extra random signal to the measure, starting at $160, 4$ s. The experimental results are shown in Figure 8. It is seen that residual r_α does not react to this fault, while residual r_β is sensitive to it. The change detector designed for the residuals r_β, is able to issue a fault alarm at $162, 4$ s, that is 2 s after the fault is really occurred.

Note that the scale of the graph of residual r_α has been intentionally changed, to show the magnitude of the residual r_α in faultless situation.

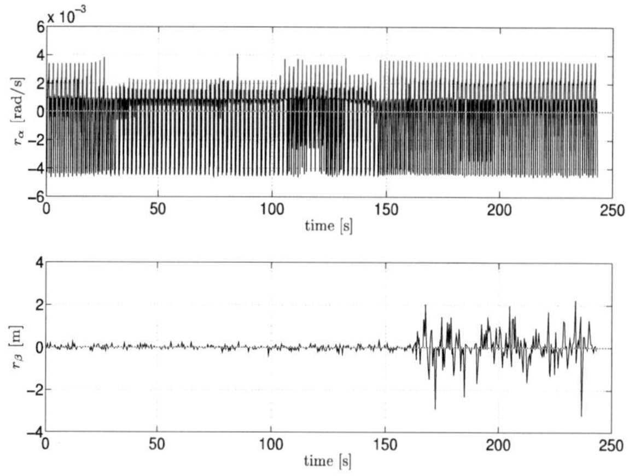

Figure 8. Laser range finder fault: residuals r_α and r_β.

Conclusion

All real control systems can malfunction and fail due to faults in their components. Logically, the chances for malfunctions increase with systems' complexity. Especially, in industrial mobile robotic systems, the complexity is permanently growing due to the continuous requirements of higher positioning accuracy and degree of automation. In such systems, several kinds of sensors need to be integrated together to ensure a more reliable guidance control system. Data fusion is required to process the navigation information collected from heterogeneous navigation sensors. Among information coming from the multi-sensor system, some can be erroneous or faulty. The reason must be either a sensor fault or a temporary inadequacy of the sensor in the work environment of industrial mobile robotic system. The existence of a sensor fault can cause undesired reactions and consequences to technical part of the navigation system, and thus aggravating the impact for man and the environment. Therefore, in the design and operation of an industrial mobile robotic system, increased attention has to be paid to reliability, safety and fault diagnosis. In order to maintain a high level of safety, performance and reliability, it is important that sensor faults are detected promptly and that the source and severity of each fault is diagnosed. The backbone of modern fault diagnosis is the model-based approach, where the mathematical model of the supervised system is used as a reference and contains the analytical redundancy. Making use of dynamic models of the system under consideration provides the most powerful approach, allowing to determine the time, size and cause of even small faults during all phases of dynamic system operations. Design and development of model-based fault diagnosis systems for industrial mobile robotic systems has been the topic of this chapter. The general and conceptual structure of a model-based fault diagnosis system, accepted by all scientific community, comprising two main stages of residual generation and residual evaluation. This chapter describes the development of a model-based fault diagnosis system, which comprises both residual generation and residual value evaluation, for an autonomous industrial mobile robot. In detail, it is shown how the proper development of a model-based sensor fault diagnosis system can improve the reliability and the safety of industrial mobile robots. In effect, the sensor fault diagnosis system permits to prevent simples navigation sensor faults and hence to reduce the risk of safety hazards. The design of the sensor fault diagnosis system was based on the structural analysis which, using graph theory, permits to determine which redundancy exists in the sensory system and thus shows the possibility to detect and handle the navigation sensor faults. Satisfactory experimental results have confirmed the developed analysis in case of industrial mobile robots.

References

[1] S. Angerer, R. Pooley, and R. Aylett, Self-reconfiguration of industrial mobile robots, *Self-Adaptive and Self-Organizing Systems (SASO), 2010 4th IEEE International Conference on*, 2010, pp. 64–73.

[2] M. Basseville and I.V. Nikiforov, Detection of abrupt changes: theory and application, *Information and System Science*, Prentice Hall, New York, 1993.

[3] G. Blaha and S.R. Gerig, Accurate navigation of industrial mobile robots, Position Location and Navigation Symposium, 1994, *IEEE*, 1994, pp. 581–585.

[4] M. Blanke, C. W. Frei, F. Kraus, R. J. Patton, and M. Staroswiecki, What is fault-tolerant control?, *Proc. of 4th IFAC Symposium on Fault Detection Supervision and Safety for Technical Processes*. Budapest, Hungary (2000), 40–51.

[5] M. Blanke, M. Kinnaert, J. Lunze, and M. Staroswiecki, *Diagnosis and fault-tolerant control*, Springer–Verlag New York, Inc., 2006.

[6] J. Borenstein, H. R. Everett, and L. Feng, *Where am I? Sensors and Methods for Mobile Robot Positioning*, University of Michigan (1996).

[7] J. Borenstein and L. Feng, Measurement and correction of systematic odometry errors in mobile robots, *IEEE Journal of Robotics and Automation* 12 (1996), no. 6, 869–880.

[8] Michael Brady, *Robot motion: planning and control*, the MIT Press, 1982.

[9] J. Chen and R. J. Patton, *Robust model-based fault diagnosis for dynamic systems*, Kluwer Academic Publishers Norwell, MA, USA, 1999.

[10] H. Chung, L. Ojeda, and J. Borenstein, Accurate mobile robot dead-reckoning with a precision-calibratedfiber-optic gyroscope, *Robotics and Automation, IEEE Transactions on* 17 (2001), no. 1, 80–84.

[11] V. Cocquempot, R. Izadi-Zamanabadi, M. Staroswiecki, and M. Blanke, Residual generation for the ship benchmark using structuralapproach, *Control'98. UKACC International Conference on (Conf. Publ. No. 455)* (1998), 1480–1485.

[12] James Crowley, Navigation for an intelligent mobile robot, *Robotics and Automation, IEEE Journal of* 1 (1985), no. 1, 31–41.

[13] P. Declerck and M. Staroswiecki, *Characterization of the canonical components of a structural graph for fault detection in large scale industrial plants*, (1991), 298–303.

[14] Guilherme N. DeSouza and Avinash C. Kak, Vision for mobile robot navigation: A survey, *Pattern Analysis and Machine Intelligence, IEEE Transactions on* 24 (2002), no. 2, 237–267.

[15] Steven X Ding, *Model-based fault diagnosis techniques: design schemes, algorithms, and tools*, Springer, 2008.

[16] P. M. Frank, *Model-based fault diagnosis*, pp. 262–269, Pergamon Press, 1992.

[17] Fredrik Gustafsson and Fredrik Gustafsson, *Adaptive filtering and change detection*, vol. 1, Wiley New York, 2000.

[18] Huosheng Hu and Dongbing Gu, Generalized predictive control of an industrial mobile robot, *IASTED International Conference on Intelligent Systems and Control*, Santa Barbara, CA, USA, 1999, pp. 28–30.

[19] ———, Landmark-based navigation of industrial mobile robots, *Industrial Robot: An International Journal* 27 (2000), no. 6, 458–467.

[20] Gianluca Ippoliti, Leopoldo Jetto, Sauro Longhi, and Andrea Monteriù, Comparative analysis of mobile robot localization methods based on proprioceptive and exteroceptive sensors, *Mobile Robots: Perception & Navigation* (2007), 215–237.

[21] Rolf Isermann, *Fault-diagnosis systems: an introduction from fault detection to fault tolerance*, Springer, 2006.

[22] R. Izadi-Zamanabadi, Structural analysis approach for fault diagnosis and disturbance decoupling, *European Journal of Control* (2003).

[23] M. Jenkin, N. Bains, J. Bruce, T. Campbell, B. Down, P. Jasiobedzki, A. Jepson, B. Majarais, E. Milios, S. B. Nickerson, J. R R Service, D. Terzopoulos, J. Tsotsos, and D. Wilkes, Ark: autonomous mobile robot for an industrial environment, *Intelligent Robots and Systems* '94. *'Advanced Robotic Systems and the Real World', IROS '94. Proceedings of the IEEE/RSJ/GI International Conference on*, vol. 2, 1994, pp. 1301–1308 vol.2.

[24] K. M. Killian, Pointing grade fiber optic gyroscope, Aerospace and Electronic Systems Magazine, *IEEE* 9 (1994), no. 7, 6–10.

[25] Józef Korbicz, *Fault diagnosis: models, artificial intelligence, applications*, Springer Verlag, 2004.

[26] Ping Li and V. Kadirkamanathan, Particle filtering based likelihood ratio approach to fault diagnosis in nonlinear stochastic systems, *Systems, Man, and Cybernetics, Part C: Applications and Reviews, IEEE Transactions on* 31 (2001), no. 3, 337–343.

[27] T. Lorentzen, M. Blanke, and H. Niemann, Structural Analysis - A Case Study of the Rømer satellite, *Proceedings of IFAC SAFEPROCESS* (2003).

[28] N. Meskin, E. Naderi, and K. Khorasani, A multiple model-based approach for fault diagnosis of jet engines, *Control Systems Technology, IEEE Transactions on* 21 (2013), no. 1, 254–262.

[29] A. Monteriù, Fault-tolerant methods for sensor fusion, *Master's thesis*, Università Politecnica delle Marche, Ancona, Italy—Technical University of Denmark, Kongens Lyngby, Denmark, 2003.

[30] A. Monteriù, Model-based fault diagnosis for mobile robotic systems, *Ph.D. thesis*, Università Politecnica delle Marche, Ancona, Italy, 2006.

[31] A. Monteriù, P. Asthana, K.P. Valavanis, and S. Longhi, Real-time model-based fault detection and isolation for UGVs, *Journal of Intelligent and Robotic Systems* 56 (2009), no. 4, 425–439.

[32] R. J. Patton, P. M. Frank, R. N. Clark, P. M. Frank, and R. Patton, *Issues of Fault Diagnosis for Dynamic Systems*, Springer, 2000.

[33] J. Petereit, T. Emter, and C.W. Frey, Safe mobile robot motion planning for waypoint sequences in a dynamic environment, *Industrial Technology (ICIT), 2013 IEEE International Conference on*, 2013, pp. 181–186.

[34] H. Roth and K. Schilling, Navigation and docking manoeuvres of mobile robots in industrial environments, *Industrial Electronics Society, 1998. IECON '98. Proceedings of the 24th Annual Conference of the IEEE*, vol. 4, 1998, pp. 2458–2462 vol.4.

[35] Alberto Rovetta, A robotic mobile platform for application in automotive production environment, *Advances in Mechanisms, Robotics and Design Education and Research* (Vijay Kumar, James Schmiedeler, S. V. Sreenivasan, and Hai-Jun Su, eds.), Mechanisms and Machine Science, vol. 14, Springer International Publishing, 2013, pp. 239–244.

[36] A. Sanchez, X. Hernandez, O. Torres, and P. Alfredo Toriz, Mobile robots navigation in industrial environments, *Computer Science (ENC), 2009 Mexican International Conference on*, 2009, pp. 155–166.

[37] Roland Siegwart and Illah Reza Nourbakhsh, *Intro to autonomous mobile robots*, MIT press, 2004.

[38] Sebastian Thrun, Dieter Fox, Wolfram Burgard, and Frank Dellaert, Robust monte carlo localization for mobile robots, *Artificial intelligence* 128 (2001), no. 1, 99–141.

[39] V. Tompa, D. Hurgoiu, C. Neamtu, and D. Popescu, Remote control and monitoring of an autonomous mobile robot, *Automation Quality and Testing Robotics (AQTR), 2012 IEEE International Conference on*, 2012, pp. 438–442.

[40] R. Zhu, Y. Zhang, and Q. Bao, A novel intelligent strategy for improving measurement precision of FOG, *IEEE Transactions on Instrumentation and Measurement* 49 (2000), no. 6, 1183–1188.

About the Editors

Dr. Fausto Pedro García Márquez earned the European Doctorate on Engineering at the School of Industrial Engineers (ETSII) of Ciudad Real, University of Castilla-La Mancha (UCLM, Spain), March 2004. He earned a degree in Engineering from University of Cartagena in Murcia, Spain, (September 1998) and Technical Engineer from the Polytechnic University School at UCLM (September 1995), and recently, he got the degree in Business Administration and Management at the Faculty of Law and Social Sciences at UCLM (December 2006). He also holds the titles of Supper Technician in Labor Risks Prevention by UCLM (July 2000) and Transport Specialist at the Polytechnic University of Madrid, Spain, (June 2001). He is Senior Lecture (with Tenure) at UCLM, and Honorary Senior Research Fellow at Bimingham University (UK). He has been distingued with the *Silver Prize* (2012) by the International Society of Management Science and Engineering Management, and the *Advancement Prize* in the Third International Conference on Management Science and Engineering Management.

Dr Mayorkinos Papaelias is currently a Lecturer and University Fellow in NDT and Condition Monitoring at the School of Metallurgy and Materials of the University of Birmingham. He is also an academic staff member of the Birmingham Centre for Railway Research and Education. Dr Papaelias received his Ph.D. in Metallurgy from the University of Birmingham in 2004. He worked as a Project Leader at TWI before returning to Birmingham University in 2006. Dr Papaelias has been involved as technical coordinator or scientific consultant in several industrial, TSB, FP6 and FP7 collaborative research projects. He is also the technical coordinator of the NIMO and OPTIMUS demonstration projects funded by the Directorate of Energy of the European Commission which deal with the development and demonstration of an integrated wind turbine condition monitoring system for diagnosis and prognosis. Dr Papaelias currently supervises a group of eight postgraduate students and three post-doctoral assistants. The research projects he is currently involved focus on railway, wind energy, concentrated solar thermal power, bridge condition monitoring, smart composites and ship-building. He is the author or co-author of more than sixty journal and conference papers in NDT and condition monitoring. Dr Papaelias also regularly authors articles for industrial magazines such as Railway Strategies and the Rail Technology Magazine. He is an acting referee for several high profile journals and international conferences. Dr Papaelias is a Member and Chartered Engineer of the Technical Chamber of Greece and Member of the International Society for Condition Monitoring. A significant part of his teaching activities at the University of Birmingham is related to non-destructive testing and condition monitoring technologies. Finally, Dr Papaelias is the Chairman of the Education Committee of the International Society for Condition Monitoring of the British Institute of Non-Destructive Testing.

INDEX

D

E

F

G

H

I

S

T

U

V

W

Y